Lasers

Lasers

Second Edition

BELA A. LENGYEL

San Fernando Valley State College,
Northridge, California

WILEY-INTERSCIENCE

a Division of John Wiley & Sons, Inc.
New York   London   Sydney   Toronto

Library of Congress Catalogue Card Number: 77-139279
ISBN 0 471 52620 7

Printed in the United States of America

10 9 8 7 6 5 4 3 2 1

Preface

This book is intended as a general introduction to an art in applied physics, that of the generation of light by stimulated emission of radiation. Being an introduction to a rather advanced subject, the exposition is based on knowledge and experience borrowed from physics and related sciences.

When writing or selecting such a book, it is important to form a clear idea about the expected degree of preparation of the person who is going to use it. This book is written for a person trained in physics at least to the extent that is expected of an engineering student graduating from one of the better schools of engineering in the United States. Specifically this means completion of two years of physics at the university level, including the study of some modern physics. The exposition of the subject is held at a level that ought not to present many difficulties to a senior majoring in physics or chemistry. I made it a point to define all new concepts and quantities in terms of those that are familiar from elementary physics. I hope that the book will be of some use to more advanced colleagues also, but it is not written for specialists in the laser field.

Some of the phenomena on which lasers are based do not fit into the framework of classical physics. More precisely said, they occur because the laws of classical physics are not suitable for the description of phenomena on the atomic and molecular scale. These must be described in terms of quantum theory, which is inevitably necessary for the full comprehension of lasers. The purpose of this book, however, is not to teach quantum mechanics, not even in an applied form. Therefore the results needed from quantum theory are stated concisely, without justification, and are used to the minimum extent necessary. Whenever possible, a classical, or semiclassical argument is given. This procedure is followed because I believe that a quantum mechanical argument presents a serious obstacle to an engineering-oriented reader and that even a reader oriented toward theoretical physics will find it easier to penetrate the quantum theoretical calculations after learning to know the subject from a more pedestrian point of view.

The selection of subject material was guided by the thought that this book ought to be general; that is, it should include exposition of all laser types. For the theoretical subjects the chosen level of presentation served as a limiting factor. An additional limiting factor was my own ability to digest material of ever-growing variety and quantity at a rate fast enough so that the parts of the manuscript that were written first would not become hopelessly obsolete at the end. Having decided to include the description of all laser types, I drew the line and abstained from entering into the discussion of any applications.

This is neither a scientific monograph nor an engineering handbook. There are many good monographs and review articles in the laser field; I bring them to the reader's attention at the proper places. To a lesser extent I also point out sources of engineering data. But I have tried to avoid cluttering the text with material that would detract from the main purpose of the book, the concise exposition of laser principles and techniques.

The selection of references was made on the basis of their usefulness to the reader. To be consistent with the educational purpose of the book, I had to resist the temptation to use the references as a means for erecting a monument to the pioneers in the laser field. I omitted citation of many outstanding contributions because there are more convenient ways of leading the reader to the information they contain. A reader interested in the historical priorities or in the original observations published will quickly find his way to the original sources by consulting the intermediate references I cite. I tried to compensate for the economy of references by inserting short historical summaries about the evolution of different aspects of the subject. The work of gathering material for the book was finished around the end of 1968; therefore very few items that appeared later found their way into the book.

Since this volume is offered to the public as the second edition of a much smaller book of conspicuously different content, an explanation about its evolution is in order. The first edition of Lasers was written during the spring of 1962. It was completed under pressure to produce, as quickly as possible, a brief comprehensive technical description of lasers at a time when no such book was available. While that book was being printed the visible helium-neon laser and the semiconductor injection lasers made their appearance. Discoveries followed one another at such a pace that in 1963 I recognized that my book was obsolete and decided to rewrite it. This time I worked more leisurely and more carefully. In about two years time I developed the subject more systematically, and in 1965 I finished a new book that included the important laser discoveries made in 1963 and 1964. This was published under the title *Introduction to Laser Physics*. It seemed at that time that it is still feasible and worthwhile to attempt a tabulation of all reported laser lines, and that it is desirable to present a selective author- and discovery-oriented bibliography in an introductory book. I straddled the fence between an introductory book and a monograph, and I included in *Introduction to Laser Physics* a chapter on nonlinear effects and one on laser applications. Further expansion of the subject matter and shifts in emphasis that took place in the technology between 1964 and 1967 suggested that drastic revisions are again necessary in order to keep my writings up-to-date. By 1967 a number of monographs appeared on lasers. They had been written mostly for the use of specialists. It seemed to me that there was still room for a general, student-oriented book. With this thought in mind, I revised the content of the book not only by adding new material but also by omitting subjects that seem less relevant now than they appeared earlier. The topics omitted include the theory of linewidth, which was a carry-over from masers, the Basov-Krokhin calculations pertaining to the operation of a gas laser with excitation transfer, and many details concerning four-level rare earth lasers which failed to attain practical significance. I also omitted the discussion of nonlinear effects and laser applications. On the other hand, the discussion of cavity modes has been expanded and an exposition of the Lamb dip and of laser amplification has been included in terms of a new and elementary formalism. Other new subjects are the semiconductor lasers excited by means other than carrier injection, the fluid lasers developed after the chelate types, and, most of all, the powerful gas lasers of ionic and molecular type together with the chemical laser types. The inclusion of the last groups of lasers required considerable expansion of the general expository material on spectroscopy.

Not only was the subject matter revised but the text of the subject matter that was retained was carefully scrutinized from the point of view of accuracy and clarity. I devoted a greater effort than before to the problem of presentation, and whenever I could do so I replaced complicated arguments found in the literature by more elementary ones.

Finally, I must answer the question, why do I call this volume *Lasers?* I do so because this title best describes its contents and because its writing was the logical continuation of the work I began in 1962 with the first edition.

The first edition of this book was written while I was a member of the technical staff of Hughes Research Laboratories. For its preparation I had the full support of my colleagues and the laboratory management.

Although I left the laboratories a long time ago and wrote the present edition entirely during my tenure at San Fernando Valley State College, I continued to receive assistance from my former colleagues at Hughes in the form of technical information, advice, and criticism. The laboratory management permitted me to use drawings and photographs that belong to the company, including many items prepared since the appearance of the first edition. In expressing my thanks for the assistance I have received from Hughes Research Laboratories, I wish to make it clear that the responsibility for the material in the book is entirely my own.

In addition to Hughes, I am indebted to Bell Telephone Laboratories; Edgerton, Germeshausen and Grier; Lincoln Laboratories of Massachusetts Institute of Technology; and Philips Research Laboratories (Eindhoven) for furnishing photographs of their apparatus and giving permission for their use. I wish to thank the American Institute of Physics for permitting the reproduction of numerous figures published in its journals.

My colleagues, Samuel Cunningham, Victor Evtuhov, Robert Hellwarth, Bernard Soffer, and Harvey Winston read various parts of the manuscript and offered numerous constructive suggestions. Mr. O. T. Sylvest, my student assistant, helped to improve the accuracy and readability of the work. To all these I express my heartfelt thanks.

My wife, Birgit Lengyel, prepared many of the illustrations, typed the entire manuscript, and assisted in the proofreading. Her participation, her unfailing support in all aspects of this enterprise, and her good natured endurance of a preoccupied, bookwriting husband is most gratefully appreciated.

Pacific Palisades, California   BELA A. LENGYEL
January 1971

Contents

Lasers

Chapter 1   Background material on radiation and atomic physics

Laser art is based on a variety of ideas and facts originating in different branches of physics and engineering. It leans heavily on the techniques of ordinary optics, it makes use of electromagnetic resonators, it involves the properties of electric discharges in gases, but most of all, it is based on those phenomena in atomic and molecular physics which are not in the framework of classical physics. Laser art is possible only because of the existence of physical phenomena which are comprehended in terms of quantum theory. It is with some justification that the Russians call lasers quantum generators of radiation.

Nevertheless, there are many practical aspects of the laser art which make it a branch of engineering. When approaching the subject of lasers from a practical point of view, it is not always possible to derive everything from first principles and yet one must be aware of the main consequences of those principles. Laser engineering is not possible without comprehension of the main features of quantum theory. The discussion of the processes involved in laser art requires knowledge of some of the basic facts of optics, atomic physics, and spectroscopy. It should be conducted with the use of the scientific terminology customary in these fields. Only an exceptional person would be familiar with all the basic laws required in this field and with the appropriate terminology, although most people actively interested in lasers presumably have encountered many of these in their studies of optics, modern physics, and atomic theory. Since the purpose of the book is to serve as an introduction, it seems appropriate to commence with a brief summary of the basic laws and relationships required and to introduce the terminology to be used in later chapters. The basic facts are summarized without an attempt at systematic exposition of physical theory. Such subjects as coherence and stimulated emission, which play a fundamental role, are discussed in some detail.

A reader well versed in modern physics may not wish to dwell on the contents of this introductory chapter; he may use it only as a source for the terminology and definition of terms employed in later chapters.

1.1 LIGHT AND THE GENERAL LAWS OF RADIATION

The subject of this book is the generation and amplification of light by stimulated emission of radiation. It therefore seems appropriate to commence by gathering, from the classical electromagnetic theory and ordinary optics, the material needed as background for the subject to be discussed. At the risk of boring the experienced reader, we begin with some elementary definitions and statements.

Our interest is in electromagnetic radiation in or near the visible region. The wavelength in the region of our main interest varies from 0.3 to 30 μm, the frequency from 10^{13} to 10^{15} Hz. The emphasis is on the fact that we are dealing with electromagnetic radiation, not that it is visible. We shall avoid all terms, so common in ordinary optics, that assess light in terms of its effects on the human eye. Consequently, we shall not speak of luminous but of *radiative* quantities, which are determined by using a detector capable of registering the transport of energy by means of electromagnetic radiation. We recapitulate the basic terms used in connection with such transport of energy.

The counterpart of luminous flux in ordinary optics is *radiative flux*. This is the rate at which radiant energy passes through a surface; it is measured in units of power, that is, in watts (joules per second) or ergs per second. The intensity of radiation incident on a surface is the *radiative flux density*. The radiative flux emitted by the unit surface area of a source is called the *radiant emittance*. The last two quantities are of the same dimension. Their common MKS unit is watts per square meter. To indicate the directional distribution of radiation of a radiating surface, we need the concept of the *radiance in a given direction*. This is the radiant flux in a given direction per unit solid angle per unit projected area of the radiator. It is usually denoted by the symbol N, and its meaning can be clarified as follows: given a radiating surface of area A and a direction at an angle θ from the surface normal, the radiative flux in a small cone of $d\Omega$ steradians around the given direction is $NA \cos \theta \, d\Omega$. When N is independent of the direction, we say that the surface radiates or scatters according to *Lambert's law*. In this case the total radiative flux from the surface is πNA. Related to N is the energy density of the radiation u, which is simply the radiative energy contained in the unit volume.

We now make use of the existence of filters and monochromators, which enable us to classify radiation according to its frequency or wavelength. All quantities pertaining to radiation may be regarded as functions of the frequency ν or the wavelength λ; their symbols are then provided with appropriate subscripts. The symbol u_ν is defined as follows: the energy density of radiation between the frequencies ν and $\nu + d\nu$ is $u_\nu\, d\nu$. The symbol u_λ refers to energy density in the wavelength interval λ to $\lambda + d\lambda$; consequently, u_ν and u_λ are related but different functions of the variables. The frequency interval ν, $\nu + d\nu$ and the wavelength interval λ, $\lambda - d\lambda$ are equivalent descriptions of the same spectral region when $d\nu/\nu = d\lambda/\lambda$. It is easily shown that $u_\nu\nu = u_\lambda\lambda$. Here ν and λ are in arbitrary units, but their product is the velocity of light.

It is usually convenient to characterize radiation by its wavelength whenever experiments or applications are concerned, but in theoretical calculations, particularly in those involving energy, frequency is a more suitable variable. When electromagnetic radiation in a cavity is in thermal equilibrium at the absolute temperature T, the distribution of radiation density according to frequency follows *Planck's law:*

$$u_\nu\, d\nu = \frac{8\pi h\nu^3}{c^3} \frac{d\nu}{e^{(h\nu/kT)} - 1}. \tag{1.1}$$

Here h is Planck's constant, k is Boltzmann's constant, and c is the velocity of light. Their numerical values are given in Appendix A, at the end of the book.

Radiation will escape through a hole cut into the walls of such a cavity at the rate of $W = uc/4$ per unit area of the hole. This is the radiative flux density at the exit of the cavity; it is the radiant emittance of the blackbody. Many solids radiate like this idealized blackbody. In fact, the spectral distribution of radiation emitted by incandescent lamps and high density arcs may be calculated in fair approximation from Planck's formula.

In experimental work, the distribution of radiative energy according to wavelength is preferred, and the radiation formula is given the form

$$W(\lambda,\, T)\, d\lambda = \frac{C_1\lambda^{-5}\, d\lambda}{\exp^{(C_2/\lambda T)} - 1}, \tag{1.2}$$

where $C_1 = 2\pi\, hc^2$ and $C_2 = hc/k$. The quantity $W(\lambda,\, T)$ is the *spectral radiant emittance.*

It follows from the aforegoing that the energy radiated from an incandescent solid is not concentrated in any frequency region. For each temperature there is a wavelength at which the spectral radiant emittance is the highest. This wavelength, λ_M, is calculable from *Wien's displacement law:*

$$\lambda_M T = a, \tag{1.3}$$

where a is a constant. The peak value of the spectral radial emittance at a given temperature is found to be proportional to the fifth power of the absolute temperature. We designate this peak by $W_M(T)$. Then

$$W_M(T) = W(\lambda_M, T) = bT^5. \tag{1.4}$$

The total radiant emittance at the absolute temperature T is given by the *Stefan-Boltzmann law* as

$$W_T = \int_0^\infty W(\lambda, T) \, d\lambda = \sigma T^4. \tag{1.5}$$

It is usually convenient in optics to deviate from the consistent use of MKS units. When the surface area is measured in square centimeters and the wavelength in ångström units (10^{-10}m), the constants introduced have the following values:

$$C_1 = 3.741 \times 10^{20} \text{ W-cm}^{-2}(\text{Å})^4, \qquad C_2 = 1.439 \times 10^8 \text{Å °K},$$

$$a = 2.898 \times 10^7 \text{Å °K} \qquad\qquad b = 1.286 \times 10^{-19} \text{ W-cm}^{-2}(\text{°K})^{-5},$$

$$\sigma = 5.679 \times 10^{-12} \text{ W-cm}^{-2}(\text{°K})^{-4}.$$

Numerical calculation of the blackbody radiation in a given spectral region is facilitated by the introduction of the variable $x = \lambda T$, because the functions $W(\lambda, T)/W_M(T)$ and $\int_0^\lambda W(\lambda', T) \, d\lambda'/W_T$ are functions of the variable x alone. These functions are tabulated in standard works [1]. The calculation of the value of $W(\lambda, T)$, or its integral over a given interval of λ, is carried out by finding the peak spectral radiant emittance $W_M(T)$, or the total radiant emittance W_T, and multiplying by the appropriate values of the tabulated functions.

A blackbody at the temperature of 5200°K has its radiation peak at 5575 Å, which is about the center of the visible spectrum, the part to which the human eye is most sensitive. Yet only about 40% of the radiation of this body falls within the visible part of the spectrum, about 6% is in the ultraviolet, and the rest is in the infrared.

Gaseous sources of light, when operated at low pressures, emit radiation consisting of groups of more or less sharp lines and possibly a continuous spectrum of lesser intensity. The frequencies of the spectral lines depend on the composition of the gas; their intensities and linewidths depend on a number of factors, such as the pressure and the temperature of the gas and the method of excitation. At low pressure the lines will be sharp, but the brightness of the gas as a lamp will be low. As the pressure increases, the brightness will increase and so will the linewidths, some extending over tens of ångströms, until at last the lines overlap and the discrete character of the spectrum disappears.

Sources of greatest brightness—greatest radiative flux in the visible region—are the high-pressure arcs and flashtubes. In order to obtain maximum brightness, flashtubes are operated at an extremely high power level which they can sustain only for short periods of time. This requires intermittent operation with a low duty cycle. They are energized by discharging large capacitors ranging from 100 to a few thousand microfarads charged between 1000 and 3000 V. Xenon tubes so activated provide a flash of the order of 1 msec with a spectral distribution approximating that of a blackbody between 6500 and 10,000°K temperature.

Light emanating from the sources discussed will be radiated in all available directions. From the flat surface of an incandescent solid it will fill a solid angle of 2π sr (not with uniform intensity, but according to Lambert's law!). To produce a parallel beam of radiation from the sources discussed so far, it is necessary to place the radiator in the focal plane of an optical system. Since the source is of finite size, the resulting beam will not be a parallel one, but will have an angular divergence equal to the angular size of the source viewed from one of the principal planes of the optical system. In order to get a sharp beam, only a small portion of an extended source may be utilized. In addition, not all energy radiated from this quasi point source will be utilized because the aperture of the optical system will act as an effective stop, eliminating a large part of the radiation. Therefore it appears that only a minute fraction of the energy of an ordinary light source may be converted to a nearly parallel beam. The higher our requirements for parallelism, the smaller this fraction becomes.

A system of mirrors and lenses can be used to direct the radiation from a source onto an object. In this manner it is possible to concentrate light at a target, and we might be tempted to try to devise an optical system that would create on a surface an image brighter than the extended source from which the light originates. In the present terminology this would mean that an image of the source is formed so that the radiance at the image is higher than at the source. A famous theorem of classical optics states that this cannot be done. More precisely, it cannot be done with Lambert-law radiators, if the media in which the object and the image are located have the same refractive index [2, p. 188].

Underlying the validity of Lambert's law and its consequences is a basic property shared by all light sources other than lasers. If light is taken from two different parts of an ordinary light source and is used to illuminate the same region of a screen, no interference effects are observed. Different portions of the same light source radiate incoherently, that is, without a fixed phase relationship. This matter will soon be examined in greater detail.

We may sum up the principal limitations of classical sources of light as follows:

Energy radiated from an intense source is distributed over a relatively broad spectral region. Powerful monochromatic sources do not exist.

The radiated energy is generally poorly collimated, and the collimation cannot be improved without sacrificing the intensity available.

Radiation from an extended source cannot be imaged with an increase in brightness.

We shall see how these limitations are overcome in the case of coherent sources.

1.2 COHERENCE OF LIGHT

The classical theory of light describes optical phenomena in terms of electromagnetic oscillations. One of the basic tools of this theory is harmonic analysis. The variation of the electromagnetic field at a point is represented as the superposition of harmonic oscillations of the form

$$E = E_i \cos (2\pi\nu_i t - \varphi_i). \tag{2.1}$$

Each oscillation has a definite amplitude E_i, frequency ν_i, and phase φ_i. The phase varies in space from point to point in a linear manner. It is useful to *think* in terms of a monochromatic radiation, which is an electromagnetic oscillation of a single frequency.

In a physical experiment one always deals with superposition of harmonic oscillations of different frequencies, but it is possible to filter radiation in such a way that for most purposes it will behave as an ideal monochromatic radiation. When this is the case, we call the radiation quasimonochromatic, or briefly monochromatic. Whether a radiation is quasimonochromatic or not depends on the experiment for which it is used.

An ideally monochromatic wave is necessarily of infinite duration; an oscillation that has the shape described by (2.1) for the finite time interval $0 < t < T$ and is zero outside of that interval may be represented as a superposition of harmonic oscillations whose frequencies are confined to a narrow region of width, approximately $1/T$ around the center frequency ν_i.

Practically monochromatic radiation is characterized by a center frequency ν_0 and a bandwidth Δ so defined that the frequency interval from $\nu_0 - \Delta/2$ to $\nu_0 + \Delta/2$ contains a large part of the energy of the radiation. In theoretical work one must perform a Fourier analysis in order to assign energy content to a frequency interval; in experimental studies it is necessary to employ an instrument with sufficient resolving power to analyze the spectral composition of the radiation.

In experimental optics it is possible to resolve radiation into its quasimonochromatic components; it is possible to measure the average intensity of radiation over a period which is long compared with ν^{-1} and Δ^{-1} and over an area whose diameter is large compared with the wavelength. Neither instantaneous nor sharply localized values can be measured; all relevant quantities must be determined in terms of the measured averages.

Theory, on the other hand, deals with amplitudes and phases. These are combined according to the rules of electromagnetic theory, but only the long-term average of the square of the resultant amplitude is subject to experimental test. Although the phase of a monochromatic wave at one point is not observable, its variation from point to point is demonstrable. Evidence for the existence of phase comes from classical interference experiments, which demonstrate that when light, emanating from a point source, is split into two beams traveling to the same final destination over two different paths the amplitudes must be added according to the well-known rules of vector addition. The direction of the vectors to be added depends on the lengths of the paths traveled.

Let the amplitudes and phases of the waves arriving over paths 1 and 2 be distinguished by appropriate subscripts. The intensity observed at a given point will be equal or proportional to (depending on the choice of units)

$$I = I_1 + I_2 + 2 \sqrt{I_1 I_2} \cos \Phi, \qquad (2.2)$$

where $I_1 = \frac{1}{2} E_1{}^2$ and $I_2 = \frac{1}{2} E_2{}^2$ are the (time average) intensities and $\Phi = \varphi_2 - \varphi_1$ is the phase difference. The latter is related to the path difference $s_2 - s_1$ as follows

$$\Phi = \frac{2\pi\nu(s_2 - s_1)}{c}. \qquad (2.3)$$

In those regions of the space in which the phase difference is 0 or an even multiple of π the intensity is large, namely $\frac{1}{2}(E_1 + E_2)^2$, whereas at those points at which the phase difference is an odd multiple of π the intensity is small, namely $\frac{1}{2}(E_1 - E_2)^2$.

A typical interference experiment is so arranged that the path difference $s_2 - s_1$ varies over a screen or over the field viewed by a telescope, and the amplitudes E_1 and E_2 are adjusted to be nearly equal. Consequently, when a monochromatic point source is used, a series of alternating light and dark bands is observed. These are the interference fringes. The light band which corresponds to zero path difference is particularly important; it is common to all frequencies. The positions of the other bands are frequency-dependent; therefore light fringes of one frequency overlap with dark ones of another. This overlap depends on the frequency difference and on the order of the fringe reckoned from the central light fringe which corresponds to zero path difference.

It is clear from ((2.2) and (2.3) that reinforcement occurs when $\nu(s_2 - s_1)/c$ is an integer, and cancellation, when it is a half integer. Therefore, the nth light fringe for wavelength λ will coincide with the nth dark fringe for wavelength $\lambda - \Delta\lambda$ when

$$s_2 - s_1 = n\lambda = (n + \tfrac{1}{2})(\lambda - \Delta\lambda).$$

Hence when

$$\frac{\Delta\lambda}{\lambda} = \left|\frac{\Delta\nu}{\nu}\right| = \frac{1}{2n + 1}, \tag{2.4}$$

the interference pattern around the nth fringe will be seriously impaired when radiation of frequency ν and $\nu + \Delta\nu$ is present in approximately equal quantities.* Since actual interference experiments are performed with a quasimonochromatic source and not an ideal monochromatic source, the number of fringes that may be clearly observed is limited by the spread of the spectrum of the source. Sometimes this fact is expressed in a different form: It is known that in a Michelson interferometer, using an ordinary spectral line as a light source, interference fringes are observed only when the path difference in the two branches is less than a few centimeters. When the path difference exceeds, say 30 cm, interference is not observed. Apparently the phase of the radiation is not preserved by the source over the length of time it takes the light to travel this distance. In an ideal monochromatic wave field, the amplitude of vibrations at any fixed point is constant, whereas the phase varies linearly with time. This is not the case in a wave field produced by a real source; the amplitude and phase undergo irregular fluctuations, the rapidity of which is related to the width $\Delta\nu$ of the spectrum. The

---

* The distinctness of the fringes is also affected by the finite extent of the illuminating source.

time interval $\Delta t = 1/\Delta \nu$ is the coherence time. During a time interval much shorter than Δt the radiation behaves like a truly monochromatic wave. This is not true for a longer time interval.

So far we have considered only the properties of light emanating from a *point source*. Such an ideal source may be approximately realized by a real source located so far from the observer that its physical dimensions are negligible compared to its distance. Now we turn to *light sources of finite extent*.

A common characteristic of all classical sources of light is the lack of coherence between light emanating from different points of the radiator. By the term "coherence" or "spatial coherence" we mean a correlation between the phases of monochromatic radiation emanating at two different points. To be precise, we ought to speak not of coherent and incoherent light but of different degrees of correlation. From a practical point of view, however, we may regard radiation emanating from two distinct sources as incoherent if we observe that the intensities of the radiation are additive. We take it as an established experimental fact that light emanating from two points of an ordinary source located well over a wavelength apart cannot be brought into interference even with extreme filtering to segregate a "monochromatic" component. We attribute this fact to a lack of correlation between the phases of distant radiators.

Related to the question of phase correlation on a light source is the question of phase correlation in a radiation field away from the source and its relation to the properties of the source as well as the geometry of the situation.

Consider more closely the electromagnetic vectors at two points, P_1 and P_2, in a wave field produced by an extended monochromatic source many wavelengths removed from both P_1 and P_2 (see Fig. 1.1). If P_1 and P_2 are so close to each other that the difference $SP_1 - SP_2$ between the paths from each source point S is small compared with the wavelength λ, then it may be expected that the fluctuations at P_1 and P_2 will be effectively the same. Furthermore, it may be expected that some correlation will exist even for greater separations of P_1 and P_2, provided that the path difference does not exceed the coherence length $c \, \Delta t = c/\Delta \nu$. By using the correlation of the electromagnetic disturbance we are led to define a region of coherence around any point in a wave field generated by an essentially monochromatic source.

The value of correlation can be tested experimentally by observing the illumination on the screen B as a function of position. The observation is carried out generally so that the illumination is measured in a region which is approximately equally distant from P_1 and P_2. The

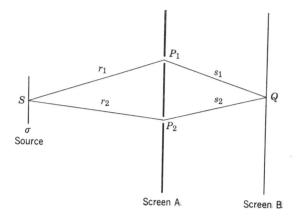

FIG. 1.1   An interference experiment with light from an extended source.

illumination is determined with small apertures around P_1 and P_2. In addition to measuring the intensities I_1 and I_2 at Q with the first and second apertures, respectively, opened, we may also measure interference effects arising from the superposition of radiation passing through these two apertures. Complete incoherence of the radiation field at P_1 and P_2 means that the intensity $I(Q)$ is

$$I(Q) = I_1 + I_2, \tag{2.5}$$

while in the case of complete coherence any value between $I_1 - I_2$ and $I_1 + I_2$ may be obtained, depending on the path difference $s_1 - s_2$. In the general case an expression of the form

$$I(Q) = I_1 + I_2 + 2\sqrt{I_1 I_2}\,\mathrm{Re}\gamma \tag{2.6}$$

is obtained. Here γ is a complex number of constant modulus $|\gamma| \leq 1$, the phase of which varies linearly with the path difference. In fact, $\mathrm{Arg}\,\gamma = \varphi_0 + 2\pi(s_1 - s_2)/\lambda$. The case $\gamma = 0$ corresponds to complete incoherence and $|\gamma| = 1$ to complete coherence, while an intermediate value of $|\gamma|$ characterizes a *partially coherent field*. As the point of observation Q is moved parallel to the line $P_1 P_2$, the intensity varies between a maximum of $I_M = I_1 + I_2 + 2\sqrt{I_1 I_2}\,|\gamma|$ and a minimum $I_m = I_1 + I_2 - 2\sqrt{I_1 I_2}\,|\gamma|$. The visibility of the interference fringes on screen B is defined as

$$v = \frac{I_M - I_m}{I_M + I_m}. \tag{2.7}$$

When the intensities I_1 and I_2 are equal, v reduces to $|\gamma|$.

The extension of the correlation concept to a polychromatic (nonmono-chromatic) field is quite straightforward but mathematically more demanding. The radiation must be represented in terms of Fourier integrals, and cross correlation must be defined as it is in the theory of stationary random processes. This analysis is carried out in the literature [2, 3]. The result is that, given two points and a time interval τ, a degree of coherence $\gamma_{12}(\tau)$, whose absolute value varies from 0 to 1 can be calculated. With this concept, the variation of the degree of coherence of a wave field generated by an extended source can be discussed. The degree of coherence of radiation between the points P_1 and P_2 can be related to the diffraction pattern of the source regarded as an aperture of specified amplitude and phase distribution. This is the substance of the *van Cittert-Zernike theorem*, which permits the calculation of the variation of $\gamma(0)$ in a plane illuminated by an extended incoherent source.

The ideal case $\gamma(0) = 1$ represents a fully coherent plane wave with the phase front coincident with the plane of observation. This, of course, cannot be achieved with a finite incoherent source. With the point P_1 held fixed and P_2 moving away from P_1, the degree of coherence $|\gamma(0)|$ decreases. Arbitrarily, the tolerance limit $|\gamma| \geq 0.88$ is set to specify the region within which the radiation is called "almost coherent." With the aid of the van Cittert-Zernike theorem it can be shown that radiation derived from a uniform, quasimonochromatic, noncoherent source of circular shape is almost coherent over a distance $d = 0.16\lambda/\alpha$, where $\alpha = \rho/r$ is the angular radius of the source as viewed from the point of observation, ρ being the radius of the circle, and r its distance from the observer (see Fig. 1.2). Therefore, an almost coherent beam of finite cross section can be obtained from a noncoherent source, but only a minute fraction of the energy radiated can be utilized in the process. In order to obtain an almost coherent beam of 1 cm diameter at 5000 Å, the source must be so far removed optically that its angular radius α is 8×10^{-6} rad. If a source of flux density w_s and surface area A radiates

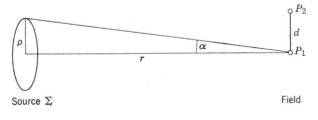

Source Σ Field

FIG. 1.2 Production of a partially coherent field.

according to Lambert's law, the flux density at a distance R from the radiator near the normal to its surface is

$$w_p = \frac{AN}{R^2} = \frac{Aw_s}{\pi R^2}.$$

Consequently, for a circular radiator of radius ρ,

$$w_p = \frac{\rho^2 w_s}{R^2}.$$

In order to obtain coherence over d cm, we must have $\alpha = \rho/R = 0.16 \, \lambda/d$. Therefore for $d = 1$ cm and $\lambda = 5 \times 10^{-5}$ cm we get $w_p = 0.64 \times 10^{-10} w_s$. In the present case the flux density in the almost coherent part of the beam is less than 10^{-10} times the flux density emitted by the source.

An almost coherent beam can be focused into a region whose dimensions are of the order of wavelength. Once an almost coherent beam is obtained, it is possible to concentrate this energy, and the degree to which this concentration is successful depends on the degree of coherence of the beam.

We can now readily appreciate some of the advantages gained by having a coherent, or almost coherent, source whose radiation is already in the form of a spherical wave or a plane wave of limited cross section. Such radiation can be concentrated by lenses and mirrors to images much brighter than the original source. Moreover, radiation emitted from a source already in the approximate form of a plane wave can be directed at a distant object with only negligible losses from diffraction effects, whereas only a small part of the radiation from a noncoherent source can be converted into an approximate plane wave.

1.3 EMISSION AND ABSORPTION OF RADIATION

It is well known that atomic systems, such as atoms, ions, and molecules, can exist in certain stationary states each of which corresponds to a definite value of the energy. The states are characterized by quantum numbers. The energy values are called the levels of the atomic system. In the case of isolated atoms, the states are described by the quantum numbers that characterize the orbits and spins of the electrons outside of the closed electronic shell. The customary nomenclature of these states makes use of symbols which indicate (to the initiated) the magnitudes and the important components of angular momenta that remain (approximately) constant [4, 5]. In the case of molecules, the description of state involves the quantum numbers associated with the rotational and vibrational motions of the molecule as well as the quantum numbers that describe the motions of the electrons.

When two or more states have the same energy, the level is called *degenerate,* and the number of states with the same energy is the *multiplicity of the level.* Frequently the word "state" is used to mean level; all states of the same energy are regarded as identical. Transitions between stationary states may occur with attendant emission or absorption of energy as radiation, or with the transfer of energy to or from another system. If the transition is radiative, the frequency of the radiation emitted or absorbed by the system is given by *Bohr's frequency relation:*

$$h\nu = E_2 - E_1, \tag{3.1}$$

where E_1 and E_2 are the energies of the states among which transition takes place and h is Planck's constant.

The level of the system with the lowest energy is the *ground level;* every other level is an *excited level.* The terms "ground state" and "excited state" are also used. An atom in the ground level can only absorb radiation. Starting with the ground level, we number the levels in increasing order of energy. When the atomic system is not in the ground level, it may change to a lower level without any external causation with the emission of radiation. This is the phenomenon of *spontaneous emission.* The probability that an atom in level n will spontaneously change to the lower level m within the unit of time is called the *spontaneous transition probability.*[*] It is denoted by A_{nm}. This quantity is characteristic of the pair of energy levels in question. In multiple levels, A_{nm} is obtained by summation over all state pairs involved. If there is a large collection of atomic systems, and N_n is the number of systems in the nth level, the total number of transitions per second from level n to level m will be approximately $N_n A_{nm}$, and the power radiated at the frequency $\nu_{nm} = (E_n - E_m)/h$ will be $N_n(E_n - E_m)A_{nm}$. Spontaneous radiation will emerge from the atoms of the assembly in a random phase, therefore the assembly of independent atoms (gas) will emit this radiation as an incoherent source.

Transitions between different atomic or molecular energy levels take place not only spontaneously but also under stimulation by electromagnetic radiation of appropriate frequency. Under certain conditions, which will soon be stated, the probability that an atomic system will change

---

[*] Strictly speaking, we are dealing with quantities that should be called rates, not probabilities. Their dimension is reciprocal time. This matter is discussed further at the end of this section.

during the unit of time from a level of index n to a level of lower energy of index m is

$$P_{nm} = A_{nm} + u_\nu B_{nm}, \qquad (3.2)$$

where u_ν is the radiation density at the frequency that corresponds to the energy difference of the levels and A_{nm} and B_{nm} are constants determined by the atomic system. In the presence of radiation of the proper frequency, the atomic system may also pass from a lower to a higher energy level. The probability of such an event (absorption) is

$$P_{mn} = u_\nu B_{mn}. \qquad (3.3)$$

Radiation emitted from an atomic system in the presence of external radiation consists of two parts. The part whose intensity is proportional to A_{nm} is the spontaneous radiation; its phase is independent of that of the external radiation. The part whose intensity is proportional to uB_{nm} is the *stimulated (induced) radiation;* its phase is the same as that of the stimulating external radiation.

In the final analysis no radiation is strictly monochromatic. For the sake of simplicity, we assume at this point that the spectral extent of each atomic line is so narrow that the distribution of energy with frequency within the line is not resolved and that we observe only the total energy emitted or absorbed.

Equations 3.2 and 3.3 are valid, and important relationships between the constants A and B hold when the radiation is isotropic, or chaotic, showing no directional preference, and when the radiation density u_ν does not vary significantly over the frequency range of the spectral line.

The relationships between the A's and B's are known as *Einstein's relations.*[*] They are usually stated in the form

$$B_{nm} = B_{mn}, \quad A_{nm} = \frac{8\pi h \nu^3}{c^3} B_{nm}. \qquad (3.4)$$

These equations are valid in vacuum for particles having only nondegenerate energy levels. When the energy levels are degenerate, Einstein's first relation takes the form

$$g_n B_{nm} = g_m B_{mn}, \qquad (3.5)$$

where g_n and g_m are the multiplicities of levels n and m, respectively. The second relation is not affected by the multiplicities. In solids, in

---

[*] For justification and further discussion of Einstein's relations see Section 1.5.

which the index of refraction η differs appreciably from unity, the second relation must be replaced by

$$A_{nm} = \frac{8\pi h \nu^3 \eta^3}{c^3} B_{nm}. \tag{3.6}$$

The reason for the appearance of the η is that the factor in front of B_{nm} arises from the counting of the radiation modes in a volume element (see Section 3.5).

We now turn to the concept of *lifetime*, which is frequently used in describing transitions between different states of an atom. The lifetime of a state is simply related to the probability of transition from that state. Let $p\,dt$ be the probability that an atom, originally in state s, will leave that state during a short time interval dt. (This interval must be so short that $p\,dt \ll 1$.) Then, for a constant p, the number of atoms in state s will decrease exponentially according to the formula $N(t) = N_0 e^{-pt}$. Hence the number of atoms leaving the state s in the time interval from t to $t + dt$ is $pN_0 e^{-pt}\,dt$. Therefore the average lifetime of the atom in state s is

$$T = \frac{1}{N_0} \int_0^\infty tpN_0 e^{-pt}\,dt = \frac{1}{p}. \tag{3.7}$$

In view of (3.7), the reciprocal of the transition probability of a process is called its lifetime. If an atomic state can be altered by several processes with lifetimes $\tau_1, \tau_2, \dots, \tau_n$ and these processes are statistically independent, then the lifetime of the state is related to the lifetimes of the processes by means of which the state can be altered by the equation

$$\frac{1}{T} = \frac{1}{\tau_1} + \frac{1}{\tau_2} + \cdots + \frac{1}{\tau_n}. \tag{3.8}$$

In a practical situation, observations are made not on a single atom but on a collection containing billions of atoms not necessarily in the same state. Given a large number N_0 of atoms, it is known that in thermal equilibrium at absolute temperature T the distribution of these atoms among the different states will follow Boltzmann's law; that is, the number of atoms in state j will be

$$N_j' = \frac{N_0 e^{-E_j/kT}}{\sum_i e^{-E_i/kT}}, \tag{3.9}$$

where E_j is the energy in state j. All states of the same level will be equally populated; therefore the number of atoms in level n is $N_n = g_n N_n'$, where N_n' refers to the population of any of the states in level n. It follows then from (3.9) that the populations of the energy levels n and m are related by the formula

$$\frac{N_n}{g_n} = \frac{N_m}{g_m} e^{-(E_n - E_m)/kT}. \tag{3.10}$$

At absolute zero all atoms will be in the ground state. Thermal equilibrium at any temperature requires that a state with a lower energy be more densely populated than a state with a higher energy.

Consider now an ensemble of atoms initially at absolute zero. This ensemble will absorb only radiation whose frequency is contained in the sequence $(E_i - E_1)/h$, where $i = 2, 3, \ldots$. If the ensemble is at equilibrium at a finite temperature T, then not only the ground state will be populated; consequently, radiation whose frequency corresponds to a transition between excited states may also be absorbed. As a matter of practical fact, it is well to remember that the first excited levels of most atoms and ions are at least 2×10^{-12} erg above the ground level, and that for $T = 500°K$ the product kT is approximately 0.07×10^{-12} erg. Therefore at moderate temperatures, generally, few atoms will occupy even the first excited level compared to the number present in the ground state, because the exponential factor in (3.10) is so small. The absorption of radiation requiring the transition from an excited level will be weak, for the number of transitions from the nth to the mth level is proportional to N_n.

As a consequence of the absorption of radiation, the equilibrium of the ensemble will be disturbed. Let us assume that monochromatic radiation is absorbed. Atoms that become excited above the first excited level by the absorption of radiation may return directly to the ground state by spontaneous or stimulated radiation, or they may follow another path and change to a lower level other than the ground level. In this manner they may cascade down on the energy scale, emitting at each step radiation different in frequency from that which originally lifted them out of the ground state. Because of the relationship (3.1) connecting energy and frequency, the radiations emitted in the cascade process, which is called fluorescence, have lower frequencies than the exciting radiation.

Consider now an ensemble that is *not* necessarily in thermal equilibrium and again designate the number of atoms per unit volume in state n by N_n. Assuming $n > m$, what is the response of the ensemble to collimated radiation of frequency ν_{nm} and density u?* The number of

---

* Actually, the density u should have the subscript ν_{nm} to indicate that we are discussing the radiation density within a spectral region surrounding the frequency ν_{nm}. To simplify notation, we drop subscripts in every instance where there is no ambiguity concerning the transition to which it refers.

downward transitions from level n to level m will be $(A_{nm} + uB_{nm})N_n$ per second and the number of upward transitions, $B_{nm}N_m$. Whenever N_n is less than N_m, which is usually the case, the incident beam will suffer a net loss of $(N_m - N_n)uB_{nm}$ quanta per second. The $A_{nm}N_n$ quanta, which are radiated spontaneously, will appear as scattered radiation. Thus a beam passing through matter in which the lower energy states are more populated than those of higher energy will always lose intensity; the material will have a positive coefficient of absorption.

An ensemble can be constructed (on paper, easily) in which N_n, the number of atoms in state n, is *greater* than N_m even though $n > m$. This ensemble is said to contain a *population inversion*. It is definitely *not* in thermodynamic equilibrium. Let us assume now that a population inversion is accomplished somehow for the pair of levels 1 and 2. This means that we have found a process that leads to $N_2 > N_1$. In this situation, the material will radiate spontaneously. It will also act as an amplifier of radiation at the proper frequency, $\nu = (E_2 - E_1)/h$; the spontaneous radiation of the same frequency will appear as amplifier noise. Laser technology deals with the practical problems related to the creation of materials in such a nonequilibrium state and with the exploitation of the amplifying capabilities for the generation of light.

In order to develop the quantitative relations that govern this amplification process, it is advisable to take a closer look at the process of absorption and to sacrifice the mathematical idealization we have made concerning the infinite sharpness of the levels and spectral lines. In an actual absorption experiment the intensity of light transmitted through a fixed layer of material is recorded while the frequency of the incident light is varied and its intensity is kept constant. The typical result of such an experiment is a curve shown in Fig. 1.3. From the assumption that the decrease of the light intensity I on passage through material

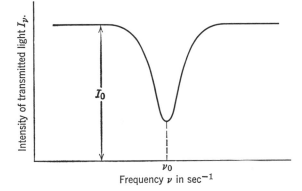

FIG. 1.3 An absorption line.

of thickness dx is proportional to $I\,dx$ it is inferred that the variation of I with depth x follows the law

$$I_\nu(x) = I_0 e^{-k(\nu)x}. \tag{3.11}$$

where $k(\nu)$ is determined from observations of the type shown in Fig. 1.3, which represents the intensity at a fixed depth in the absorbing material. The frequency ν_0 is the center of the absorption line. When x is measured in centimeters, $k(\nu)$ is expressed in reciprocal centimeters. From these observations we may obtain $k(\nu)$ as a function of frequency, and when this is done we will have a curve similar to that shown in Fig. 1.4. The total width of the curve, at the place where $k(\nu)$ has fallen to one half of its peak value k_0, is the width of the absorption line and is denoted by $\Delta\nu$. Frequently, this quantity is called the "halfwidth." This does not mean one-half the width of the curve, but the full width at half peak value.

An important relationship links the total area under the curve in Fig. 1.4 with the Einstein coefficients and the populations of the states that are responsible for the absorption centered around ν_0. This relationship was derived by Füchtbauer and Ladenburg in the early 1920's.*

Consider a parallel beam of light of frequency between ν and $\nu + d\nu$ and intensity I_ν traveling in the positive x-direction through a layer of atoms bounded by the planes x and $x + dx$. Let the velocity of light in this medium be $v = c/\eta$, where η is the index of refraction. The phase front will travel through the slab of thickness dx in the time $dt = dx/v$. Suppose there are N_1 atoms/cm³ in level 1, of which $dN_{1\nu}$ are capable of absorbing in the frequency range ν to $\nu + d\nu$, and N_2 atoms/cm³ in

---

* The present proof follows the method of Mitchell and Zemansky [6]. However, our definition of the Einstein coefficients and theirs differ by a factor of $c/4\pi$.

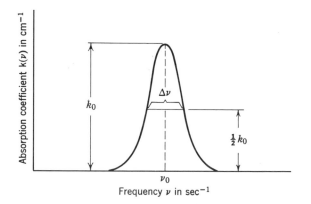

FIG. 1.4 Variation of absorption coefficient with frequency in an absorption line.

level 2, of which $dN_{2\nu}$ are capable of emitting in the same range. Then, as the phase front advances from x to $x + dx$, the decrease of energy in the beam is

$$-d(I_\nu \, d\nu) = h\nu(B_{12} \, dN_{1\nu} - B_{21} \, dN_{2\nu})I_\nu \frac{dx}{v}. \qquad (3.12)$$

Here we have made use of the fact that radiation emitted by stimulation is coherent with the stimulating radiation; therefore it will reinforce the beam. Radiation emitted spontaneously does not contribute significantly to the beam because it is not collimated. From (3.12) it follows that

$$-\frac{1}{I_\nu}\frac{dI}{dx}\,d\nu = \frac{h\nu\eta}{c}\,(B_{12}\,dN_{1\nu} - B_{21}\,dN_{2\nu}). \qquad (3.13)$$

We now recognize the left-hand member as $k(\nu)\,d\nu$, and we obtain by integration over the entire line centered around ν_0

$$\int k(\nu)\,d\nu = \frac{h\nu_0\eta}{c}\,(B_{12}N_1 - B_{21}N_2). \qquad (3.14)$$

Here B_{12} may be eliminated by means of (3.5) and B_{21} may be expressed in terms of A_{21} by making use of (3.6). In this manner we get the *Füchtbauer-Ladenburg formula*

$$\int k(\nu)\,d\nu = \frac{c^2 A_{21}}{8\pi\nu_0{}^2\eta^2}\frac{g_2}{g_1}\left(N_1 - \frac{g_1}{g_2}N_2\right). \qquad (3.15)$$

This is a basic formula, which we write as

$$\int k(\nu)\,d\nu = \kappa\left(N_1 - \frac{g_1}{g_2}N_2\right). \qquad (3.16)$$

The constant κ may be written in various forms, one of which is

$$\kappa = \frac{\lambda^2 A_{21}}{8\pi\eta^2}\frac{g_2}{g_1}. \qquad (3.17)$$

Another useful form favored by many authors is

$$\kappa = \frac{c^2}{8\pi\eta^2\nu^2 t_2}\frac{g_2}{g_1}. \qquad (3.18)$$

where $t_2 = 1/A_{21}$ is called the lifetime of atoms in level 2.

The situation most frequently encountered is such that the material is unexcited, so that few atoms are found in states other than the ground state. In this case essentially all absorption takes place as a result of a transition from the ground state. As long as the intensity of the absorbed radiation is not excessive the number of atoms in the terminal state is negligible. Denoting the absorption of the totally unexcited material by $k(\nu)_0$, we have:

$$\int k(\nu)_0 \, d\nu = \kappa N_0, \tag{3.19}$$

where N_0 is the total number of atoms per unit volume.

It is interesting to note that, under the conditions described, the integral of the absorption coefficient is simply proportional to the number of atoms present, and that its value is independent of line shape. The physical meaning of the constant κ is revealed in (3.19). It is the *integrated absorption cross section* per atom for the line in question, the quantity $\sigma(\nu) = k(\nu)_0/N_0$ is the *absorption cross section* per atom.

It is reasonable to assume that the lineshape itself does not depend on the distribution of the atoms among the energy levels, that is, the function $k(\nu)$ may be written as the product of a function of ν and a function of N_1 and N_2. Taking into consideration that (3.16) must hold, we write

$$k(\nu) = \kappa g(\nu, \nu_0)\left(N_1 - \frac{g_1}{g_2} N_2\right), \tag{3.20}$$

where ν_0 is the center frequency of the line, and $g(\nu, \nu_0)$ is a line shape function that is different from 0 only in a small region around ν_0. It is so normalized that

$$\int g(\nu, \nu_0) \, d\nu = 1, \tag{3.21}$$

when the integration is extended over the entire region in which $g \neq 0$.

The peak value of the absorption coefficient is proportional to $g(\nu_0, \nu_0)$, the peak value of the lineshape curve. Other things being equal, it will be inversely proportional to the spread of the line. The peak value of absorption in the unexcited material is of particular importance. It is denoted by k_0 and is related to the other variables as follows

$$k_0 = \kappa N_0 g(\nu_0, \nu_0). \tag{3.22}$$

For simplicity we introduce the notation $g(0)$ for $g(\nu_0, \nu_0)$. It is reasonable to do this because g is usually a function of $\nu - \nu_0$.

When the material is in thermal equilibrium, the distribution of atoms among the levels is described by (3.10). For any positive value of the absolute temperature we get

$$\frac{N_n}{g_n} < \frac{N_m}{g_m}, \tag{3.23}$$

whenever the inequality $E_m < E_n$ holds.

The nonequilibrium situation, in which the inequality (3.23) is reversed, is frequently referred to in the literature as a state of *negative temperature*. A negative value of T is calculated from the distribution of atoms among the energy levels by means of Boltzmann's formula (3.10). The idea is applicable only to a pair of levels, and it arises from the use of this formula in connection with a pair of levels in a system not in thermal equilibrium. Temperature in this connection does not have its customary meanings: $kT/2$ is not the average energy of the system per degree of freedom, and nothing can be inferred from the value of T about the distribution of the population in states other than for the pair from which this negative value of T was calculated. The use of the term "negative temperature" does not facilitate the understanding of nonequilibrium phenomena. It is best to avoid the use of this term and speak of population inversion instead. The term negative temperature is introduced here merely to provide a connection with the language of the pertinent literature. The expression, "negative temperature is established for levels n and m," sometimes encountered, means nothing more than that the inequality (3.23) is reversed while $E_m < E_n$.

When population inversion takes place for levels n and m, formula (3.16) gives a negative value for the integrated absorption coefficient. We have a condition of negative absorption, that is, we have amplification. Negative absorption, or amplification, is the consequence of the excess of stimulated radiation over absorbed radiation. In a material that is in the condition of negative absorption for a frequency region, an incident light wave will grow according to the law (3.11), which in this case represents an exponential growth at the rate of $\alpha = -k(\nu)$.*

The amplification rate α is calculated from (3.20), which may be given the form.

$$\alpha(\nu) = \kappa g(\nu, \nu_0) N, \tag{3.24}$$

where

$$N = \frac{g_1}{g_2} N_2 - N_1 \tag{3.25}$$

is the numerical measure of the population inversion.† More useful still is the *relative population inversion* $n = N/N_0$, which is -1 for a totally

---

* In complex materials, typically in semiconductors, the intensity of a light beam may be diminished by processes other than the one considered here. Light might be scattered, for example. When such additional loss mechanisms are present, population inversion may not always lead to amplification because the gain due to the excess of stimulated emission over absorption may be cancelled by losses of other kinds. In this case $N_2/g_2 > N_1/g_1$ is necessary but not sufficient for negative absorption.

† Many authors use $N_2 - g_2 N_1/g_1$ as the measure of population inversion. Their practice requires that the constant κ introduced in (3.17) be defined without the factor g_2/g_1. In the light of the physical meaning of κ, we prefer the definition of N as in (3.25).

unexited material and 0 for the material which neither absorbs nor amplifies. We then obtain

$$\alpha(\nu) = \kappa N_0 g(\nu, \nu_0)n = k(\nu)_0 n, \tag{3.26}$$

hence

$$\int \alpha(\nu) \, d\nu = \kappa N_0 n. \tag{3.27}$$

Thus the amplification rate and the integrated amplification rate are now expressed in terms of the relative population inversion and the measurable absorptive properties of the material in its unexcited state.

1.4 SHAPE AND WIDTH OF SPECTRAL LINES

Einstein's coefficients, introduced in Section 1.3, are determined by the structure of the atom. As we have seen, they determine the total rate of emission and absorption integrated over the entire spectral line. The rate of emission or absorption in a narrow spectral range between ν and $\nu + d\nu$ is $k(\nu)d\nu$, a quantity related to A_{21} through (3.15). Here a narrow spectral range is one whose extent in frequency is small compared with the width of the spectral line. In laser technology we deal with amplification in fine spectral regions, which are generally much narrower than the width of the spectral line as observed in a gas or in a crystal. Since the population distribution and the Einstein coefficient determine only the integral of $k(\nu)$, the peak value of the absorption or amplification depends on the width and the shape of the line. It is therefore desirable to take a close look at the shape of a spectral line as observed in emission or absorption from an aggregate of atoms, such as a gas.

The natural, or intrinsic, linewidth of an atomic line is extremely small. This is the linewidth that would be observed from atoms at rest, without interaction with one another. There is a theoretical limit for linewidth under such circumstances, but this may be disregarded in most instances, because it is small compared with the broadening effects of other causes which are invariably present. The two major factors of line broadening are the frequency variations resulting from the thermal motion of the atoms and those resulting from the interruption of absorption or emission of radiation by atomic collisions.

The thermal motion of the atoms is the cause of the *Doppler-broadening* whose frequency dependence is calculated as follows:

The probability that a fixed (say, x) component of the velocity of an atom in a gas at absolute temperature T is between v_x and $v_x + \Delta v_x$ is proportional to $[\exp - (mv_x^2/2kT)] \, \Delta v_x$. The Doppler shift in frequency is related to the relative velocity v_x toward the observer according to the equation

$$\frac{\nu - \nu_0}{\nu_0} = \frac{v_x}{c},$$ (4.1)

where c is the velocity of light. Therefore the Doppler effect gives rise to the following Gaussian frequency distribution

$$P(\nu)\, d\nu = P_0 e^{-\beta(\nu-\nu_0)^2/\nu_0^2}\, d\nu,$$ (4.2)

where $\beta = mc^2/2kT$. Here $P(\nu)\, d\nu$ is proportional to the spectral emittance of the gas.

The constant P_0 is determined from the requirement that the integral of the probability distribution $P(\nu)$ over all frequencies must be one. Therefore

$$P_0 = \frac{c}{\nu_0}\left(\frac{m}{2\pi kT}\right)^{\frac{1}{2}}.$$ (4.3)

The width of the distribution (4.2) at half power is

$$\Delta\nu = 2\frac{\nu_0}{c}\left(\frac{2kT \log 2}{m}\right)^{\frac{1}{2}}.$$ (4.4)

Here m is the mass of the molecule. One may introduce the molecular weight $M = N_0 m$ and the gas constant $R = kN_0$ by multiplying the atomic quantities by Avogadro's number. By substitution of the proper numerical values the following formula is obtained for the Doppler broadening of spectral lines:

$$\Delta\nu = 7.162 \times 10^{-7}\left(\frac{T}{M}\right)^{\frac{1}{2}}\nu_0.$$ (4.5)

It is to be noted that this linewidth, for a given line, depends only on the temperature of the gas.

The second major cause of line broadening in a gas is the collision of radiating particles (atoms or molecules) with one another and the consequent interruption of the radiative process. A finite wavetrain is never purely monochromatic; the spectrum of a wavetrain is spread in inverse proportion to the length of the train in the time domain. As an atomic collision interrupts either the emission or the absorption of radiation, the long wavetrain, which otherwise would be present, becomes truncated. After the collision the process is restarted without memory of the phase of the radiation before the collision. The result of frequent collisions is the presence of many truncated radiative or absorptive processes. The linewidth of the radiation of this aggregate is, of course, greater than that of an individual uninterrupted process.

The lineshape, that is, the distribution of frequencies, must be computed statistically.

The original classical computation of this kind was carried out around the turn of the century by H. A. Lorentz, who showed that when the frequency of collisions is small, compared with the undisturbed frequency ν_0, the following expression describes the *frequency distribution of the collision-broadened (Lorentz) line*

$$g(\nu) = \frac{\Delta\nu}{2\pi} \frac{1}{(\nu - \nu_0)^2 + (\Delta\nu/2)^2}. \tag{4.6}$$

Here ν_0 is the center frequency, and $\Delta\nu$ is the width between the half-power points of the curve. The factor $\Delta\nu/2\pi$ assures normalization according to the area under the curve

$$\int_{-\infty}^{+\infty} g(\nu)\, d\nu = 1. \tag{4.7}$$

The linewidth $\Delta\nu$ is related to the average time τ, which elapses between consecutive interrupting collisions:

$$\Delta\nu = \frac{1}{\pi\tau}. \tag{4.8}$$

Since the frequency of collisions is proportional to the density of the gas, the Lorentz linewidth is proportional to the density.

Although both Doppler and collision broadening result in bell-shaped curves for the distribution of frequencies, these curves are quite different. The difference is illustrated in Fig. 1.5, which shows Gaussian– and Lorentz–type curves of the same linewidth plotted on the same scale. The peak values of these curves are related to the linewidths as follows:
for the Gaussian curve

$$g(0)_G = \frac{2}{\Delta\nu}\left(\frac{\log 2}{\pi}\right)^{\frac{1}{2}} = \frac{0.939}{\Delta\nu}, \tag{4.9}$$

for the Lorentz curve

$$g(0)_L = \frac{2}{\pi\Delta\nu} = \frac{0.637}{\Delta\nu}. \tag{4.10}$$

The peak of the Gaussian curve exceeds that of the Lorentz curve by almost 50%.

In an actual situation, factors producing both types of broadening may be present at the same time. The combination of these factors leads to more complex lineshapes for which we refer to the literature [6, 7]. Frequently, one of the factors predominates; in that case calculations based on that factor alone will lead to approximately correct results.

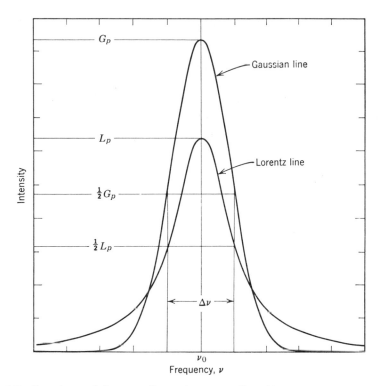

FIG. 1.5   Gaussian and Lorentz lines of common linewidth. G_p and L_p denote the peak intensities.

The spectral lines of ions in a crystal lattice are broadened by the variations of the electric field in which the ions are located. This is so because the degenerate energy levels of the ions are split by the crystal field, and the fluctuations of this field are often large enough to obliterate the fine structure. Some broadening is caused by the fact that different ions are located in different (average) fields because of their positions with respect to imperfections and strains in the crystal. Another cause of the line broadening is the thermal motion of the lattice elements. Lattice vibrations subject the ions to statistically varying fields and thus cause a temperature-dependent line broadening [8].

1.5   ABSORPTION AND STIMULATED EMISSION

In Section 1.3 we introduced laws governing the transition of atomic systems from one state to the other with the emission or absorption of radiation. Here we will re-examine the laws already stated and show their relationship to physical theory. In the present section we shall try to provide support for the statements the reader has been asked to accept on faith, and we shall examine some essential details omitted earlier. A reader whose interest is confined to the technical aspects of lasers may leave this section unread, until he becomes puzzled about the nature of the stimulated radiation.

Our first objective is to show that the existence of stimulated emission of radiation and the validity of Einstein's relations are consequences of Planck's law of blackbody radiation (1.1), Bohr's frequency relation (3.1), and Boltzmann's law (3.9), which governs the distribution of particles in thermal equilibrium with a heat reservoir.

The discussion follows the line of Einstein's original argument [9]. We assume the presence of a collection of identical atoms in a closed cavity, the walls of which are maintained at a constant (absolute) temperature T. The atoms are characterized by the existence of a sequence of energy levels $E_1, E_2, \ldots$, among which transitions may take place with emission or absorption of radiation of the frequency determined by Bohr's condition.

We know from experience that whenever atoms exist with energy more than the minimum possible, spontaneous emission of radiation will eventually take place. We also know that radiation may be absorbed, and the rate of absorption of energy is proportional to the intensity of the proper spectral component of the radiation. Analogy with the behavior of driven macroscopic harmonic oscillators and the application of Bohr's correspondence principle then suggest that a process of stimulated emission also ought to occur, and that its rate ought to be proportional to the spectral radiative flux density. Thus we are led to postulate that transitions from a higher level 2 to a lower level 1 occur at the rate of

$$P_{21} = A_{21} + u_\nu B_{21}, \qquad (5.1)$$

where u_ν is the radiation density at the frequency

$$\nu = \frac{E_2 - E_1}{h}, \qquad (5.2)$$

and where A_{21} and B_{21} are constants characteristic of the atom. The "up" transitions from level 1 to level 2 occur at the rate of

$$P_{12} = u_\nu B_{12}. \qquad (5.3)$$

Having made these assumptions, we seek to determine the constants A_{21}, B_{12}, and B_{21}.

In thermal equilibrium the number of atoms in each state remains the same (except for small fluctuations). Therefore, if N'_j denotes the number of atoms in the jth state in thermal equilibrium, then

$$N'_1 P_{12} = N'_2 P_{21}. \tag{5.4}$$

According to Boltzmann's formula (3.9) and (5.2),

$$\frac{N'_2}{N'_1} = \frac{e^{(-E_2/kT)}}{e^{(-E_1/kT)}} = e^{-h\nu/kT}. \tag{5.5}$$

In place of N'_2/N'_1 we introduce the ratio P_{12}/P_{21} and obtain from (5.1) and (5.3) that

$$\frac{u_\nu B_{12}}{A_{21} + u_\nu B_{21}} = e^{-h\nu/kT}. \tag{5.6}$$

Solving for u_ν, we get

$$u_\nu = \frac{A_{21}}{B_{12}} \frac{1}{e^{h\nu/kT} - B_{21}/B_{12}}. \tag{5.7}$$

Thermal equilibrium is maintained within the cavity by means of radiation whose spectral distribution is governed by Planck's law (1.1). Therefore the energy density u_ν given by (5.7) must be consistent with Planck's law for any value of T. This is possible only when

$$B_{21} = B_{12}$$

and

$$A_{21} = \frac{8\pi h\nu^3}{c^3} B_{12}. \tag{5.8}$$

We have thus derived Einstein's relations (3.4) for the nondegenerate case. The general formulas involving the multiplicities g_1 and g_2 are easily obtained by applying Boltzmann's law in its more general form (3.10).

In the course of the above derivation it was assumed that the radiation field to which the atoms are subjected is of the type found in a black cavity. First, it is chaotic, showing no directional or local preference. Second, the variation of u_ν with ν is slow so that, over the spectral linewidth of an atomic transition, u_ν may be regarded as constant. It is not possible to remove entirely the above restrictions for the validity of Einstein's relations. It is not true, for example, that the rate of stimulated emission in a resonant cavity is proportional to u_ν at any point in that cavity.

In a favorable situation it is possible to determine A_{21} experimentally by measurements of the intensities of spectral lines and by measurements of the rate of decay of fluorescence. The B coefficients are accessible by means of absorption measurements.

The calculation of these coefficients from basic principles is among the difficult problems of quantum mechanics. The B coefficients relate to the rate of change of state of an atom subjected to an external electromagnetic field. They reflect the perturbation of the atom by an external action which varies with time; therefore their calculation is the object of the time-dependent perturbation theory. This theory is developed in essentially all textbooks on quantum theory [10, 11, 12]; therefore only a summary of the method and the pertinent results are given here.

The atom is characterized by Schrödinger's equation. In the unperturbed condition the atom is in one of the stationary states described by the wavefunction

$$\Psi = \psi_n(\mathbf{r})e^{-i\omega_n t}. \tag{5.9}$$

Associated with this wave function is the energy $E_n = \hbar\omega_n$.* It is an eigenvalue of the time-independent Schrödinger equation. The integer n is representative of all quantum numbers required to describe the state, and the variable $\mathbf{r}$ is representative of all position vectors required to describe the particles of the system.

Perturbation is the result of an additional term H' in the Hamiltonian of the system. It is assumed to be 0 up to the time $t = 0$. Therefore, until $t = 0$, the system has the wave function of (5.9); its energy is E_n. The consequence of the perturbation is that for $t > 0$, the wave function of the system is no longer given by (5.9). It may be written in the form

$$\Psi = \sum_m c_m(t)\psi_m(\mathbf{r})e^{-i\omega_m t}. \tag{5.10}$$

Perturbation theory enables us to calculate the coefficients $c_m(t)$ in terms of the perturbing Hamiltonian and the wave functions ψ_m. The essential intermediaries in the calculation are the matrix elements

$$H_{nm}(t) = \int \psi_n e^{i\omega_n t} H' \psi_m e^{-i\omega_m t} \, dv, \tag{5.11}$$

where $H'(\mathbf{r},t)$ is the operator associated with the perturbation, and the integration is extended over the entire configuration space. It is shown in quantum mechanics that the following relation holds:

$$\frac{dc_m}{dt} = -\frac{i}{\hbar} H_{nm}(t), \quad m = 1,2, \ldots . \tag{5.12}$$

* It is convenient in this section to use the angular frequency $\omega = 2\pi\nu$ and the modified Planck's constant "h-bar": $\hbar = h/2\pi$.

It is one of the fundamental principles in quantum theory that the quantities $|c_1(t)|^2$, $|c_2(t)|^2$, etc., give the probability that an energy measurement on the system (5.10) yields the values E_1, E_2, etc. Thus the rate of change of $|c_m(t)|^2$ gives the rate at which the atom is changing from the original state n to the final state m.

To calculate the transition rates, we must carry out the integration in (5.11) and then use the calculated matrix elements to integrate (5.12). Since $c_n(0) = 1$, and $c_m(0) = 0$ for $m \neq n$, we have

$$c_m(t) = -\frac{i}{\hbar} \int_0^t H_{nm}(t') \, dt', \qquad (5.13)$$

for $m \neq n$.

The matrix element depends on the perturbing field and on the wave functions of the initial and the final states. The simplest field is that of a plane-polarized, monochromatic plane wave whose vector potential is of the form

$$A = A_0 \mathbf{u} \cos (\omega t - \mathbf{k}.\mathbf{r}). \qquad (5.14)$$

The unit vector $\mathbf{u}$ specifies the direction of the electric polarization. The propagation vector $\mathbf{k}$ is perpendicular to $\mathbf{u}$; its magnitude is $2\pi/\lambda$.

We assume that we are dealing with an atomic system whose dimensions are much smaller than the wavelength of the incident radiation. Let us place the origin at the center of the atom. The quantity $\mathbf{k}.\mathbf{r}$ will be smaller than one, when $\mathbf{r}$ is confined to the region accessible to the electrons. Then the evaluation of the matrix elements and the subsequent calculation of the functions $|c_m(t)|^2$ may be accomplished by expansion according to powers of $\mathbf{k}.\mathbf{r}$, and the following results are obtained [10, 11, 12]:

1. A significant perturbation of the atom takes place only when ω, the angular frequency of the incident wave, is nearly equal to $\omega_m - \omega_n$, or to $\omega_n - \omega_m$. In the neighborhood of $\omega_{nm} = \omega_m - \omega_n$ the time-dependent factor of $|c_n(t)|^2$ is of the form

$$\left[\frac{\sin \frac{1}{2}(\omega - \omega_{nm})t}{\frac{1}{2}(\omega - \omega_{nm})} \right]^2. \qquad (5.15)$$

A similar expression describes the time-dependent factor for $\omega \approx -\omega_{nm} = \omega_{mn}$. These expressions show the resonant character of the emission and absorption, which corresponds to the behavior of a classical oscillator in a driving harmonic field.

2. In the first order of approximation (in terms of powers of $\mathbf{k}.\mathbf{r}$) the time-independent factor in the integral (5.11) represents the interaction of the electric field derived from (5.14) with the electric dipole

moment of the atom. Higher order approximations account for the interaction of the magnetic field with the magnetic dipole moment of the
atom and the interaction of the electric field with the quadrupole
moment.

First-order calculations result in

$$H_{nm} = \mathbf{E}.\mathbf{u}(n,m), \qquad (5.16)$$

where the electric dipole matrix $\mathbf{u}$ is the quantum mechanical analog of
the classical electric dipole moment

$$\mathbf{u}_c = \Sigma e_i \mathbf{r}_i \qquad (5.17)$$

summed over all charged particles of the atom. The components of $\mathbf{u}$ are
given by

$$\begin{aligned}
\mathbf{u}_x(n,m) &= e\int \psi_n^* x \psi_m \, dv, \\
\mathbf{u}_y(n,m) &= e\int \psi_n^* y \psi_m \, dv, \\
\mathbf{u}_z(n,m) &= e\int \psi_n^* z \psi_m \, dv.
\end{aligned} \qquad (5.18)$$

Ultimately, in this order of approximation the transition rate is proportional to

$$|\mathbf{E}|^2 |\mathbf{u}(n,m)|^2 \cos^2 \theta,$$

where θ is the angle included between $\mathbf{E}$ and $\mathbf{u}$. It must be averaged
when the atoms are free to orient themselves with respect to the field.
Thus we conclude that the rate of stimulated emission of radiation is
proportional (in first approximation) to the square of the electric field
of the incident radiation. We note, without proof, that the emitted radiation is coherent (in phase) with the stimulating radiation.

In the simple situations that prevail in a single plane wave the square
of the electric field is proportional to the radiation density u. It is then
permissible to say that the rate of stimulated emission is proportional
to u. In a blackbody cavity the radiation field is chaotic, and the rate
at which stimulated emission is obtained is again proportional to the
radiation density.

The derivation of Einstein's formula from perturbation theory
is completed by passing from a monochromatic field to a radiation field
with a broad spectrum, a step which requires integration of (5.15) in
the frequency domain and averaging over all spatial directions. The
result is

$$B_{21} = \frac{8\pi^3}{3h^2} |\mu(2,1)|^2. \qquad (5.19)$$

The rate of spontaneous emission may then be calculated from (5.8). It is

$$A_{21} = \frac{64\pi^4 \nu^3}{3hc^3} |\mu(2,1)|^2. \tag{5.20}$$

Although the calculation of the dipole matrix elements seems formidable, in most instances it is easy to tell whether they are 0.

As a consequence of the symmetry properties of the integrals (5.18), most combinations of states lead to a zero matrix element. Such transitions are called *forbidden transitions*. The term "forbidden" means that a transition among the states concerned does not take place as a result of the interaction of the electric dipole moment of the atom with the radiation field. The allowed transitions are specified by the *selection rules*, which help to sort out the pairs of states capable of yielding a nonzero matrix element for electric dipole radiation.

Atomic energy levels are generally so labeled and organized in tables that it is relatively easy to pick out the transitions permitted by the selection rules.

Selection rules will be discussed in greater detail in Section 1.5, after the classification and nomenclature of atomic energy levels has been introduced. It should be noted now that transitions forbidden by the selection rules may occur by means of mechanisms other than electric dipole radiation. They occur at a much slower rate than the permitted transitions.

As a result of the selection rules, an atom may get into an excited state from which it will have difficulty returning to the ground state. A state from which all transitions to lower energy states are forbidden is *metastable;* an atom entering such a state will generally remain in that state much longer than it would in an ordinary excited state, from which an easy escape is possible. A metastable state has a long lifetime; therefore, when atoms are stirred into transitions by collisions or radiation they tend to accumulate in these metastable states.

1.6 SPECTROSCOPIC NOMENCLATURE AND UNITS

Since the emission and absorption of light is accompanied by the rearrangement of an atomic system, the spectral lines are described as transitions between specific stationary states. The stationary states, in turn, are characterized by the wave function ψ of the atomic system.

As a simplification, which is permissible in most cases, one may consider the electrons in an atom individually and consider the emission of a spectral line as the result of the rearrangement of the configuration of a single electron with respect to the rest of the atom.

The electrons in an atom are characterized by three *orbital quantum numbers* n, l, and m, and the *spin quantum number* s. The orbital quantum numbers are integers; s is $\pm\frac{1}{2}$. The first, n, governs the radial distribution of the wave function; it corresponds to the principal quantum number of Bohr's theory. In the case of atoms with relatively simple structure, the value of n is the primary determinant of the energy of the electron in question. The quantum number l varies from 0 to $n-1$; it determines the orbital angular momentum, whose largest component is $l\hbar$. For brevity we shall say that the orbital angular momentum is $l\hbar$ with apologies to the expert in quantum mechanics who knows that the *magnitude* of this vector is $\sqrt{l(l+1)}\,\hbar$. The quantum number m describes the orientation of the angular momentum vector with respect to an external field; it may assume the values $-l$ to $+l$; that is, for a fixed value of l a total of $2l+1$ values of m are possible. With different quantum numbers there are associated different wave functions, that is, different states of the electron. When several states have the same energy, the level is said to be degenerate, as we have already noted. The degeneracy of a level may be removed by the application of an external field, by the field of the other electrons of the same atom, or by that of neighboring atoms. This removal of degeneracy comes about most often by the appearance of a small energy difference associated with the reversal of the spin of an electron.

When an atom contains many electrons, the electrons that form a closed shell may be disregarded, and the energy differences associated with transitions in the atom may be calculated by considering only the electrons outside of the closed shell. Thus in the case of the alkali metals only one electron, and in the case of the alkali earths only two, need to be considered.

An electron is called an s, p, d, or f electron if its azimuthal quantum number is $l = 0$, 1, 2, or 3, respectively. For larger values of l the letters of the alphabet are used in their natural order, starting with f for $l = 3$. The notation $3p$ indicates an electron with $n = 3$, $l = 1$.

In describing the state of a multielectron atom it is well to remember that the electrons within the atom are interchangeable, and that no two electrons can have the same quantum numbers (including s). The ground state of Li, for example, is described by the symbol $1s^2 2s$, which means that there are two electrons (with opposing spins) in the $1s$ state, and one in the $2s$ state. The sum of the orbital angular momenta of these electrons is 0, and the sum of the spin angular momenta is $\frac{1}{2}$, giving the total angular momentum $\frac{1}{2}$. All these momenta are in the units of $\hbar = h/2\pi$.

The total angular momenta of most atoms and ions of interest to us may be obtained by first adding vectorially the orbital angular momenta of the individual electrons and combining the spin angular momenta separately. The orbital angular momenta combine into a vector $\mathbf{L}$, whose magnitude L is an integer; the spins combine into a vector characterized by $\mathbf{S}$, whose magnitude is an integer for an even number of electrons and a half integer for an odd number. The total angular momentum $\mathbf{J}$ of the atom may then be obtained by vector addition of $\mathbf{L}$ and $\mathbf{S}$.* When the vectors $\mathbf{L}$ and $\mathbf{S}$ are constants of motion (invariants) we speak of $L - S$ *coupling.* Alternatively, it is called *Russell-Saunders coupling.*

It is shown in quantum theory that for each fixed value of S there are $2S + 1$ different possible spin configurations. Configurations with $S = 0$ are called singlets, those with $S = \frac{1}{2}$ are doublets, etc. For an atom or ion with two electrons, S is either 0 or 1; therefore, such an atom will have singlet and triplet states. The collection of states with common values of J, L, and S is called a *term.* In general, a term will contain a number of states differing in the orientation of their momentum vectors. The multiplicity of a term is $2J + 1$.

The following nomenclature evolved historically for the designation of atomic terms in the Russell-Saunders momentum-coupling scheme. The symbol characterizing the term is of the form $^{2S+1}X_J$, where the letter X stands for S, P, D, F, G, etc., depending on the value of the orbital angular momentum L. The letter S is used as a symbol for the terms with $L = 0$ and is not to be confused with the spin quantum number. The letters P, D, F, etc., are used as symbols to designate terms with $L = 1, 2, 3$, etc. The superscript preceeding the basic symbol is determined by the spin quantum number; it indicates whether the state is a singlet, doublet, or triplet, etc. Different elements of the multiplet are distinguished by the value of J, the total angular momentum; this is added as a subscript to the right. When necessary, the configuration of the excited electron is also given; it precedes the letter symbol. Thus the ground state of Li has the symbol $2s^2S_{1/2}$; the symbols of some of its excited states are $3s^2S_{1/2}$, $2p^2P_{1/2}$, $2p^2P_{3/2}$. The last two states differ by the opposite orientation of the spin of the excited $2p$ electron with respect to the orbital momentum. The total angular momentum J varies in integral steps from $|L - S|$ to $L + S$. Hence, when $L \neq 0$, then $S = 1$ leads to three different values of J, namely $L - 1$, L, and $L + 1$, but when $L = 0$ then

---

* We are using the symbols $\mathbf{L}$, $\mathbf{S}$, and $\mathbf{J}$ to denote the angular momentum vectors whose maximal components are $L\hbar$, $S\hbar$, and $J\hbar$.

$J = 1$ is the only value possible. Nevertheless, such a state is still called a triplet state, since there are three independent wave functions that belong to the common energy of a $^3S_1$ state.

Note that the notation based on the Russell-Saunders coupling scheme is applicable only when the interaction of the orbital angular momenta $\mathbf{l}_i$ of the individual electrons is so strong from one electron to another that these momenta combine to give a resultant $\mathbf{L} = \Sigma \mathbf{l}_i$, which is a constant of the motion. Similarly, the individual spins $\mathbf{s}_i$ must so combine to form the constant $\mathbf{S} = \Sigma \mathbf{s}_i$. The general laws of dynamics assure the constancy of $\mathbf{J} = \mathbf{L} + \mathbf{S}$ only; the constancy of $\mathbf{L}$ and $\mathbf{S}$ separately follows only when the spin-orbit interaction is small compared to the spin—spin and orbit—orbit interactions. This type of situation prevails for a large number of elements; its presence is recognized spectroscopically by noting that the splitting of the levels of a multiplet is small compared to the energy differences of levels having the same electron configuration but different values of L.

The reverse of the case described above is one in which there is a considerable interaction between the orbital angular momentum $\mathbf{l}_i$ and the spin angular momentum $\mathbf{s}_i$ *of the same electron.* Then each $\mathbf{l}_i$ combines with the corresponding $\mathbf{s}_i$ to form a $\mathbf{j}_i$, the total angular momentum of an individual electron. The $\mathbf{j}_i$'s are loosely coupled to each other, that is, they are approximate constants of motion. (Naturally, $\mathbf{J} = \mathbf{j}_1 + \mathbf{j}_2 + \ldots + \mathbf{j}_n$ is always a constant.) When this is the case, we speak of *j-j coupling.* Its presence is recognized by the observation of the Zeeman splitting of spectral lines, which follows laws different from those applicable in the Russell-Saunders case.

Pure *j-j* coupling occurs very seldom. Instead, we find atoms like neon, in which electrons in inner shells obey the rules of Russell-Saunders coupling, while the total angular momentum $\mathbf{j}$ of an external electron with a higher quantum number n may be weakly coupled to the resultant $\mathbf{J}_c$ of the core electrons.

So far the discussion of atomic and ionic energy levels has been based on an atomic system free of external influence. When an ion is located in a crystal lattice, the electric and magnetic fields prevailing at the site of the ion may exert a profound influence on the energy level structure of the ion. This is the case for a chromium ion in ruby, which is of great importance for laser technology. In ruby the Cr^{3+} ion is surrounded by an approximately octahedral field of oxygen ions. The crystal field, that is, the electrostatic field resulting from the presence of the O^{2-} ions, splits the originally degenerate levels of Cr^{3+}. The calculation of the splitting of ionic levels in a crystal field is based on the theory of group representations. It involves the irreducible representations of the symmetry group applicable to the crystal in question. The number of components into which a multiple level may be split is determined by group theory; the split levels are designated by the appropriate symbol of group theory with the spectroscopic symbol of the original level suppressed or abbreviated. The actual shifts in energy levels are determined by the crystal field parameters. Their calculation is among the most complicated tasks of quantum theory.

A useful concept in spectroscopy is the *parity* of a state. It is defined as even or odd, depending on the parity of the number Σl_i, where the summation over the (scalar) quantum numbers is extended over all electrons of the atom.

The selection rules for electric dipole radiation mentioned in Section 1.5 may now be stated precisely.

The general selection rules are the following:

1. Transitions must change the parity.
2. $\Delta J = 0$ or ± 1, but transition from $J = 0$ to $J = 0$ is excluded.

In atoms for which Russell-Saunders coupling is applicable the following additional selection rules hold:

3. $\Delta L = 0$ or ± 1.
4. $\Delta S = 0$.

When the state of only one electron changes, the parity rule requires a change in L. In this case the third rule becomes $\Delta L = \pm 1$. Transitions in which the states of two electrons change at the same time are considerably less probable than transitions involving a single electron.

In order to facilitate the application of the selection rules, terms of *odd parity* are usually provided with an upper index o. For the same reason it is practical to list or plot the energy levels of singlets, triplets, etc., in separate groups. Such groupings are used in the energy level diagrams of elements in Chapter 9.

The selection rules were derived by considering electric dipole radiation only. Other radiative mechanisms are less effective. In general, they lead to less frequent transitions than electric dipole radiation. When the electric dipole transition is forbidden, however, secondary mechanisms lead to transition rates that are generally several orders of magnitude slower than the rates of permitted dipole transitions. Thus transitions forbidden by the selection rules will occur, but they will occur relatively rarely. The selection rules are most rigidly in force for elements at the beginning of the periodic table. They lose their effectiveness in complex atoms and in a strong interaction of an atom with another, as it takes place in a collision or in a crystal lattice.

It is, in principle, most desirable to adopt a single system of units and to use it consistently and exclusively. This procedure is cumbersome to follow in all details in the laser field, which encompasses atomic quantities as well as those of power engineering. Although we use rationalized mks units for input and output power calculations, on the laboratory and atomic scale we shall give preference to the cgs system, thus insuring compatibility with our sources of reference. Some deviations from the cgs system are traditional in spectroscopy. The wavelength of visible radiation is customarily expressed in ångström units ($1\text{Å} = 10^{-10}$m), whereas in the infrared region the micrometer (1μm $= 10^{-6}$m) is the preferred unit. No one measures atomic energy levels in ergs or joules. They are expressed and tabulated either in electron volts (1.602×10^{-12} erg) or in reciprocal centimeters. The use of the reciprocal centimeter as a unit of energy originates from Bohr's frequency relationship, which may be written in the form:

$$\frac{1}{\lambda} = \frac{E_2 - E_1}{hc}. \tag{6.1}$$

The quantity E/hc has the dimension of reciprocal length. It is frequently referred to as the energy, although correctly it should be called the *wave number*. The tabulation of energy levels in reciprocal centimeters enables us to obtain by direct subtraction of two tabulated entries the reciprocal of the wavelength (*in vacuo*) corresponding to a transition between the levels. One electron volt is equivalent to 8066 cm^{-1} (for conversion see Appendix B).

Wavelengths in the visible and the near-infrared region are measured in air, and usually the value in air is quoted. For this reason, the reciprocals of the wave numbers obtained by subtracting tabulated entries must be corrected to secure agreement with the measured values (see Appendix C). The correction is determined by the deviation of η, the refractive index of air, from 1. Its magnitude can be gauged by the fact that $\eta - 1$ varies from 277×10^{-6} to 274×10^{-6}, as the wavelength varies from 0.6 to 1.0 μm.

REFERENCES

1. *American Institute of Physics Handbook,* McGraw-Hill, New York, 1963.
2. M. Born and E. Wolf, *Principles of Optics,* Pergamon, New York, 1959.
3. M. J. Beran and G. B. Parrent, *Theory of Partial Coherence,* Prentice Hall, Englewood Cliffs, N. J., 1964.
4. G. Herzberg, *Atomic Spectra and Atomic Structure,* Dover, New York, 1944.
5. A. Beiser, *Concepts of Modern Physics,* McGraw-Hill, New York, 1967.

6. A. C. G. Mitchell and M. W. Zemansky, *Resonance Radiation and Excited Atoms,* Cambridge University Press, England, 1934 (1961).
7. G. Birnbaum, *Optical Masers,* Academic Press, New York, 1964.
8. G. H. Dieke, Spectroscopic observations on maser materials, *Advances in Quantum Electronics,* J. R. Singer, Ed., Columbia University Press, New York, 1961, pp. 164–186.
9. A. Einstein, Zur Quantentheorie der Strahlung, *Phys. Z.,* **18,** 121–128 (1917).
10. E. C. Kemble, *Fundamental Principles of Quantum Mechanics,* McGraw-Hill, New York, 1937 (esp. Section 54).
11. A. A. Vuylsteke, *Elements of Maser Theory,* Van Nostrand, Princeton, N. J., 1960, esp. Chapter 4.
12. R. L. White, *Basic Quantum Mechanics,* McGraw-Hill, New York, 1966. esp. Chapter 11.

Chapter 2   Survey of
masers and lasers

2.1   THE DEVELOPMENT OF MASERS

Masers and lasers are devices that amplify or generate radiation by means of the stimulated emission process. Their names originate from the initials of Microwave Amplification by Stimulated Emission of Radiation, which was the objective achieved by the older device, the maser. The word "laser" was coined when the technology shifted from microwaves to light.

The common element of these devices is a medium made to amplify in a narrow frequency region by population inversion achieved for a pair of energy levels. The main problem to be solved in the construction of these devices is the creation and maintainance of population inversion. All naturally occurring processes tend to move a system toward thermal equilibrium, and therefore they tend to destroy population inversion and counteract the processes contrived by man to create it.

Speculation about the use of stimulated emission for amplification began around 1950. There are controversies concerning the priorities of early proposals, but the first successful scheme was conceived, and the first maser built, by C. H. Townes and his students [1] at Columbia University in 1954. It utilized the inversion of population between two molecular levels of ammonia to amplify radiation of frequency 23,870 MHz, which has a wavelength of about 1.25 cm. The population inversion was accomplished by separating, in an inhomogeneous electrical field, the ammonia molecules which happened to be in the upper state from those in the lower state. The separated "good" molecules were led into a resonant cavity where they became available for amplification of a signal of the proper frequency. The main purpose of this device was to serve as a low-noise amplifier, but with the addition of feedback, it was also used as a high-precision signal generator.

Soon after the appearance of the ammonia maser, methods other than particle selection were introduced for the creation of population inversion. These methods are called excitation (or pumping) because they make the material amplifying by raising a substantial fraction of its atoms or molecules to an energy level above the ground level. Excitation is accomplished by transferring energy from an outside source to enhance the population of certain favored energy levels.

The most direct method for such selective excitation is called *optical pumping*. It consists of irradiation with a signal of higher frequency than the one to be amplified. The process utilizes at least three levels of the active material. The ground level (1), from which the exciting radiation raises the atoms to the top level (3), and an intermediate level (2), which is not directly populated because the exciting radiation does not contain a spectral component capable of raising atoms directly from the ground level to level 2. Then a population inversion may result between levels 2 and 3. Such an amplifier is called a three-level maser or laser. It has many variants. Other intermediate levels may be involved, and spontaneous transitions usually take an important part in the operation of the excitation—de-excitation cycle. Three- and four-level schemes of amplification will be discussed in detail in connection with solid lasers. We note here that the optical excitation scheme for masers was proposed in 1955, independently and almost simultaneously, by Basov and Prokhorov in the USSR and by Bloembergen in the United States. By 1957 several solid three-level masers were built utilizing paramagnetic ions embedded in host crystals.

Multilevel solid masers are excited by means of a high-frequency microwave generator. They provide amplification at a lower frequency. One of the popular materials for such masers is ruby. The chromium ions in the ruby constitute the active element. The separation of the relevant energy levels of chromium, which provide amplification at around 9300 MHz, are affected by an external magnetic field. Therefore it is possible to tune these amplifiers by varying this field.

Particle sorting and three-level pumping are not the only means of establishing population inversion in masers. During the late fifties attention was given to a number of pulsed or intermittent methods of excitation involving the application of incident radiation for a short period, after which the system was available for amplification. These schemes never attained much practical significance.

Many masers were constructed for applications in radio astronomy and as components of radar receivers. These were mostly of the ruby-type and served as preamplifiers of systems designed to receive very weak signals.

Maser theory and technology is the subject of several textbooks and review articles. Among the elementary books, we suggest those of Singer [2] and Troup [3]. Advanced topics of this subject are discussed in the books of Siegman [4] and Yariv [5] and in the review article of Weber [6]. The author's view on the controversial history of the invention of masers and lasers is expounded in a historical review article [7], which contains numerous references to early original publications on this subject.

2.2 THE DEVELOPMENT OF LASERS

While the maser technology was being established, the desirability of extending stimulated emission techniques to the infrared– and optical regions was recognized by many active in the maser field. The challenge and the difficulties that lay ahead in 1958 were surveyed in the classical article of Schawlow and Townes [8], which marks the beginning of an era of highly competitive search for laser materials and excitation processes. They wrote:

Maser techniques give the attractive promise of coherent amplification at the high (optical) frequencies and of generation of very monochromatic radiation.

As one attempts to extend maser operation toward the very short wavelengths, a number of new aspects and problems arise, which require a quantitative reorientation of theoretical discussions and considerable modification of experimental techniques used.

The new problems and the new aspects of old problems were mainly the following:

1. In the optical region the confinement of the working material has to be accomplished in a cavity permitting a large number of electromagnetic oscillations (modes) in the frequency range of the spectral line. In contrast, the maser cavity could be constructed so that only one oscillatory mode was permitted.

2. Spontaneous emission of radiation in the optical region is greatly enhanced over stimulated emission because of the presence of the factor ν^3 in (3.4) of Chapter 1. Since the change in frequencies from the ammonia maser to a laser in the visible region represents a change by a factor of over 10^4, the factor of enhancement of the ratio of spontaneous to stimulated radiation is greater than 10^{12}. As a consequence, in the optical region stimulated emission is submerged in the incoherent spontaneous radiation until a high level of radiation is reached. Another

serious disadvantage of the high spontaneous emission rate is that it requires a rapid rate of supply of excitation in order to maintain an adequate population at the starting level of the laser.

3. In the microwave region optical pumping can be accomplished by using radiation from tunable, monochromatic signal generators. In the visible and the infrared region such sources are lacking. We already noted in Section 1.1 that powerful, monochromatic sources of light did not exist prior to the advent of lasers.

4. Energy differences required for the generation of visible radiation are large compared to kT for convenient laboratory temperatures, while the energy differences utilized in masers are small. In one sense, this factor is helpful, because it makes refrigeration less critical for lasers. On the other hand, the large energy differences make the Boltzmann factors more unfavorable, and the rates at which energy is fed into lasers is much higher than it is in masers.

Some explanation is in order concerning the role of the cavity modes. As was immediately recognized by Schawlow and Townes, it is necessary for the success of the laser that most of the available optical energy be channeled into a very few modes. They wrote:

If many modes, rather than a single one, are present in the cavity, a rather large background of noise can occur. . . . If many nearby modes are present, a very small change in cavity dimensions or other characteristics may produce a shift of the oscillations from one mode to another, with a concomitant variation in frequency.

Since the number of modes per unit frequency interval could not be substantially reduced for a large cavity, Schawlow and Townes concluded that it is desirable to make all but a few of the modes of a multimode cavity lossy in order to suppress oscillations in all these modes. This conclusion led to the employment of the Fabry-Perot interferometer, a semiopen structure with two mirrors facing each other at the end. It is now the standard configuration of lasers.

Schawlow and Townes did not complete the invention of lasers because they could not find a material and the means of exciting it to the required degree of population inversion. This was accomplished in 1960 by Maiman at Hughes Research Laboratories. The discovery, or invention, was the outgrowth of a long study on the fluorescent properties of ruby. Maiman irradiated pink ruby crystals with light from a xenon flashlamp, measured the characteristics of the red fluorescent radiation emitted by the ruby, and studied the distribution of the chromium ions among its energy levels as a function of irradiation. With increasing intensity of irradiation, he noted significant decrease in the population of the ground state and concluded that it is possible to reach a level of irradiation at which population inversion takes place with the ruby becoming amplifying at 6943 Å.

The final experiments were conducted with small ruby cubes and cylinders about 1 cm in diameter whose endfaces were parallel and coated with a reflective layer, leaving a small hole for the exit of radiation. When the irradiation through the sides of the ruby was increased beyond a threshold level, Maiman [9] observed that

(a) the lifetime of the fluorescense was shortened,
(b) the linewidth of the emitted radiation was drastically reduced, and
(c) the emitted radiation became extremely intense and highly directional.

From these facts, Maiman concluded that the emitted intense, red radiation was produced mostly by stimulated emission. Subsequent measurements of Maiman [10] and a group of investigators at Bell Telephone Laboratories [11] confirmed this conclusion. They revealed the following additional properties of the radiation emitted from the ruby excited above a threshold level:

1. The radiation is emitted *coherently* over a significant area of the ruby surface.
2. The intensity of the radiation is subject to very rapid oscillations

Maiman's discovery surprised the scientific community engaged in maser research because the extension of the maser art into the optical region had been discussed at length and the opinion had been formed that ruby was a poor prospect for a maser material. It had been expected that the active element of a solid laser would be a "four-level" material, that is, an atom or ion in which the terminal level of the laser transition was a level above the ground level, not the ground level itself, as in ruby. It was also thought that gases excited optically, or by the passage of an electric discharge, are even more suitable laser materials than solids. Semiconductors had also been recognized early as potential laser materials, and a variety of schemes had been proposed for their excitation.

After the discovery of the first laser the discoveries of expected laser types followed in rapid succession. Four-level solid lasers, utilizing ions of uranium and rare-earth metals, made their first appearance before the end of 1960. Their number multiplied rapidly during the following two years. Among these lasers the neodymium laser achieved practical significance because it could be operated at room temperature, and it could deliver power at a rate comparable with that of the ruby. The construction of these solid lasers is similar to that of the ruby laser. They are optically excited by radiation from flashlamps. Such excitation is made possible by the presence of relatively broad absorption bands in the solids. A combination of these broad absorption bands with lower, sharp energy levels is essential for all solid lasers.

Before the end of 1960 Javan and his associates [12] at Bell Telephone Laboratories announced the successful operation of a helium—neon laser. This laser obtained its excitation from an electric discharge through the gas. The laser light was obtained from neon atoms, whose population inversion was maintained in a steady state by means of an energy-exchange cycle in which the helium atoms participated. The helium—neon laser operated continuously and steadily, emitting an even more monochromatic radiation than the solid lasers. Although the original operation of this laser was confined to about five lines in the infrared region, around 1.1 μm, within two more years many other lines were obtained from the same laser, including the popular 6328 Å line. Many other gas lasers were discovered during the years 1963 to 1967. Most of these operate at low power level, but gas lasers with high power output have also been constructed using ionic and molecular transitions.

Semiconductor lasers were first developed in 1962. In their most common form, the injection laser, coherent light is emitted from the junction; that is, the thin transition region separating the p-type and the n-type semiconductor. The excitation of this type of laser is accomplished by the application of an electrical field which injects current carriers into the junction. A large fraction of the electrical energy expended in the semiconductor crystal is thus directly converted into stimulated radiation. The lasers so constructed operate with the highest efficiency, but their active regions are extremely thin, usually of the order of a few microns. Other methods are also available for the excitation of semiconductors. The interesting history of the discovery of semiconductor lasers is reviewed briefly at the end of Chapter 7.

Liquid and chemical lasers soon followed. The first of these lasers was developed early in 1963 but did not attain the same practical importance as the other types.

The search for lasers began as an extension of the maser field. Originally, lasers were called *optical masers,* and many of the early books and papers carry that term in their title. The term "laser" was used only colloquially at first. Eventually it displaced the more cumbersome optical maser.

2.3 COMMON SOLID LASERS*

The ruby laser is not only the first but also the most important of the solid laser types. Its working element is the pink ruby crystal, excited by irradiation from a flashlamp and provided on opposite faces with reflecting surfaces. The original configuration of such a laser is shown in Fig. 2.1. A ruby cylinder of about 1 cm in diameter and 2 to 10 cm long is surrounded by the coils of a flashlamp. The end faces of the crystal are ground and polished parallel to each other and are provided with a reflective coating which permits only a small fraction of the incident light to pass through.

When the flashlamp is triggered, it emits a green and blue flash of short duration. The chromium ions in the ruby absorb this light through their broad absorption bands, thus raising many ions from their ground state into several broad energy levels above ground. From these levels the ions spontaneously change to a lower sharp level, where they may accumulate to such an extent that this level may become more densely occupied than the ground level. When this occurs, the ruby becomes amplifying at around 6943 Å.

A greatly simplified energy level structure of a three-level laser material, such as ruby, is shown in Fig. 2.2. The ground state is denoted by index 1. Excitation is supplied to the solid by radiation of frequencies which produce absorption into the broad band 3. Most of the absorbed energy is transferred by fast, radiationless transitions into the intermediate sharp level 2. The energy difference is given up to the crystal lattice

---

* The scientifically correct title of this section ought to be Lasers Based on Ions in Solid Materials.

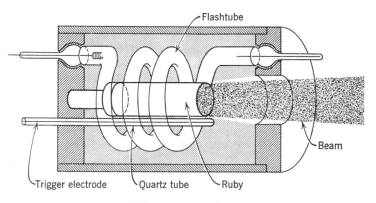

FIG. 2.1 Ruby laser.

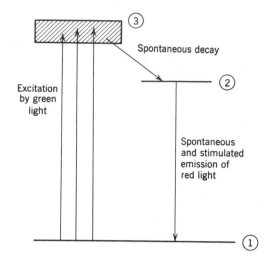

FIG. 2.2 Simplified energy-level diagram of chromium ions in ruby.

as heat. The emission of radiation, associated with the spontaneous return from level 2 to ground level, is ordinary fluorescence. Such fluorescence will take place even at a low level of excitation. When the exciting radiation is sufficiently intense, it is possible to obtain more atoms at level 2 than are left at ground level. The spontaneously emitted photons traveling through the crystal will stimulate additional radiation, and emission thus induced is superposed on the spontaneous emission. Stimulated emission will also take place when the population of the ground level is larger than that of level 2, but absorption will exceed stimulated emission and the net result will be a loss in the number of photons.

It is easier to establish population inversion in a material that has one additional level participating in the fluorescence cycle. The cycle in such a four-level material is shown in Fig. 2.3. Excitation and radiation transfer proceed as in the case of a three-level material, but there is now an additional normally unoccupied level above the ground level, and the transitions which produce the laser light terminate on this level instead on the ground level. Therefore laser action may begin as soon as there is a significant occupancy of the initial level, which in Fig. 2.3 is denoted by index 3. A pulsed four-level laser need not operate with the energy waste characteristic of ruby, from which no laser output is obtained for the energy invested in removing, by excitation, one-half of the atoms from the ground level.

Most ordinary solid lasers are of the four-level type, the ruby, having only three levels, being rather exceptional. These solid lasers consist of active atoms or ions of a transition metal, rare-earth metal, or actinide incorporated in a hard, ionic crystal or glass. In the case of ruby the host crystal is sapphire; the favorite host crystals for other elements are various garnets, usually yttrium aluminum garnet, tungstates, mostly $CaWO_4$, and fluorides. Special glasses are frequently used in technical lasers, especially with Nd^{3+} as the active element. Next to ruby neodymium is the most commonly used solid laser material. It is capable of producing powerful infrared radiation in several wavelength regions, 1.06 μm being the most frequently used. Almost all rare-earth elements may be used in such solid lasers. They provide coherent radiation at a number of wavelengths between 0.6 and 2.6 μm.

Lasers of the ordinary solid type have certain properties in common that are not shared by semiconductor lasers. The bulk of their material is the host which does not directly participate in the laser cycle. The active material is present in generally low concentration, 1% or less. The frequency of the emitted radiation is characteristic of the individual ion modified by its environment. Consequently, the spectral content of the laser output is well defined by the materials and the temperature. Lasers of this type are usually made in the form of rods, not too different in size from the ruby rods already described. Their excitation is accomplished by optical pumping, that is, by irradiation from another light source.

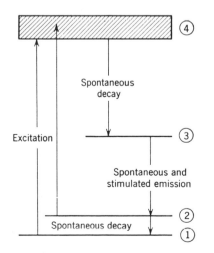

FIG. 2.3 Simplified energy-level diagram of a four-level laser.

Laser action is usually understood to imply the generation of light by means of a laser. It is possible only if the material can be excited to provide amplification for some wavelength region and if, in addition, a minimum feedback is established in the region of space that contains the amplifying material. This feedback is accomplished by the partially reflective mirrors. The laser will operate as a light generator when amplification on passage of light back and forth is more than sufficient to compensate for the losses at the end faces and for the losses that might occur as a result of scattering on crystal imperfections. The mathematical form of this statement is known as the *threshold condition*. It relates the minimum gain, that is, amplification per unit length in the laser material to the design characteristics of the laser. The exposition of this condition is one of the first analytical tasks to be undertaken in Chapter 3.

The top levels in Figs. 2.2 and 2.3 are shown as broad bands in contrast with the other levels. This breadth at the top level is a practical necessity because there is not enough energy available from ordinary sources of radiation in a narrow band. If a laser were to be used to excite a second laser, a material with a narrow top level would be acceptable. Under ordinary circumstances powerful flashlamps are pushed to the limit of their capabilities to provide sufficient excitation for materials, such as ruby, that are capable of utilizing incident radiation from 3800 to about 6100 Å.

Solid-state lasers generally operate intermittently. The reasons for this are primarily technical. First, it is difficult to provide a sufficiently powerful source of exciting light capable of continuous operation; second, a great deal of heat evolves within the laser which must be dissipated. Ordinary ruby lasers are excited for periods of a few milliseconds, the length of the period being determined by the duration of the exciting flash.

Construction of continuously operating solid lasers is largely an engineering problem. Quite a number of such lasers have been constructed. In order to attain continuous operation it is necessary to increase the efficiency of the excitation process and to improve the cooling of the active material.

Whether operated by pulses or in a continuous wave regime, the output of a solid laser undergoes rapid pulsations. The intensity fluctuations consist of many irregular spikes of about 1-μsec duration, but regular pulsations may be obtained under highly controlled circumstances. Figure 2.4 shows the trace of the output of a typical ruby laser along with a trace of the exciting radiation. Fluorescence begins immediately after irradiation starts, but stimulated emission starts, in this case, about

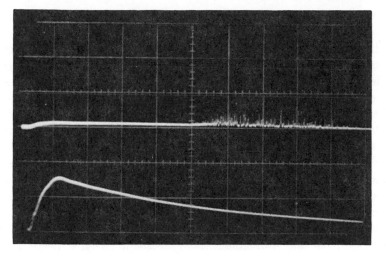

FIG. 2.4 Intensity of radiation from ruby versus time: light emitted from ruby, upper trace; light emitted from flashlamp, lower trace; time scale, 0.1 msec/cm.

0.5 msec later. The sharp spikes correspond to the rapid pulsations of the intensity of stimulated emission. They are shown on an expanded time scale in Fig. 2.5. The irregularity of these pulsations and their apparent lack of reproducibility indicate that they are caused or influenced by a number of factors. These pulsations, and methods for controlling them, are discussed in detail in Chapters 4 and 6.

A variation of temperature has a profound effect on several aspects of fluorescence. It affects the time of decay of fluorescence as well as the width and the position of the fluorescent line. All these variations are reflected in laser performance. In general, an increase in temperature shortens the fluorescent decay time and broadens the line. Both effects tend to increase the intensity of the incident pumping light required

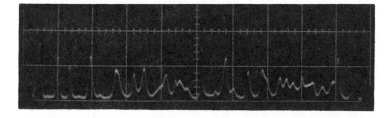

FIG. 2.5 Pulsations of ruby on an expanded time scale: time scale 5 μsec/cm.

to reach the threshold of oscillations. A shift of the peak fluorescent frequency causes a corresponding shift in the spectrum of the laser output. In addition, an increase of temperature may significantly increase the population in the terminal level of a four-level laser, thus providing another threshold-inhibiting factor. Whether or not this factor is relevant depends on the separation of the ground level from the terminal level of the laser. In any case, laser action becomes always more difficult to achieve at a higher temperature. In many solids it has been achieved only at liquid nitrogen temperature or lower.

Threshold energy is sometimes used as a figure of merit to describe a solid laser. This quantity is the least electrical input energy required to obtain laser action. It is not a characteristic of the material alone, and comparisons of threshold energies are physically meaningful only if applied to similar configurations employing similar pumping sources. It follows from the aforegoing that threshold energies decrease with a decrease of the temperature.

Solid lasers are most useful for the generation of a powerful pulse of radiative energy lasting for a millisecond or less and delivering an energy between 0.1 and 100 J. Although the radiation generated by solid lasers is more monochromatic than the radiation obtainable from ordinary intense light sources, its spectral width is much broader than that of the radiation obtainable from gas lasers. Modulation of the amplitude or frequency of the radiation generated by solid lasers is generally difficult, but solid-state lasers readily lend themselves to generation of extremely short pulses of the higest peak intensity by means of giant-pulse techniques which are discussed in Chapter 6. Peak intensities of many million watts may be generated with pulse lengths of the order of a microsecond.

2.4 GAS LASERS

Before the invention of the first successful laser it was the consensus among would-be inventors that the working material of a laser would probably be a gas excited by irradiation from a spectral lamp. With one minor exception, this did not turn out to be the case. An optically excited gas laser was indeed constructed in 1962, long after other, more practical lasers made their debut. Gould and his co-workers constructed a cesium-vapor laser, excited by radiation from a helium lamp. This laser owes its existence to a chance coincidence of spectral lines. It is a freak, or a monument to prodigious amounts of effort and federal money expended for an objective that became almost meaningless before it was accomplished.

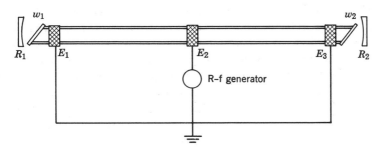

FIG. 2.6 Schematic diagram of a gas laser with radio-frequency excitation. R_1 and R_2 are curved mirrors coated on the side facing the discharge tube; E_1, E_2, and E_3 are electrodes on the outside of the discharge tube; w_1 and w_2 are flat glass windows oriented at Brewster's angle.

Practical gas lasers derive their excitation from a variety of complex processes that take place in an electric discharge, principally from excitation by electron impact and transfer of excitation between colliding atoms and molecules. The essential features of the typical gas laser are shown schematically in Fig. 2.6. The gas is confined in a glass or quartz tube typically 50 cm long and of 0.5-cm inside diameter. The discharge may be excited by radio-frequency current, using electrodes placed on the tube externally, or, alternatively, internal electrodes may be provided, in which case the discharge may be powered by direct, or low-frequency alternating current. Feedback is provided by external mirrors the orientation of which must be carefully aligned with the axis of the tube. One of these mirrors is made partially transparent to provide an exit port for the laser. The mirrors may be plane, but for reasons to be discussed later, it is advantageous to use spherical mirrors. The glass envelope is terminated by flat exit ports oriented at Brewster's angle to eliminate reflections at these ports. Naturally they are reflection-free for only one polarization, and the laser is designed to operate in that polarization. It is not necessary to provide a shield or reflector on the sides of the tubes for optical reasons, but shields are usually provided for the sake of safety.

The basic source of energy in an electric discharge is the acceleration of the electrons by the field between the electrodes. The excitation of an atomic or molecular species in a gas laser derives from collisions with electrons and in some instances with atoms of another species so constituted that a transfer of excitation energy may take place on collision. As the atoms, excited by one means or another, cascade down the energy scale, a certain stationary nonequilibrium situation will be established in which the number of atoms in each energy state remains constant. This situation requires that the rate at which the atoms arrive in a state, due to all causes, be equal to the rate at which they leave that state. The number of atoms in each state adjusts itself to establish a balance. Those states from which escape is slow will accumulate a large number of atoms. In particular, crowding will occur in the so-called metastable states, which are higher in energy than the ground state but from which radiative transitions to lower levels are forbidden by the selection rules of quantum mechanics.

Whether population inversion is achieved depends on the rate of excitation and on the decay rates of all levels involved in the cascading process. Several processes contribute to decay of a single level: radiative processes, collisions with electrons, and collisions with other atoms including those which constitute the walls of the vessel containing the gas. In addition to these phenomena one must consider the possibility of the resonance trapping of radiation, that is, the absorption of radiation by originally unexcited atoms.

The frequency of occurrence of these phenomena depends not only on the composition of the gas but also on pressure and the geometrical form of the container. Although the detailed discussion of these factors must be postponed, it is clear that their effects are complex and inter-related, and that laser action can be achieved only in rather exceptional circumstances.

The best-known gas laser has a mixture of He and Ne as its working material. Such a laser was first constructed in 1960 by Javan, Bennett, and Herriott [12] at Bell Telephone Laboratories. It was originally designed to operate in the near-infrared region, emitting radiation at several wavelengths around 1.1 μm as a result of transitions between groups of neon levels designated by $2s$ and $2p$, respectively.

The functioning of this laser is explained with the aid of the partial energy-level diagram of Fig. 2.7, which shows some of the lowest energy levels of helium and of neon. The 2^3S state of He is metastable; a direct radiative transition to singlet ground state is forbidden, but a helium atom can arrive in this state by an electron collision process. When helium atoms in the 2^3S state collide with neon atoms in the ground state, the excitation may be transferred to the neon atoms which then end up in one of the $2s$ states, the highest of which lies only about 300 cm^{-1} below the 2^3S level of helium. Radiative transitions may then take place from the four $2s$ levels to the ten $2p$ levels. The $2p$ levels may be less populated than the $2s$ levels because there is no direct transfer to them from a helium level. Whether an inversion will actually take place depends on the relative abundance of the He and Ne atoms

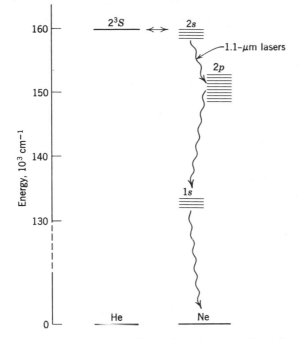

FIG. 2.7   Energy levels of He and Ne involved in the cycle of the 1.1-μm laser.

in the mixture and on the electron temperature. The $2p$ levels, fortunately, do not get overcrowded because they are readily drained by transitions to the lower $1s$ levels. The success of this scheme naturally depends on the achievement of the correct ratios of excitation and decay rates which are determined by the gas pressure, the power expended in the discharge, and the radius of the discharge tube, which enters into the matter because of the collision of excited atoms with the walls.

Another fortuitous near-coincidence of a metastable helium level with a group of neon levels is responsible for the popular, visible, 6328-Å laser and for an infrared laser that emits radiation at 3.39 μm. The selective transfer of excitation from helium to neon atoms makes the achievement of negative absorption in neon easier, but it is not absolutely necessary. Negative absorption can be achieved in pure neon as well as in other pure noble gases, provided that the electron density in the discharge is maintained at a proper level and provided gas pressure and tube dimensions are chosen so as to prevent the overpopulation of the terminal levels. The neon lasers described are typical of the atomic gas lasers, except for the transfer from helium, which is absent in other gases. The discharge in these lasers is a glow discharge with a current of the order of 10 mA. The light emitted is characteristic of the spectrum of the complete atom. These lasers are generally continuously operated. Although they usually provide only a modest power output of around 1 mW, the output is highly precise in spectral content. If properly operated, the atomic gas laser is the most monochromatic light source available. Atomic gas lasers are excellent sources of steady, coherent radiation of moderate intensity.

By means of high current arc discharges one may obtain stimulated emission from the ions of noble gases. Such ionic lasers are operated with power supplies capable of delivering short current pulses of the order of 100 Å at a rather high voltage (10 kV). The peak output of such a laser is of the order of 1 W. The spectral output is not as well defined as that of the atomic gas lasers.

Lasers may be based on the molecular spectra of several substances, the best known of these being carbon dioxide, nitrogen, and water vapor. Stimulated emission can be obtained from pure CO_2, but the powerful CO_2 lasers usually contain a mixture of three gases: nitrogen, helium, and carbon dioxide. The nitrogen molecules are excited by electron collision to a level with a long lifetime, and the excited nitrogen molecules may give up their energy when they collide with the carbon dioxide molecules. Their role is similar to that of the helium atoms in the He-Ne laser. The role of helium gas, in this case, is quite different. It helps to remove the energy of the CO_2 molecules in the terminal level of the laser transition, and thus exercises a desirable influence on the population difference between the levels involved in laser action. Although, in the case of atomic lasers, it is generally necessary to use a gas discharge with a small diameter, this is not true for molecular lasers. Carbon dioxide lasers can be scaled up to several meters in length and over 10 cm in diameter. Several hundred watts of power are available in continuous operation at 10.6 μm wavelength from such large lasers. Nitrogen gas is a useful energy donor for a number of molecules, including CO, N_2O, and CS_2. Acting as a laser material by itself, N_2 is the source of coherent radiation in several wavelength bands, most notably those in the vicinity of 0.87, 0.89, 1.05, and 1.23 μm. Atomic and ionic transitions of nitrogen also provide a variety of laser lines. Water vapor produces molecular lasers at many wavelengths in the far infrared region. Gas lasers have been constructed, using as active materials the atoms or ions of the halogens and of the atmospheric gases. Laser oscillations have been obtained from many other elements in vaporized state, the most notable among these being mercury.

More than a hundred separate spectral lines have been observed in stimulated emission from some elements and some compounds. The total number of gas lasers is over 2000. Although most of the lines observed have not led to the development of practical lasers, the number and variety of gas lasers is considerable. In technical importance they have overtaken the other laser types.

2.5 SEMICONDUCTOR LASERS

Semiconductor lasers differ in appearance, size, and method of excitation from the crystalline and glass solid lasers discussed in Section 2.3. The only features these lasers have in common with the other types previously discussed are the use of stimulated emission and population inversion for the creation of an amplifying medium.

Semiconductors differ from the other materials considered in their energy level structure. In ionic crystals and in ions located in glasses individual ions possess their own energy levels which might be modified by the fields of neighboring atoms. Emission and absorption of light occurs as transitions take place among these levels. The energy level structure of a semiconductor, however, is a property of the entire crystal. The rather complex theory of semiconductors may be briefly (and inadequately) summed up by stating that the energy levels of a pure semiconductor crystal are grouped in bands, each band representing a vast number of closely spaced levels. There are gaps between these bands—ranges of forbidden energy. At a low temperature only the lowest bands of the semiconductor are filled with electrons and filled completely. The highest filled band is called the *valence band;* the lowest empty band the *conduction band.* Excitation of such a crystal represents the transfer of an electron from the valence band into the conduction band and the creation of a hole in the valence band. Excitation occurs, for example, when light is absorbed in the semiconductor. Conversely, when an electron in the conduction band unites with a hole in the valence band, emission of radiation takes place.

Addition of certain impurities (donors) to the semiconductor produces excess electrons, which are essentially in the conduction band. When this is done, we have what is called an *n-type semiconductor.* Other impurities (acceptors) create holes in the valence band. The resulting material is called a *p-type semiconductor.*

Light emission can be obtained at the interface, or junction, of *n*- and *p*-type semiconductors if an external electric potential is applied which drives both the electrons and the holes into the junction. This process is called *carrier injection.*

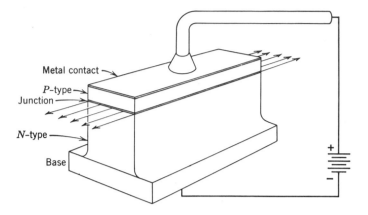

FIG. 2.8 Structure of a semiconductor junction laser.

The structure of a semiconductor junction laser is illustrated in Fig. 2.8. It consists of a single crystal wafer of a III-V compound, typically GaAs, whose linear dimensions are of the order of 1 mm. The p-n junction is made by diffusing an acceptor element (e.g., Zn) in one side of a donor-doped (e.g., Te) crystal. The area of the junction is of the order of 10^{-4} cm^2. The thickness of the junction, from which all light originates, is a few μm.

The electric field is impressed in the form of a pulse of about 1-μsec duration; the electromotive force required is approximately 1.5 V. The current in the junction is a highly nonlinear function of the impressed voltage; it increases rapidly as the voltage approaches the potential difference of the band gap. The emission of light is noticeable even at relatively low current densities. The radiation emitted under such circumstances is called *recombination radiation* because it arises when holes and electrons recombine with the emission of energy in the form of photons.

When a junction diode is prepared for laser operation, its front and back surfaces are cut perpendicular to the plane of the junction and parallel to each other. These end surfaces form the terminating mirrors of the laser. The index of refraction of the semiconducting crystal is usually so high that it is not necessary to increase the reflectivity of the end surfaces by coating. The side surfaces are usually offset by a small angle or are roughly finished to avoid regeneration of radiation in an undesired direction.

As the current through the semiconductor junction is increased, the emitted light intensity varies linearly until a threshold is reached at several thousands amperes per square centimeter. At this threshold the intensity increases rapidly, the radiation pattern becomes highly directional, and the spectral width of the emitted radiation is narrowed. These phenomena are characteristic of the onset of stimulated emission.

The operation of semiconductor lasers is generally intermittent because of the large heat dissipation in the junction. The properties of the semiconductors involved are greatly temperature-dependent. For reasons explained in Chapter 7, semiconductor junction lasers are operated cooled to liquid nitrogen temperature (77°K). The output of a semiconductor laser has a spectral width several orders of magnitude greater than the output of a ruby or rare-earth laser.

The GaAs junction laser operates between 8400 and 8500 Å, depending on the temperature. Other similar compounds—GaP, InAs, InP, InSb, and their mixtures—provide a wide variety of laser materials. By adjusting the composition of the mixture it is possible to "tune" the laser output over a frequency range covering the far-red region of the spectrum and the near-infrared region below 1.0 μm. Variations in the distribution of donor– and acceptor atoms also lead to frequency variations. Thus the output frequency of semiconductor lasers is never as precisely defined as that of other laser types. The output energy of a single pulse is much smaller than that obtainable from ruby and neodymium lasers. On the other hand, some semiconductor lasers may be pulsed 100 to 1000 times per second. The conversion efficiency of semiconductor lasers is the highest of all lasers. Efficiencies as high as 70% have been observed. Because excitation is achieved by the passage of current through the laser, it is possible to modulate the laser output by modulating the exciting current.

Carrier injection is not the only method for the excitation of semiconductors, and the III-V compounds are not the only laser materials. It is possible to excite semiconductor lasers by electron beams as well as by optical pumping. With the addition of these methods, and with the inclusion of such semiconductor materials as ZnO, ZnS, CdS, PbS, PbSe, and PbTe, the wavelength range of semiconductor lasers extends from 0.3 to 30 μm. These lasers form the subject matter of Chapter 7.

2.6 LIQUID LASERS

Liquid lasers derive their excitation from irradiation by means of a flash lamp. The configurations of the exciting source and the active material are similar to those used for solid lasers. A noteworthy difference between the appearance of solid and liquid lasers is due to the large thermal expansivity of the liquid. The liquid laser material cannot be confined in a fixed glass or quartz container. Provision must be made for an expansion volume, or the length of the laser must be made variable.

One type of liquid laser consists of a solution of a rare-earth chelate. This is a metallo-organic compound in which a metal ion is surrounded by oxygen atoms arranged at the vertices of a regular octahedron or cube. The oxygens atoms themselves are members of a ketone group in an organic compound. This compound is in an ionic form so that the positive charge on the metal ion is compensated by negative charges on the organic ions. The absorption of the incident light takes place in the organic part of the molecule. The resulting excitation is then in part transferred to the metal ion, and thus a level of the metal ion is populated. Stimulated emission takes place from this level when a lower, empty level is available. In order for this process to lead to population inversion, it is necessary that the organic ligands have metastable levels near suitable levels of the metal ion so that an excitation transfer can take place. A number of organic ligands may be used, including benzoylacetonate, dibenzoylmethide, and benzoyltrifluoroacetonate. The most suitable metal ion is Eu^{3+}.

Chelate laser materials have very large absorbance in the spectral region of the exciting light. This high absorbance is a serious disadvantage from the point of view of laser construction because the pumping light is greatly attenuated over a distance of a millimeter, therefore only very thin tubes of the liquid can be excited effectively. Most solvents cause a partial dissociation of the chelate. This is, again, a deleterious effect, since different ions will be surrounded in the solution by different types of environment and will therefore have somewhat different spectral characteristics. Finally, the solvent renders the energy transfer process quite inefficient by providing alternative paths for the dissipation of the energy stored in the metastable ligand. As a result of these adverse factors, energy utilization in chelate lasers is much less efficient than in solid lasers. The energy required for threshold excitation is strongly temperature-dependent. Several hundred joules are required for very tiny lasers.

A more efficient liquid laser material consists of the solution of neodymium ions in an acid that does not contain hydrogen. When hydrogen is present, molecular vibrations of hydrogen drain the excitation off the metal ions. A suitable hydrogen-free solvent for Nd_2O_3 is a mixture of $SeOCl_2$ and $SnCl_4$. This highly corrosive liquid provides a laser material over a large concentration range of Nd ions. The exciting light is absorbed directly by the Nd ions as in solid Nd lasers. Laser action takes place at $\lambda = 1.056$ μm. The efficiency of the process is about as good as it is in ruby, and the threshold excitation is comparable with the requirements of good four-level lasers.

The most recent, and the most interesting, of the fluid laser types are the fluorescent organic dyes excited by irradiation from solid lasers or special, fast flashlamps. These dye lasers are tuneable, convenient sources of coherent visible radiation.

REFERENCES

1. J. P. Gordon, H. J. Zeiger, and C. H. Townes, The maser—New type of amplifier, frequency standard, and spectrometer, *Phys. Rev.*, **99**, 1264–1274 (1955).
2. J. R. Singer, *Masers*, Wiley, New York, 1959.
3. G. Troup, *Masers and Lasers*, Methuen, London, 2nd ed., 1963.
4. A. E. Siegman, *Microwave Solid State Masers*, McGraw-Hill, New York, 1964.
5. A. Yariv, *Quantum Electronics*, Wiley, New York, 1967.
6. J. Weber, Masers, *Rev. Mod. Phys.*, **31**, 681–710 (1959).
7. B. A. Lengyel, Evolution of masers and lasers. *Am. J. Phys.*, **34**, 903–913 (1966).
8. A. L. Schawlow and C. H. Townes, Infrared and optical masers, *Phys. Rev.*, **112**, 1940–1949 (1958).
9. T. H. Maiman, Stimulated optical radiation in ruby. *Nature*, **187**, 493–494 (1960).
10. T. H. Maiman, R. H. Hoskins, I. J. D'Haenens, C. K. Asawa, and V. Evtuhov, Stimulated emission in fluorescent solids II. Spectroscopy and stimulated emission in ruby, *Phys. Rev.*, **123**, 1151 1157 (1961).
11. R. J. Collins, D. F. Nelson, A. L. Schawlow, W. Bond, C. G. B. Garrett, and W. Kaiser, Coherence, narrowing, directionality and relaxation oscillations in the light emission from ruby, *Phys. Rev. Letters*, **5**, 303–305 (1960).
12. A. Javan, W. R. Bennett, Jr., and D. R. Herriott, Population inversion and continuous optical maser oscillation in a gas discharge containing a He-Ne mixture, *Phys. Rev. Letters*, **6**, 106–110 (1961).

Chapter 3   Excitation and oscillation problems in laser theory

The excitation of the laser material to the extent required for laser operation presents analytical problems that are common to a variety of laser types. These problems are discussed in this chapter, together with analytical problems that arise from the fact that lasers are resonant structures with characteristic frequencies and modes of oscillation that represent a steady-state distribution of the electromagnetic field in the laser. Although the problems of excitation and those of mode structure are to some extent interdependent, we focus our attention first on the laser material, neglecting the detailed, frequency-dependent effect of its boundaries. Thus the boundaries of the laser will enter into the developments of the first three sections only insofar as they are the main causes of radiation loss which must be compensated for by amplification. In the later sections, attention is focused on the confinement of the radiation in the laser with definite boundaries and on the frequency selectivity resulting from the resonant structure. The technical consequences of the existence and nature of the resonant structure are explored in the last sections of the chapter.

The discussion is so oriented that its results are primarily applicable to ruby lasers and gas lasers. Some attention is given to four-level lasers, such as neodymium. Problems specific to the excitation and mode structure of semiconductor lasers are reserved for discussion in the chapter dealing with those lasers.

3.1   THE THRESHOLD CONDITION

The laser, as a device, consists of a pair of parallel mirrors between which is a piece of material made amplifying in a limited frequency region. This device is represented schematically in Fig. 3.1, in which

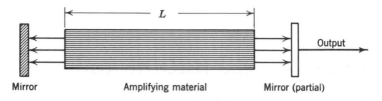

FIG. 3.1 Schematic diagram of a laser.

the reflectors are shown detached from the active material. The length of the active material is denoted by L, the optical distance between the reflectors by L'. When the mirrors are deposited on the end surfaces of the active material, then $L' = \eta L$, where η is the index of refraction. This pertains to the case of ordinary ruby lasers. For gas lasers the refractive index of the active material is very nearly one, but the mirrors are usually outside of the active medium so that $L' > L$.

At least one of the reflectors is partially transmitting, therefore its reflection coefficient is noticeably less than one. The reflection coefficient, or reflectivity, r, is defined as the fraction of the light intensity reflected. Thus at each reflection the fraction $(1 - r)$ is not returned into the laser. If the reflection coefficients of the mirrors are r_1 and r_2, respectively, the energy of a wave in one full passage back and forth diminishes due to incomplete reflection by the factor $r_1 r_2$. Generally, there are other sources of loss present in addition to the reflection losses although, in the case of the ruby laser, the reflection losses are probably the most important. In gas lasers the diffraction losses may be quite significant. In any case let us denote by $e^{-2\gamma}$ the fraction of the intensity remaining after a full round-trip passage through the laser. Alternatively, this is the fraction of photons remaining in the laser after one complete round trip. The number γ is positive and may be regarded as a measure of loss in a single passage. When all losses other than reflection losses may be neglected, we have $e^{-2\gamma} = r_1 r_2$, and therefore $\gamma = -\frac{1}{2} \log r_1 r_2$.

Oscillations may be sustained in the laser if the amplification of the radiation through the active material is sufficient to compensate for the fraction of the energy lost due to all causes. In each passage through the laser the intensity of the radiation is increased by a factor $e^{\alpha L}$ by virtue of the amplification in the material (see Section 1.3). Therefore, after taking the loss factor $e^{-\gamma}$ into consideration, the intensity changes from 1 to $F = e^{(\alpha L - \gamma)}$. When F is not less than 1, oscillations will build up, starting from a small disturbance; when F is less than 1, they will die out. Clearly, if the situation $\alpha L > \gamma$ is somehow brought about, the intensity of radiation of the proper frequency will build up rapidly until it becomes so large that the stimulated transitions will deplete the upper level and reduce the value of α. This is a dynamic situation that in most solid lasers gives rise to pulsations. If the level of excitation is such that αL is less than γ for all frequencies, the intensity of radiation does not build up at any frequency.

The threshold of laser oscillations is attained when the peak value α_m of the amplification curve satisfies the equation

$$\alpha_m L = \gamma. \tag{1.1}$$

This equation is called the *threshold condition*.

The amplification within the laser material is a function of the frequency ν and of the relative population inversion n in the laser material. It was shown in Section 1.3 that

$$\alpha(\nu) = k(\nu)_0 n, \tag{1.2}$$

where $k(\nu)_0$ is the absorption in the unexcited laser material and

$$n = \frac{1}{N_0} \left(\frac{g_1}{g_2} N_2 - N_1 \right). \tag{1.3}$$

Thus a laser of a given length and mirror reflectivity will operate only if the population inversion is large enough to insure that

$$\alpha_m = n k_0 \geq \frac{\gamma}{L}. \tag{1.4}$$

When the inequality holds, the laser will operate in the frequency interval in which $\alpha(\nu)$ is above γ/L. This interval is shown in Fig. 3.2 between ν_1 and ν_2.

An explicit expression for the population inversion at threshold involves the spontaneous transition rate between the laser levels and the shape of the spectral line as well as the design parameters (γ and L) of the laser. Using (3.22) in Chapter 1, we obtain from (1.4)

$$\frac{g_1}{g_2} N_2 - N_1 = \frac{\gamma}{L \kappa g(0)}. \tag{1.5}$$

A general conclusion may be drawn from (1.5). Since the intensity of the exitation determines the population inversion, and this in turn has to exceed the minimal value $\gamma/k_0 L$, a certain trade-off is possible between the reflection coefficient that determines γ and the active length L of the laser. Any deterioration of the reflector must be compensated

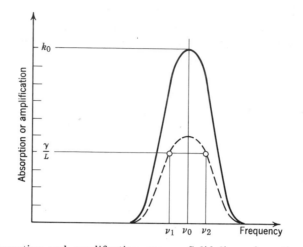

FIG. 3.2 Absorption and amplification curves. *Solid line:* absorption of the un-excited material $k(\nu)0$; *broken line:* amplification for $n = 0.5$.

for by increased length, or a penalty is paid in the form of increased threshold of excitation. It is also significant that the onset of oscillations can be prevented by lowering r and thereby increasing γ.

Let $\tau = L'/c$ be the time for a single passage of a photon through the laser. Then photons will travel back and forth at the rate of $1/\tau$. Given a large number of photons P_0, it follows from the definition of γ that after m passages their number will be $P_0 e^{-\gamma m}$. Hence photons traveling axially in the unexcited laser decay in time according to the law

$$P = P_0 e^{-\gamma t/\tau}. \tag{1.6}$$

The average lifetime of a photon* in the laser is therefore

$$t_p = \frac{\tau}{\gamma} = \frac{L'}{c\gamma}. \tag{1.7}$$

On combining (1.5) and (1.7) with (3.18) in Chapter 1, we get

$$\frac{g_1}{g_2} N_2 - N_1 = \frac{8\pi\eta^2\nu^2}{c^3 g(0)} \frac{g_1}{g_2} \frac{L'}{L} \frac{t_2}{t_p}. \tag{1.8}$$

In the special case when the mirrors terminate the active material,

---

* The photon lifetime t_p is related to the Q of the laser as a resonator by the formula $Q = 2\pi\nu t_p$. For the definition of Q see p. 83.

$L' = L\eta$, therefore (1.8) reduces to

$$\frac{g_1}{g_2} N_2 - N_1 = \frac{8\pi\eta^3\nu^2}{c^3 g(0)} \frac{g_1}{g_2} \frac{t_2}{t_p}. \tag{1.9}$$

This last equation clearly demonstrates that the population inversion required for threshold is proportional to the ratio of two lifetimes: one characteristic of the active material (t_2), the other of the laser construction (t_p).

Equation 1.9 specifies the minimum inversion necessary for laser oscillation. It is known as the *Schawlow-Townes condition* and is most frequently stated in terms of other parameters than those employed here. Following the original practice of Schawlow and Townes, most authors prefer to describe the lineshape in terms of the linewidth $\Delta\nu$, which is inversely proportional to $g(0)$. Then for a Lorentz-type line it follows with the aid of (4.10) in Chapter 1 that

$$\frac{g_1}{g_2} N_2 - N_1 = \frac{4\pi^2\eta^3\nu^2 \Delta\nu}{c^3} \frac{g_1}{g_2} \frac{t_2}{t_1}, \tag{1.10}$$

or, on introducing the wavelength in the material, $\lambda = c/\eta\nu$,

$$\frac{g_1}{g_2} N_2 - N_1 = \frac{4\pi^2 \Delta\nu}{\lambda^3 \nu} \frac{g_1}{g_2} \frac{t_2}{t_p}. \tag{1.11}$$

Similar formulas can be found for Gaussian lines. It is to be noted that our formulas are based on $\Delta\nu$, the full width of the line between half-power prints, whereas Schawlow and Townes* use the half-width $\delta\nu = \Delta\nu/2$.

When k_0, the peak absorption coefficient of the unexcited material, has been determined experimentally, the threshold population inversion can be calculated directly from (1.4). Then

$$\frac{g_1}{g_2} N_2 - N_1 = \frac{N_0\gamma}{k_0 L}. \tag{1.12}$$

It is instructive to calculate this for a typical case. The measured value of k_0 for pink ruby is 0.28 cm^{-1}; then for a ruby rod 10 cm long, coated with mirrors of reflectivities $r_1 = 1.00$ and $r_2 = 0.96$, we have $\gamma = 0.02$, hence

$$n = \frac{0.02}{0.28 \times 10} = 0.0072.$$

* See p. 59, Chapter 2 [8].

The terminal laser level in ruby has the multiplicity $g_1 = 4$; the initial level is a combination of two nearby levels each of which has a multiplicity 2. These levels are so closely coupled that for many purposes they may be considered a single level of multiplicity $g_2 = 4$. With this approximation we get for ruby

$$n = \frac{N_2 - N_1}{N_0}.$$

Thus, in the example just calculated threshold is reached when the population in the upper level exceeds that of the ground level of 0.7%. Since the concentration of Cr in pink ruby is 1.6×10^{19} atoms/cm³, and since relatively few atoms are in level 3 at any one time, the levels 1 and 2 each contain about 8×10^{18} atoms/cm³; their population difference in the above example is around 5.6×10^{16} atoms/cm³.

3.2 OPTICAL EXCITATION OF THE THREE-LEVEL LASER

The work of Maiman, which culminated in the successful construction of the ruby laser, began with the study of the transition rates in ruby and continued with calculations concerning the excitation rate necessary to produce the population inversion required for laser action [1, 2, 3]. We shall review the significant features of these calculations because the method employed in these is applicable to many lasers.

The simplified energy-level model of the ruby is shown in Fig. 3.3,

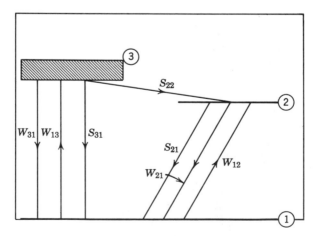

FIG. 3.3 Energy-level diagram for a three-level fluorescent solid. Stimulated transitions are indicated by W, spontaneous ones by S.

which includes only three of the many energy levels of ruby. Transition rates among these levels are indicated by the letters S and W, the former identifying spontaneous, the latter stimulated, transitions. The principal difference between these types is that the S rates are independent of the radiation density, the W rates are proportional to it. Thus the rate W_{13} is the rate at which atoms, or ions in the ground state, are excited to level 3 by means of the incident pumping radiation absorbed in the ruby. The transition from the absorption band of the ruby, – level 3, to the starting level of the laser, – level 2, takes place without radiation, but transitions between level 2 and the ground level may occur in several ways. The spontaneous transition rate S_{21} includes the spontaneous emission rate A_{21} plus the rate S'_{21} of radiationless transitions.

The occupation of levels in such a three-level system with a total of N_0 atoms is governed by the *rate equations*

$$\frac{dN_3}{dt} = W_{13}N_1 - (W_{31} + S_{31} + S_{32})N_3,$$

$$\frac{dN_2}{dt} = W_{12}N_1 - (W_{21} + S_{21})N_2 + S_{32}N_3, \tag{2.1}$$

$$N_0 = N_1 + N_2 + N_3.$$

We calculate the population ratio N_1/N_2 in a stationary state, that is, when the time derivatives are zero. In this case

$$W_{13}N_1 = (W_{31} + S_{31} + S_{32})N_3,$$

$$W_{12}N_1 = (W_{21} + S_{21})N_2 - S_{32}N_3. \tag{2.2}$$

On eliminating N_3 we get

$$\frac{N_2}{N_1} = \left(\frac{W_{13}S_{32}}{W_{31} + S_{31} + S_{32}} + W_{12}\right)(S_{21} + W_{21})^{-1}. \tag{2.3}$$

The preceding calculations are applicable to any three-level system. The special properties of the ruby permit certain simplifications. It is instructive to learn how these properties were discovered and how they affect the success of laser operation.

Before he attempted the construction of lasers, Maiman [1] measured the ratio of fluorescent quanta emitted in the R line to the number of quanta absorbed from the exciting beam. The result was around 70%, which indicated that most of the ions went through the fluorescent cycle via level 2 instead of returning directly from level 3 to level 1 via the spontaneous transition, whose rate is S_{31}. This can happen only if S_{32} is much larger than S_{31} so that the ions will pass to level 2 before many have a chance to return to level 1 directly. Moreover, the spontaneous transition from level 2 to level 1 must be mostly radiative so that S_{21} is not much greater than A_{21}. The measurements of Maiman and others [1, 3] on the intensity and the rate of decay of the fluorescence of ruby show that $S_{32} \approx 100\ S_{31}$, so that S_{31} may be omitted from (2.3). The transition rate S_{21} is about 330/sec at 300°K and about 230/sec at 77°K and below. Maiman's estimate for S_{32} is 2×10^7/sec. This value is large compared to all transition rates, including the W_{13} attainable with high pumping intensities. As a result of the dominance of S_{32} the first term in the brackets of (2.3) is approximately equal to W_{13}, and the following approximate equation results:

$$\frac{N_2}{N_1} = \frac{W_{13} + W_{12}}{S_{21} + W_{21}};$$ (2.4)

hence

$$\frac{N_2 - N_1}{N_2 + N_1} = \frac{W_{13} + W_{12} - S_{21} - W_{21}}{W_{13} + W_{12} + S_{21} + W_{21}}.$$ (2.5)

Two further simplifications may now be made. As a result of the high value of S_{32} the occupation of level 3 will be small compared with the other two levels. Thus $N_1 + N_2 \approx N_0$. Moreover, Einstein's relation implies that $W_{12} = W_{21}$. Therefore

$$\frac{N_2 - N_1}{N_0} \approx \frac{W_{13} - S_{21}}{W_{13} + S_{21} + 2W_{12}}.$$ (2.6)

In the case of ruby the multiplicities g_1 and g_2 are equal, therefore the left-hand side of (2.6) is the relative population inversion n. The condition for reaching $n = 0$ is therefore $W_{13} = S_{21}$. Physically, this means that the rate at which atoms are raised to level 3 is equal to the spontaneous return rate from level 2 to level 1. This rate is not sufficient to reach threshold.

Until the threshold is exceeded, the level of radiation is low, therefore $W_{12} \approx 0$. With this additional approximation we have from (1.12) and (2.6)

$$\frac{W_{13}}{S_{21}} \approx \frac{1 + \gamma/k_0 L}{1 - \gamma/k_0 L}$$ (2.7)

as the condition for maintaining a steady state at threshold. Thus the excitation rate for threshold is of the form

$$W_{13} = \Gamma S_{21},$$ (2.8)

where

$$\Gamma = \frac{1 + \gamma/k_0L}{1 - \gamma/k_0L}. \tag{2.9}$$

The condition for zero absorption is obtained by setting $\Gamma = 1$.

As we have already noted, the return rate S_{21} includes the spontaneous radiative return rate A_{21} and the rate of radiationless relaxation S'_{21}. The latter is dependent on the temperature. This dependence accounts for the fact that the measured value of S_{21} in pink ruby is 330 sec^{-1} at room temperature, but only 230 sec^{-1} at 77°K. The temperature dependence of the quantities entering (2.9) explains why the excitation rate required for the operation of a ruby laser increases with the temperature.

Let us now estimate the pumping radiation intensity required for the excitation of a ruby to transparency $n = 0$. The excitation rate W_{13} can be related to the intensity I of the pumping radiation within the ruby. If, for the sake of simplicity, the pumping radiation is assumed to have the character of a plane wave, the rate at which its energy is absorbed in a volume element of cross sectional area A and thickness dx is $Ik_pA\ dx$, where k_p is the rate of absorption at the pump frequency. This quantity is the product of the number of active atoms in the ground state times the absorption coefficient per atom; that is, $k_p = N_1\sigma_p$. The rate at which the active atoms gain energy is $W_{13}N_1Vh\nu_p$, where V is the volume and N_1, the number of atoms per volume element capable of absorbing the pumping radiation. Equating the energy loss in the beam to the energy gain in the material, we have

$$IN_1\sigma_p = W_{13}N_1h\nu_p. \tag{2.10}$$

Hence,

$$I = \frac{W_{13}h\nu_p}{\sigma_p}. \tag{2.11}$$

In order to obtain an estimate of the required irradiation we assume, as Maiman did, that the excitation is accomplished mainly by the green band of the xenon flashtube whose frequency is centered around 5.4×10^{14} sec^{-1}. The energy of one quantum is then about 3.6×10^{-12} erg. The measured average cross section of atomic absorption of ruby in the green band is $\sigma_p = 10^{-19}$ cm^2. Then, for $W_{13} = S_{21} = 330$ cm^{-1}, we have $I = 330$ sec$^{-1} \times 3.6 \times 10^{-12}$ erg $\times 10^{10}$ cm$^{-2} = 1200$ W/cm^2. Thus irradiation at the rate of 1200 W/cm^2, applied to a ruby slab from one side, would maintain the ruby in transparent condition for the R line.

Illumination by means of a plane wave is a mathematical artifice. A configuration more likely to be realized in practice is one in which the material is isotropically illuminated over most of its surface. Moreover, the illuminating source is likely to be high-pressure discharge with a more or less continuous spectral output resembling the spectral distribution of the blackbody radiator. It is therefore realistic to consider the laser crystal to be immersed in an isotropic blackbody radiation and to determine the minimum temperature of the blackbody source which may produce population inversion in a three-level fluorescent solid.

The rate of excitation W_{13} is $u(\nu)B_{13}$. If we identify $u(\nu)$ with the spectral energy density of the blackbody radiation and make use of Einstein's relations, we obtain

$$W_{13} = \frac{g_3}{g_1} \frac{A_{31}}{e^{h\nu/kT} - 1}, \qquad (2.12)$$

where $\nu = \nu_{13}$. The condition $W_{13} = S_{21}\Gamma$ can then be solved for the minimum source temperature T_s which is capable of producing adequate illumination in the green absorption band of ruby to maintain the required amplification. The result is

$$T_s = \frac{E_3 - E_1}{k \log (1 + g_3 A_{31}/g_1 S_{21}\Gamma)}. \qquad (2.13)$$

We may now make a rough estimate of the source temperature required to produce zero absorption in ruby using the following numerical data of Maiman [1, 3]:

$$E_3 - E_1 = 3.6 \times 10^{-12} \text{ erg},$$
$$A_{31} = 3 \times 10^5 \text{ sec}^{-1},$$
$$S_{21} = 330 \text{ sec}^{-1} \text{ (at } 300°K).$$

The relevant levels have the following multiplicities: $g_1 = 4$, $g_3 = 12$. From these data $T_s = 3300°K$. The actual temperature is considerably higher than this because many adverse complicating factors were neglected. Nevertheless, this computation, which follows the line of reasoning employed by Maiman [2], indicates the rough magnitude of the temperature involved.

It should be stressed that the above calculations do not imply that the ruby laser operates in a steady-state condition. On the contrary, the laser material must be raised from its unexcited condition to the threshold; then laser oscillations begin in a dynamic manner. As the threshold is exceeded, the radiation density grows rapidly, and the resulting stimulated emission rate soon overtakes the rate at which excitation is supplied. Then, as the population inversion is depleted, it may fall below threshold and the laser is momentarily extinguished. The output of an ordinary ruby laser normally consists of such rapid pulsations.

The method of rate equations can only give a rough estimate of the power requirements of the ruby laser. It is more applicable to gas and four-level solid lasers, which may be operated in a continuous regime with essentially constant output power.

3.3 OPTICAL EXCITATION OF FOUR-LEVEL LASERS

The analysis of the kinetics of an optically excited four-level solid laser requires equations involving the populations of all four levels and in principle involves transitions of several types from every level to every other level. As a practical matter, equations containing only the dominant transition rates describe the physical situation adequately. The most important transitions in a four-level solid laser are shown in Fig. 3.4. The pumping rate is indicated by W_{14}, nonradiative rates by S. The spontaneous and stimulated transition rates at the laser frequency are denoted by A_{32} and W_{32}, respectively. The material and levels involved are generally chosen so that the S rates are high in comparison to the other rates even at the highest level of irradiation. The starting level of the laser transition, level 3, is such that direct transitions from this level to the ground state are forbidden, and transitions to level 2 occur only at a moderate rate, that is, at a rate small compared to the S rates. The rate equations for the occupation numbers N_1, N_2, N_3, and N_4 can then be written as in the three-level case. These equations and their mathematical solutions are available in the literature [2, 4]. The general solution is of limited value; the specific computations

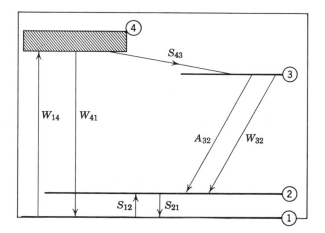

FIG. 3.4   Principal transitions in a four-level solid laser.

usually exploit the special situation which renders certain transitions dominant over all others. We shall forego the exposition of the general mathematical apparatus and confine the discussion to the main physical principles applicable and to their quantitative consequences.

In a successful four-level laser most of the atoms are in levels 3 and 1. This is achieved only when the transition rates S_{43} and S_{21} are high so that levels 4 and 2 drain rapidly. Level 2 is most frequently a sublevel of the ground level of the ion. It is split off from the ground level by the action of the crystal field. When this is the case, it is essential that the laser operate at a low temperature so that $kT \ll E_2 - E_1$, otherwise the terminal laser level will be populated by atoms excited by the available thermal energy.

In the simplest limiting case, when N_2 and N_4 are neglected in comparison to N_1 and N_3, the laser is approximately a two-level system. Then (before the onset of laser action)

$$\frac{dN_3}{dt} = W_{14}N_1 - A_{32}N_3. \tag{3.1}$$

Therefore in a stationary state

$$\frac{N_3}{N_1} = \frac{W_{14}}{A_{32}}. \tag{3.2}$$

Even though the last two equations are obtained from a very rough model, they indicate that under some circumstances the threshold may be reached with moderate excitation. The relation between population inversion and threshold is the same in the three— and four-level case. Thus when the laser operates from level 3 to level 2, which is assumed to be practically empty, the threshold condition (1.9) takes the form

$$N_3 = \frac{8\pi\nu^2\eta^3 t_3}{c^3 g(0) t_p}, \tag{3.3}$$

where $t_3 = A_{32}^{-1}$ is the decay rate of level 3 and t_p the photon lifetime.* The rate at which energy is supplied by the pump is $P = W_{14} N_1 h\nu_p V$. With the aid of (3.2), the optical power required to reach threshold is calculated to be

$$P = A_{32}N_3Vh\nu_p = \frac{8\pi\nu^2\eta^3 Vh\nu_p}{c^3 g(0) t_p}. \tag{3.4}$$

This expression does not depend on the transition rate A_{32}, but is directly proportional to the linewidth Δ because $g(0)$, the peak of the lineshape, is inversely proportional to Δ.

---

*It is assumed that the mirrors are in contact with the active material.

From (3.4) it is clear that under the conditions assumed it will require less power to excite materials with a narrow fluorescent linewidth and that it will be easier to provide excitation for lasers operating in the infrared than in the visible.

A typical and useful four-level laser has as its active material Nd^{3+} ions in crystals such as $CaWO_4$. The terminal laser level is about 2000 cm^{-1} above the ground level and that level is normally empty at room temperature, since kT at room temperature corresponds to only about 200 cm^{-1}. It has, in fact, been observed that such a laser may be excited to threshold by irradiation of an order of magnitude less than required for a similar ruby.

3.4 STANDING WAVES IN A LASER

The laser consists of a vast number of atomic amplifiers placed between two partially reflecting mirrors which cause radiation to travel back and forth through the amplifying medium and permit a fraction of the radiation to emerge as output. Up to this point we have discussed the properties of the amplifying medium as if it were unbounded. Now we shall take into account the effects of the mirrors whose presence and shape determine the structure of the electromagnetic field within the laser.

In the most elementary analysis of structure the mirrors are assumed to be plane, their dimensions to be very large compared to the wavelength, and the field is assumed to consist of plane waves of uniform amplitude traveling in a direction perpendicular to the mirrors. This oversimplified picture leads to the recognition of the existence of standing waves between the mirrors. Moreover, it reveals the fact that waves reflected several times will meet in reinforcement only if a certain simple relation is satisfied between wavelength (or frequency) and the length of the laser.

In the simplest case, when the refractive index η does not vary between the mirrors, the condition of reinforcement is that the distance of the mirrors is an integral multiple of the half-wavelength, $\frac{1}{2}\lambda/\eta$, within the laser. Thus

$$n\lambda = 2L\eta \qquad (4.1)$$

or, in terms of frequency,

$$\frac{\nu}{c} = \frac{n}{2L\eta} . \qquad (4.2)$$

When the optical path between the mirrors is inhomogeneous, it is of advantage to introduce the *optical distance*

$$L' = \int_0^L \eta \, dz. \qquad (4.3)$$

It is then found that (4.1) and (4.2) remain valid with L' substituted for ηL. The proof of this proposition is based on the fact that complete reinforcement takes place when the total phase change in a round-trip passage through the laser is an integral multiple of 2π. Aside from the phase change of π on each mirror, the phase change through the laser is $2 \int_0^L k \, dz$, where $k = 2\pi\eta/\lambda$. Straightforward calculation then leads to

$$\frac{\nu}{c} = \frac{n}{2L'}. \tag{4.4}$$

Plane waves of a frequency satisfying (4.4) and directed along the laser axis (perpendicular to the mirrors) may be called the *axial modes* of the laser. They are analogous to the free oscillating modes of a damped harmonic oscillator.

This plane-wave analysis is applicable as a rough approximation to the simplest situation that may occur in a gas laser with plane mirrors. In gas lasers, the diameters of the mirrors are large compared to the wavelength, and the distance between the mirrors is large compared to the mirror diameter. (Typical orders of magnitude for these quantities are 10^{-6}, 10^{-2}, and 1 m.) There is no optical boundary on the sides, therefore rays directed at an angle to the laser axis wander off and are lost. With this elementary, and not entirely justified, picture we can concentrate on the quasiplane waves that travel in the axial direction.

Since the threshold condition is met only in a restricted frequency range, laser oscillations will occur only for a few discreet frequencies ν_n which satisfy (4.4) and which lie in this restricted range. The frequency difference of two such consecutive axial modes is

$$\nu_{n+1} - \nu_n = \frac{c}{2L'}. \tag{4.5}$$

These modes are equally spaced in frequency. In a typical gas laser whose length is 1 m and whose index of refraction is 1 this frequency spacing is 1.5×10^8 sec^{-1} = 150 MHz.

Consecutive modes of this type are thus so closely spaced that ordinarily several of them will lie within the range where the excitation exceeds the threshold. The output of the laser will then consist of several lines, separated from each other by the frequency difference $c/2L'$. These lines have a finite linewidth, determined by the losses of the laser as an electromagnetic resonator. The interrelation of the factors that determine the spectrum of the laser output is shown in Fig. 3.5.

Although the uniform plane-wave theory correctly predicts the fre-

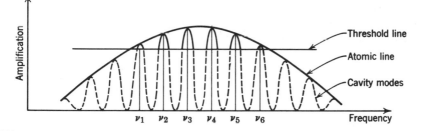

FIG. 3.5 Spectral factors in a laser. Six cavity resonances are shown in the frequency region of laser operation. The laser output frequencies are ν_1 to ν_6.

quency separation of the principal resonances in a laser, it is deficient for several reasons. The assumption of uniform plane waves of finite cross section is inconsistent with the basic laws of optics and electromagnetism. What happens on the sides cannot be ignored, especially when one is forced to consider radiation that does not progress strictly axially. Off-axis radiation is mostly reflected when it arrives at the side surface of a solid laser because the laser material has a large refractive index. Therefore the structure of the radiation field within the solid laser resembles that of a field in a closed metal cavity with some coupling to the outside. The structure of gas lasers is open, and here diffraction effects, combined with focusing effects produced by curved mirrors, dominate. Both cavity theory and the diffraction theory of Fabry-Perot structures show the existence of less symmetrical modes than those already considered, but the principal modes in these theories have the frequency separations calculated from the plane-wave theory.

3.5 MODES OF OSCILLATION OF A CLOSED OPTICAL CAVITY

A solid laser, such as the ruby laser, resembles a long cylindrical cavity with conducting walls and parallel end surfaces. The resemblance is incomplete, of course. Some radiation escapes from the ruby cylinder, so it is not a completely reflecting enclosure. The fact that some radiation escapes as output through one of the end surfaces does not create much of a problem. The escape of this radiation may be regarded as a perturbation or as a loss in an otherwise loss-free system. It may be taken into account in the same manner as damping in the theory of an otherwise harmonic oscillator. Radiation may also leave through the side surfaces, provided total reflection does not take place.

Radiation escapes only when the ray strikes the side surface at a relatively small angle of incidence because ruby has a high refractive index. A ray that travels in a direction in which it is not totally reflected at the sides is of no interest in laser theory because only a very small fraction of the radiative energy will develop into such rays. The bulk of stimulated radiation will develop in fields which propagate nearly, or exactly, in the axial direction. Therefore for the purpose of describing the modes of the stimulated radiation we make use of a cavity with conductive walls as the first approximation.

The reader may already have noticed with dismay that we are still referring to rays, even though our declared objective is to explore the radiation field in a cavity where the ray concept is not appropriate. To obtain the electromagnetic field configuration in a cavity one must solve Maxwell's equations with the appropriate boundary conditions. Neither the ray theory, nor the scalar wave theory is appropriate in this case.

A derivation of the modes and characteristic frequencies of cylindrical resonators is out of place here. This subject is developed in texts on microwave transmission lines [5]. It is useful, however, to summarize some of the results available in this discipline because they provide orientation concerning laser mode problems.

Let $\omega = 2\pi\nu$ denote the angular frequency of a periodic electromagnetic field and let $k = \omega\eta/c$ be the corresponding wave number in a medium of refractive index η. Periodic solutions of Maxwell's equations in a rectangular cavity may be represented as sums of terms derived from vector potentials whose dependence on coordinates and time is of the form

$$\exp[i(k_1 x + k_2 y + k_3 z - \omega t)], \qquad (5.1)$$

where

$$k_1^2 + k_2^2 + k_3^2 = k^2. \qquad (5.2)$$

Boundary conditions at the conductive walls of the cavity further restrict the values of k_1, k_2, and k_3 because the tangential component of the electric field must vanish at the walls. If the coordinate system and the dimensions of the cavity are chosen as indicated in Fig. 3.6, expressions of the type (5.1) combine to produce sinusoidal variation of the electric field with the coordinates. The boundary conditions then require that $k_1 a = l\pi$, $k_2 b = m\pi$, and $k_3 L = n\pi$, where l, m, and n are integers.

In the customary configurations, the sides a and b are nearly equal, whereas L is much longer. The z-axis is taken along the longest dimension. It is called the longitudinal axis, and a plane $z = $ constant is called a transverse plane. It is known from the theory of microwave

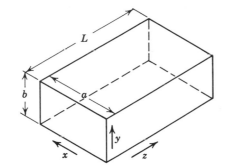

FIG. 3.6 Rectangular electromagnetic resonator.

transmission lines that the characteristic oscillations of a perfectly conducting hollow cylindrical cavity are either of the *transverse electric*
(TE) type or the *transverse magnetic* (TM) type. The meaning of these
terms is as follows: A field configuration is called transverse electric
if the electric vector everywhere in the cavity is perpendicular to the
cylinder axis, that is, when $E_z = 0$. Similarly, $H_z - 0$ characterizes
the TM fields. All possible oscillations in the cavity can be represented
as sums of oscillations of TE and TM types. A uniform plane wave
is of the *transverse electromagnetic* type because both the electric and
the magnetic vector are perpendicular to the direction of propagation.
A transverse electromagnetic (TEM) wave may also exist in a coaxial
transmission line, but not in a hollow transmission line or in a hollow
resonator* [5].

The different modes of oscillation in a rectangular cavity are identified
by their type and by the indices l, m, and n. One of the simplest TE
modes is characterized by the following variation of its electric field
components

$$E_x = 0, \qquad E_y = \sin\frac{\pi x}{a}\sin\frac{n\pi z}{L}, \qquad E_z = 0.$$

The transverse field distribution of this mode is referred to as the TE_{10}
mode, indicating that $l = 1$ and $m = 0$. The entire resonator mode is
called the TE_{10n} mode. The general TE_{lmn} mode polarized along the
y-axis has the field configuration

$$E_x = 0, \qquad E_y = \sin\frac{l\pi x}{a}\cos\frac{m\pi y}{b}\sin\frac{n\pi z}{L}, \qquad E_z = 0.$$

* By a hollow resonator or transmission line we mean a boundary configuration
whose cross section is a simple closed curve.

The transverse field distribution of the first of these modes (TE_{10}) contains no nodal lines of the electric field in the cross-sectional plane. In the general case (TE_{lm}) there are $l-1$ nodal lines parallel to the y-axis and m nodal lines parallel to the x-axis. The situation is similar for TE waves polarized in the x-direction and not very different for the TM waves. For each type of oscillation the simplest transverse field distribution contains no nodal line, but as one proceeds toward higher-order modes (higher values of l and m) the number of nodal lines increases and a net of two orthogonal sets of nodal lines develops.

When the resonator is a right circular cylinder of radius r and length L, the complex exponential dependence of the field in (5.1) is replaced by the cylinder functions. In place of (5.2) one obtains

$$\kappa^2 + k_3{}^2 = k^2, \tag{5.3}$$

where k and k_3 have the same meaning as before. The constant κ governs the variation of the electromagnetic field in the cross-sectional plane. For TM waves the electric field is proportional to $J_l(\kappa\rho)\exp(il\varphi)$, for TE waves to $J_l'(\kappa\rho)\exp(il\varphi)$. The boundary conditions require that $J_l(\kappa r) = 0$ for TM waves and $J_l'(\kappa r) = 0$ for TE waves. Combining all boundary conditions into a single formula, we obtain

$$\left(\frac{p_{lm}}{r}\right)^2 + \left(\frac{n\pi}{L}\right)^2 = k^2, \tag{5.4}$$

where p_{lm} is the mth root of the Bessel function of order l in the case of TM waves and the mth root of the derivative of the same function in the case of TE waves. The number l indicates the number of radial nodal lines; $m-1$ is the number of nodal circles within the cross section and $n-1$ is the number of nodal planes between the end surfaces.

In microwave theory the emphasis is on finding the lowest order oscillating modes of a resonator; that is, the least values of k (and ν) consistent with the dimensions of the cavity. In laser theory one is searching for nearly axial modes in a frequency range that permits a vast number of modes because it lies orders of magnitude above the frequencies of the lowest order modes. By nearly axial modes we mean those modes whose transverse field configurations contain few, if any, nodal lines. For these modes l and m are small integers. The number n, on the other hand, is very large, varying in practical situations between 10^4 and 10^6. The numbers p_{lm}, introduced above, are between 2 and 20 when l and m are restricted to small integers. Therefore the first term on the left of (5.4) is very small compared to the second so that the frequency calculated by neglecting p_{lm} altogether is almost the correct frequency.

In fact, neglecting p_{lm} is equivalent to introducing the plane wave approximation of Section 3.4 because for $p_{lm} = 0$, (5.4) reduces to $n\pi/L = k$, which is equivalent to (4.2).

It is shown in electromagnetic theory that the cavity modes are orthogonal to each other in mathematical terminology. Physically, this means that electromagnetic oscillations may be excited in any one of these modes without exciting other modes. Ideally, there is no interaction between oscillations in different modes. This isolation is true only in the idealized case when the cavity walls are infinitely conducting and possess the exact geometrical shape postulated. Deviations from these idealized conditions, or inclusion of polarizable material in the cavity, may result in the coupling of the modes, that is, in transfer of energy from one mode of oscillation to another.

In the laser the situation is so arranged that the ideal conditions are maintained only for relatively few modes. The cavity is made deliberately lossy for modes of unwanted kinds. The modes favored by the experimenter are those which have the greatest axial symmetry, that is, the least value of l and m. Given l and m, we have a fixed transverse mode configuration. There still may be a number of values of n for which (5.4) gives a frequency within the range where the laser is amplifying. The frequencies of these oscillations, which belong to a fixed transverse configuration, are (approximately) equidistant, the differences between adjacent frequencies being equal to $c/2L\eta$, the value calculated from the standing-wave theory in Section 3.4. This result is obtained as follows: Let l and m be fixed, and let k_0 denote the wave number that satisfies

$$\left(\frac{p_{lm}}{r}\right)^2 + \left(\frac{n\pi}{L}\right)^2 = k_0{}^2 \tag{5.5}$$

for some value of n. Furthermore, let k_1 satisfy

$$\left(\frac{p_{lm}}{r}\right)^2 + \left(\frac{(n+1)\pi}{L}\right)^2 = k_1{}^2. \tag{5.6}$$

After subtracting (5.5) from (5.6) we obtain

$$k_1{}^2 - k_0{}^2 = \frac{(2n+1)\pi^2}{L^2}, \tag{5.7}$$

which may be put in the form

$$k_1 - k_0 = \frac{(2n+1)\pi^2}{(k_1 + k_0)L^2}. \tag{5.8}$$

As noted before, the difference of the k's is small; both are nearly equal to $n\pi/L$. Then, replacing $\frac{1}{2}(k_1 + k_0)$ by $n\pi/L$, and neglecting 1 next to $2n$, we get

$$k_1 - k_0 \approx \frac{\pi}{L}. \tag{5.9}$$

Now we introduce the frequencies by means of the equation $k = 2\pi\nu\eta/c$. The frequency difference of adjacent modes then becomes

$$\nu_1 - \nu_0 = \frac{c}{2L\eta}. \tag{5.10}$$

This result does not depend on n and on the mode type, but it is valid only when modes of the same transverse configuration are compared. The frequency difference between modes of different types involves the differences of their p's.

Actually, the solid laser is not a box with conducting walls. It is a dielectric resonator with boundary conditions appropriate to the air-dielectric interface. The application of these boundary conditions at the surface of a cylinder leads to somewhat more complicated electromagnetic field distributions than those obtained in a cylindrical cavity with conductive walls [5]. Nevertheless, the equation that determines the resonant frequencies is always of the form of (5.3). The essential difference between the dielectric cylinder and the conductive cavity is that in case of the latter the possible values of κ are determined by the solutions of the equations $J_l(\kappa r) = 0$ and $J_l'(\kappa r) = 0$; the equations that determine the κ's in the dielectric case are more complicated. Although the values of κ calculated in this case are different, this does not matter very much as far as the frequencies of the laser are concerned. We have already found that the frequency spacing of similar modes is independent of the κ's. The frequency differences of different modes are, of course, affected, but this fact is of no practical significance.

We have repeatedly made the statement that the number of possible oscillations in a laser structure is very large. It is now proper to explain precisely what we mean by such a statement and to give some quantitative information concerning the density of modes. When discussing the number of possible electromagnetic oscillations in a cavity, one always has to make reference to a frequency range because each cavity has a lowest characteristic frequency and then an unlimited sequence of other characteristic frequencies. The question is, how many of them will fall in a prescribed frequency range? It is shown in many elementary textbooks that given a cubical enclosure of side a with perfectly reflecting walls, filled with a material of refractive index η, the number of electro-magnetic oscillations of frequency not exceeding ν is given by the equation

$$P(\nu) = \frac{8\pi\nu^3\eta^3 a^3}{3c^3};$$ (5.11)

hence the mode density per unit volume per unit frequency interval is

$$p(\nu) = \frac{1}{a^3}\frac{dP}{d\nu} = \frac{8\pi\nu^2\eta^3}{c^3} = \frac{8\pi}{\lambda^3\nu}.$$ (5.12)

This expression plays an important role in connecting spontaneous and stimulated emission rates. Einstein's second relation introduced in Section 1.3 may be written in the form

$$A_{nm} = h\nu p(\nu)B_{nm}.$$ (5.13)

The frequency of the ruby radiation is $\nu = 4.3 \times 10^{14}$ sec^{-1}. The corresponding wavelength in ruby ($\eta = 1.76$) is $\lambda = 4.0 \times 10^{-5}$ cm; therefore the mode density in this spectral region, calculated from (5.12), is approximately 0.9 sec/cm^3. Thus counting all possible modes in a 1-cm cube of ruby, these are less than 1 Hz apart in frequency in the relevant spectral range. A linewidth of 0.1 Å corresponds to a relative frequency spread of about one part in 70,000, and therefore to a frequency spread of about 6×10^9 Hz. Clearly, such a range of frequency contains a tremendous number of modes. Only a very small fraction of these modes lies in a narrow cone around the longitudinal axis of the laser, and the laser acts as a generator only for these favored modes.

The material introduced in this section serves primarily as orientation concerning modes of oscillation and their frequency distribution in a cylindrical structure that resembles the laser geometrically. A laser is actually more complicated than the idealized structures considered here, and the physics of laser oscillations cannot be meaningfully discussed without involving the loss rates of the possible modes of oscillation. The fact that the sides of the laser are not reflecting makes a significant difference, especially in the case of gas lasers. Although the cavity theory provides a fair approximation for solid lasers, an entirely new approach is necessary for the calculation of the modes of gas lasers because the latter are largely open structures. This new approach will be developed in Section 3.7 after the general consequences of mode structure on laser operation are explored.

3.6 RELATION OF MODE STRUCTURE TO LASER OUTPUT

The essential general characteristics of a laser are the following: It is an amplifying structure within a narrow frequency range determined by the threshold condition. The amplification is frequency-dependent, and the shape of the amplification curve is determined by the shape of the spectral line. It is also a resonant structure with many modes of oscillation, whose characteristic frequencies lie close together. In general, the region of amplification of the laser material contains a vast number of characteristic frequencies of the resonant structure.

The laser, as a resonant structure is characterized by its modes of oscillation which are typified by the stable field configurations of a closed cavity. Each mode of the laser may be geometrically described by the electric field incident on one of the terminal surfaces and by the number of transverse nodal surfaces between the mirrors of the laser. The physical characterization of a mode involves, in addition to the geometrical parameters, the specification of the damping rate of the mode.

In the geometrical description we are concerned primarily with the distribution of amplitude, phase, and polarization over a typical cross section of the laser, or over one of its mirrors. This distribution in the transverse plane, or on similar surface, is called the *transverse mode distribution*. The number of transverse nodal surfaces is proportional to the length of the laser. This number, together with the transverse mode distribution, determines the characteristic frequency of the resonant oscillation. Many different modes have the same resonant frequency. So, in general, the laser is a highly degenerate electromagnetic resonator.

As in all degenerate eigenvalue-eigenfunction problems in mathematical physics, only the eigenvalues (the characteristic frequencies) are uniquely determined. The eigenfunctions (modes) are not unique because all linear combinations of eigenfunctions that belong to the same eigenvalue are equally valid eigenfunctions. In lasers, for example, we may combine two modes, one with horizontal, the other with vertical polarization and obtain new modes of the same frequency. In a laser of circular or square cross section transverse modes differing only in the role of the x- and y-axes have equal resonant frequencies and may be combined into new modes.

In a perfect electromagnetic cavity each oscillating mode is independent of every other oscillating mode (orthogonality) and each oscillation is loss-free, that is, undamped. In the actual physical situation a certain amount of coupling will exist between different modes and each mode will have its own finite rate of damping. This damping rate is customarily described by the Q of the oscillation.* The loss rate is proportional to $1/Q$. In a laser the loss rate varies greatly from one mode to another because of the difference from mode to mode in the rate of escape of radiation. Axial modes, that is, modes with a high degree of symmetry about the laser axis, have much higher Q's than other modes.

Each mode excited within the laser makes its contribution to the laser beam. Therefore, if modes of different frequency are excited, the laser output will not be monochromatic. The directional distribution of the emitted radiation is characteristic of each mode and the coupling of the mode to the outside. The axial modes radiate in a symmetric beam concentrated around the laser axis. Off-axis modes radiate at an angle to this axis. A cylindrical laser of perfect symmetry produces a far-field radiation pattern consisting of a central spot surrounded by a number of rings. The central spot consists of the radiation emitted by the axial modes and off-axis modes account for the rings.† When it is desired to obtain the maximum power in a concentrated, monochromatic beam, the off-axis modes are deliberately suppressed.

The fact that the laser is a resonant structure makes it necessary to re-examine the laws of absorption and stimulated emission of radiation as they apply to atomic systems within a resonant cavity. Let the modes of this cavity be indexed somehow, and let i be the index that identifies a mode. Let furthermore n_i denote the number of photons in the mode i. Then the quantum theory of radiation provides the following result [6]:

The rate at which an atom in this cavity absorbs a photon from mode i with an increase of internal energy from E_1 to E_2 is proportional to $n_i\delta(E_2 - E_1 - h\nu_i)$.‡ The rate at which the radiation field gains a photon in the mode i with a corresponding loss of energy by the atom is proportional to $(n_i + 1)\delta(E_2 - E_1 - h\nu_i)$.

This theory accounts for spontaneous emission, since emission takes place into each available mode with equal probability when $n_i = 0$ for all i's. The spontaneous emission rate is proportional to the total number of available modes. In the case of a large cavity, this number is given by (5.12). It follows readily that the ratio of spontaneous– to stimulated emission rate must be equal to the mode density times the energy of

---

* The definition of Q is that the energy stored in a free-running oscillator with a resonant angular frequency ω_0 varies with time as $\exp(-\omega_0 t/Q)$.
† Diffraction rings may also arise, but that is another matter.
‡ Here $\delta(x)$ denotes Dirac's delta function or a similar positive function concentrated around $x = 0$ and normalized so that its integral is one.

a quantum $(h\nu)$, a result in agreement with Einstein's relation (3.6) in Chapter 1.

The essential result of the theory needed here is that *stimulated emission into each mode takes place at a rate proportional to the radiation density in that mode*. As a consequence, the laser will operate with each mode feeding separately on the available population inversion. The number of photons in any particular mode will grow if the threshold condition is satisfied for that particular mode. Because different modes differ in resonant frequency and loss factor, only selected modes will reach threshold. First, the resonant frequency of the mode must lie within the region where the amplification curve is positive. Second, the threshold condition must be satisfied for the particular mode. The general threshold condition (1.4), $\alpha = \gamma/L$, may be expressed in terms of the photon lifetime t_p as follows

$$\alpha = \frac{1}{ct_p}\frac{L'}{L},\tag{6.1}$$

where L' is the optical distance between the mirrors. Each mode has its own photon lifetime t_{pi}, which is related to the Q of the mode as follows:

$$Q_i = 2\pi\nu_i t_{pi}.\tag{6.2}$$

Therefore the threshold condition for the ith mode is*

$$\alpha(\nu) = \frac{2\pi\nu}{Q_i c}\frac{L'}{L} = \frac{2\pi}{\lambda Q_i}\frac{L'}{L}.\tag{6.3}$$

In order to concentrate most of the available energy into a few modes, the laser is designed so as to make only a few Q_i's large. Laser operation will then take place only at a few selected frequencies. This situation is illustrated in Fig. 3.5 (p. 75), where the atomic linewidth is much broader than the linewidth of the axial modes shown in the figure.

The output frequencies in this figure are shown to be equally spaced, as we would expect them to be from the conclusions reached in Sections 3.4 and 3.5. In these sections it was shown that the high-order axial modes of a passive cavity are equidistant in frequency. The frequency difference of an adjacent mode was calculated as $c/2L\eta$. A close examination of the spectrum of a multimode laser shows that the actual operating frequencies are displaced toward the center frequency of the atomic

---

* The subscripts were omitted from ν and λ because these quantities vary little within the linewidth.

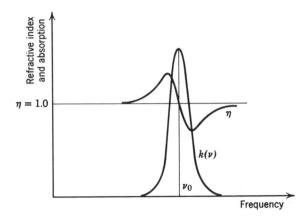

FIG. 3.7 Graphic representation of the dispersion relation.

line and that the constancy of the frequency differences is only approximate. The reason for these anomalies is the fact that the laser is *not* a passive cavity. When significant absorption or amplification takes place in the cavity, the refractive index η is no longer a constant. It varies rapidly in the vicinity of the peak absorption or amplification. This variation is well known from the studies of dispersion and is entirely analogous to the variation of the phase of the response as the driving frequency of an oscillator is swept past the frequency of resonance. The graphical relationship between the variation of $\eta - 1$ and $k(\nu)$ is shown in Fig. 3.7. This figure shows an approximately linear variation of $\eta - 1$ near the peak of the absorption curve. When absorption is replaced by amplification, the sign of the $\eta - 1$ curve is reversed. In this case the curve $\eta(\nu)$ rises in the immediate neighborhood of ν_0 at a rate inversely proportional to the linewidth. The mathematical relationship connecting the index of refraction with absorption is the Kramers-Kronig relation. With its aid Bennett [7] has shown that in the case of a Lorentz-type atomic line of center frequency ν_a and linewidth $\Delta\nu_a$, the actual oscillating frequency of a mode is

$$\nu' = \frac{\nu_c\,\Delta\nu_a + \nu_a\,\Delta\nu_c}{\Delta\nu_a + \Delta\nu_c},\tag{6.4}$$

where ν_c is the peak mode frequency and $\Delta\nu_c$ the linewidth of the mode in the passive cavity. When the peaks of the atomic and cavity lines coincide, then $\nu' = \nu_c = \nu_a$, otherwise ν' lies between ν_c and ν_a.

In any case

$$\nu' - \nu_c = \frac{(\nu_a - \nu_c)\,\Delta\nu_c}{\Delta\nu_a + \Delta\nu_c}, \tag{6.5}$$

and, since $\Delta\nu_a \gg \Delta\nu_c$, we have approximately

$$\nu' - \nu_c = (\nu_a - \nu_c)\frac{\Delta\nu_c}{\Delta\nu_a}. \tag{6.6}$$

This relation holds approximately even when the lineshape deviates from the Lorentzian form. The shift of the laser output from the passive cavity mode toward the center of the atomic line is called *mode pulling*.

3.7 DIFFRACTION THEORY OF THE PLANE FABRY-PEROT INTERFEROMETER

The cavity theory serves as a basis for orientation and provides adequate solutions for solid-state lasers with reasonably high refractive indices. It is not applicable to gas lasers, which are open structures consisting of a pair of plane or curved mirrors at the ends of an amplifying column. In a typical case a pair of plane circular mirrors 2 cm in diameter may be located at a distance of 1 m from each other. In a situation of this type, diffraction loss may not be negligible, in fact, it may be an important factor determining the distribution of energy in the interferometer during oscillation. For laser oscillations to occur the total loss in power from scattering, diffractive spillover, and incomplete reflection on the mirrors must be balanced with power gained by travel through the active medium. In the presence of decoupled or orthogonal modes of oscillation the threshold condition must be satisfied for each mode in which oscillations are to occur.

The parallel partially transparent mirrors of the laser form a Fabry-Perot interferometer. When such an instrument is operated as a passive device with uniform plane waves continuously supplied from the outside, the internal fields may also be essentially uniform plane waves. In a laser, however, where power is supplied only from within the interferometer, the loss of power from the "edges" of the wave by diffraction will cause a marked departure from uniformity.

What, then, are the modes of the Fabry-Perot interferometer? These modes may be defined and discussed in terms of *self-reproducing field configurations* over the surfaces of the reflectors. A field configuration is called self-reproducing or a *transverse mode* if, after propagation from one reflector to the other and back, the field returns to the same phase

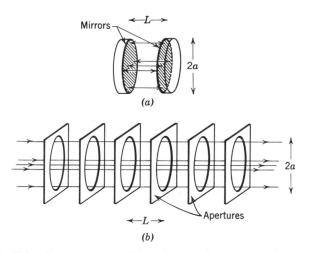

FIG. 3.8   Fabry-Perot resonator (a) and equivalent transmission system (b).

and amplitude pattern; that is, the function representing the complex amplitude over the reflector is multiplied by a fixed complex number which gives the total phase shift and the loss of the round trip. For every such transverse mode there is a sequence of longitudinal modes for which the round-trip phase shift is an integral multiple of 2π.

It is instructive to look at the problem of finding stable, or self-reproducing, modes in a Fabry-Perot resonator in the following manner: As far as diffraction is concerned, the Fabry-Perot resonator consisting of two parallel mirrors is equivalent to an infinite sequence of apertures spaced at distances equal to the distance of the mirrors. Such equivalent structures are illustrated in Fig. 3.8. In order to account for the reflection losses at the mirrors, we may imagine that each aperture contains an attenuating filter to reduce the amplitude of the radiation by the required amount.*

In the transmission line, or "beam waveguide", representation just introduced, a self-reproducing mode is an optical field distribution which passes through the transmission line in a stationary manner. It is attenuated but not changed in any other way. A uniform plane wave evidently does not do this, since it changes by diffraction as illustrated at the left side of Fig. 3.9. Several diffraction lobes arise on passage through the first aperture. Only the main lobe enters the second aperture, but its intensity distribution is no longer uniform, nor are the equiphase

---

* When the mirrors are unequal, we have an infinite sequence of pairs of unequal apertures, but this does not change the principle involved.

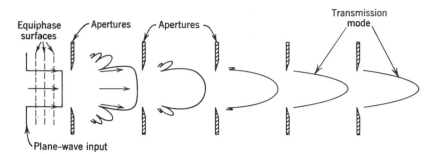

FIG. 3.9 Formation of a self-reproducing mode in a transmission line with plane-wave input.

surfaces plane after passage through the first aperture. It is intuitively clear that passages through many apertures will modify the phase and amplitude distribution in such a manner that the distribution is less and less affected by the apertures. Ultimately, then, if any light passes through at all, its phase and amplitude must be so distributed that its form is unaffected by these apertures. This suspected development is illustrated on the right side of the figure. Calculations show that this intuitive reasoning leads to the correct result. They also show that a symmetric distribution is not the only self-reproducing mode. There are many (mathematically, an infinite number) other modes. These are solutions of an integral equation whose eigenvalues determine the diffraction losses in these modes. The reader interested in the mathematical aspects of the mode theory is referred to the excellent and very readable article by Fox and Li [8], in which the foundation of the subject was developed, and to a number of comprehensive review articles [9—12] which may serve as introductions to the rather extensive literature on the subject. Here we confine ourselves to a brief statement of the method and a summary of its results.

The basis of the calculation is the Huygens-Fresnel principle which enables one to calculate the distribution of amplitude and phase of the radiation in one aperture from the amplitude and phase distribution in the preceding aperture. Let the complex function $u_1(x_1, y_1)$ describe the amplitude and phase in one aperture. Then, according to the Huygens-Fresnel principle, the function $u_2(x_2, y_2)$ that describes the amplitude and phase distribution in the next aperture is given by the Kirchhoff integral

$$u_2(x_2, y_2) = \frac{ik}{4\pi} \int_{\sigma_1} (1 + \cos \theta) \frac{e^{ikR}}{R} u_1(x_1, y_1)\, dx_1\, dy_1. \qquad (7.1)$$

Here $k = 2\pi/\lambda$, R is the distance between the points $P_1(x_1, y_1, 0)$ and $P_2(x_2, y_2, L)$, and θ the angle included between the line P_1P_2 and the z-axis. The integration is extended over the entire aperture σ_1.

A distribution $v(x, y)$ is called *self-reproducing* if setting u_1 equal to v results in $u_2 = \gamma v$, where γ is a complex constant. This means that the original distribution is reproduced, except for a uniform shift in phase and a uniform reduction in amplitude. If we write $\gamma = \exp(\alpha + i\beta)$, the number α will be a measure of the reduction in amplitude and β a measure of the phase shift. The determination of the self-reproducing distribution thus requires the solution of the homogeneous integral equation

$$v(P) = \gamma \int_{\sigma_1} K(P, P')\, v(P')\, dP', \qquad (7.2)$$

whose kernel is given by

$$K = \frac{ik}{4\pi} \frac{e^{ikR}}{R} (1 + \cos \theta). \qquad (7.3)$$

This kernel is symmetric since R and $\cos \theta$ are symmetric in the coordinates (x_1, y_1) and (x_2, y_2). The solution of the integral equation requires the introduction of coordinates appropriate to the symmetry of the aperture and expansion of the kernel, taking into account the orders of magnitude of the wavelength, the aperture dimensions, and the aperture distance L.

There are no complete analytic solutions available for the integral equation described, but considerable information is available concerning the nature of the solutions. In the case of rectangular apertures, which were studied in great detail—more for their mathematical convenience than for their physical significance—the variables may be separated and the solution represented in a product form $v(x, y) = v_1(x)v_2(y)$. For circular mirrors the solution takes the form $v(r, \phi) = R_n(r)e^{-in\varphi}$, where n is an integer. The solutions are classified in analogy with the modes of waveguides and transmission lines by their nodal lines, that is, the lines that divide the aperture into regions of opposite phase. Some of these modes, applicable to lasers with square and circular mirrors, are schematically represented in Fig. 3.10. The dominant, or TEM$_{00}$, modes have no nodal lines; there is no phase reversal over the aperture. In the case of the TEM$_{10}$ mode, however, the aperture is divided into two antisymmetric halves, as shown in the figure. For circular mirrors the TEM$_{mn}$ mode has m nodal lines through the origin ($e^{im\varphi}$ dependence) and n concentric nodal circles over the aperture. Only vertical polarization is shown in the figures, but horizontal polarization is naturally also

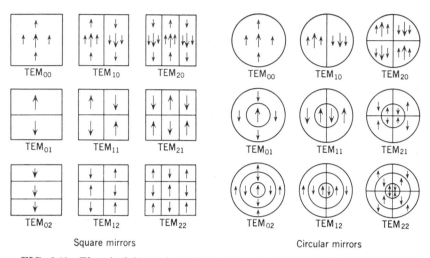

Square mirrors   Circular mirrors

FIG. 3.10 Electric field configurations for interferometers with plane mirrors.

possible and so are all linear combinations of horizontally and vertically polarized modes.

Fox and Li [8] calculated numerically the most important of these modes for a variety of mirror types, including the plane circular mirrors to which we confine ourselves here. The solution was obtained by iteration starting with a distribution of the appropriate symmetry. The amplitude and phase distribution of the dominant TEM_{00} mode was obtained by starting with the trial function $u_1 = 1$ and calculating the subsequent distributions by the repeated application of (7.1), renormalizing the peak amplitude to one before the next iteration. Results of this calculation are shown in Fig. 3.11. The actual dimensions of the mirrors are not important; the only relevant parameter is the *Fresnel number* $N = a^2/\lambda L$, where a is the radius of the circular mirror apertures and L the distance between the mirrors.

The role of the Fresnel number is based on the similarity law of diffraction [13]. This law asserts that the diffraction patterns caused by two objects (or apertures) will be geometrically similar if the quantity $x^2/L\lambda$ is the same for both objects. Here x is an arbitrary linear dimension of the object and L the distance to the screen. Physically, this means that both objects, when viewed from their screens, contain the same number of Fresnel zones. The undulations seen on the curves of Fig. 3.11 are related to the number of Fresnel zones in the diffraction problem. Fox and Li also obtained several interesting nonsymmetric distributions and for several geometries calculated the diffraction losses

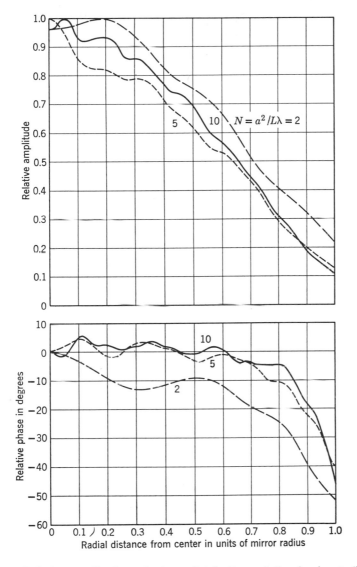

FIG. 3.11 Relative amplitude and phase distributions of the dominant (TEM$_{00}$) mode for circular plane mirrors. (Reproduced from the *Bell System Technical Journal* with the permission of the American Telephone and Telegraph Company.).

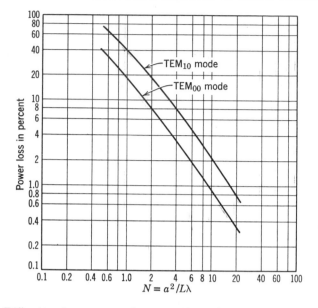

FIG. 3.12 Diffraction loss per transit versus $N = a^2/L\lambda$ for circular plane mirrors. (Reproduced from the *Bell System Technical Journal* with the permission of the American Telephone and Telegraph Company.)

associated with their modes as functions of the number N. In Fig. 3.12 we reproduce their diffraction-loss data pertaining to the interferometer with plane circular mirrors. From this figure it appears that the loss due to diffraction in a single passage in the TEM_{00} mode is about 0.9% for $N = 10$, a value comparable to the usual loss caused by incomplete reflection in a gas laser. The Fresnel number $N = 10$ is applicable to a laser 1 m long and 7 mm in diameter, with a wavelength of 1.15 μm. It is interesting to note that the adjustment of phase and amplitude distribution over the reflectors causes the diffraction loss to decrease in comparison to its value for a uniformly illuminated aperture.

The variation of diffraction loss from mode to mode is of interest because, if there is not much difference in loss between the modes, a laser excited well above threshold will oscillate in several modes simultaneously. Although losses associated with incomplete reflection at the mirrors may be larger than the diffraction losses, the former are constant for all modes; therefore the differences from mode to mode in the sum of all losses arise mainly from differences in diffraction losses. The threshold condition, as we know, involves the total gain along the path and the sum of all losses in a two-way passage. Kotik and Newstein [14] pointed out that in a laterally extended active Fabry-Perot interferometer an oblique mode may be excited before an axial one because the oblique ray passes through a longer amplifying path than an axial ray, and the losses of the modes are nearly identical. The maximum transverse dimension of the mirrors must be limited in terms of the mirror distance, mirror reflectivity, and wavelength in order that the axial mode be favored.

The frequency separation of different modes is of interest in connection with gas lasers since they possess an extremely high resolution. In calculating the mode separations we must proceed with the self-reproducing configurations and not with the cavity calculations of Section 3.5. Waveguide theory warns us not to assume that the velocity of phase propagation of the interferometer modes is equal to the velocity of light. Rather, it is to be expected that this velocity approaches that of light for large values of $N = a^2/\lambda L$. Fox and Li calculated the phase shift in one passage relative to the geometrical phase shift, that is, $2\pi L/\lambda$. Their results show the existence of a differential phase shift which depends on the transverse mode and on the Fresnel number. The frequency difference of consecutive modes of the same type is the same as in the plane-wave case (4.5).

The radiation pattern of the laser is in principle calculable from the phase and amplitude distributions of the self-reproducing configurations. In practice the laser will seldom oscillate in a single-mode type, and the observed radiation pattern is used to identify the different oscillating modes that are actually present.

3.8 LASERS WITH SPHERICAL MIRRORS

A plane parallel interferometer formed by two parallel plane mirrors is not the best multimode resonator. Considerable improvement may be obtained when two similar concave spherical reflectors are used in a confocal arrangement, that is, with the center of one sphere on the other reflector. A schematic representation of such an interferometer is shown in Fig. 3.13.

Fox and Li [8] calculated the first few modes of a *confocal spherical interferometer* using the technique already described. They observed the remarkable properties of the confocal system as contrasted with the plane one: In the confocal system the field is more concentrated near the axis of the reflector and falls to a lower value at the edge than in the plane system. The amplitude distribution is smooth; the ripples of Fig. 3.11 are absent. The surface of the reflector is a phase front of the wave. The losses are orders of magnitude lower in confocal systems than in comparable plane systems. The phase shifts per transit for each

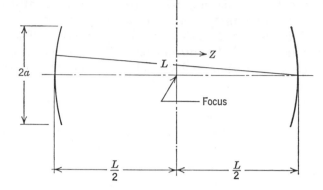

FIG. 3.13   Confocal spherical interferometer.

configuration are independent of N and are multiples of $\pi/2$. This is shown in Fig. 3.14. The frequency difference of consecutive modes of the same type in the confocal interferometer are the same as in the plane interferometer.

Boyd and Gordon [15] solved the integral equation pertaining to the confocal case. Their calculations show that for the fundamental mode the distribution of amplitude in the central part of the reflector is nearly Gaussian, and that the surface of the reflector is an equiphase surface; this is not so in the case of the plane reflectors. The distance

$$w_s = \left(\frac{L\lambda}{\pi}\right)^{\frac{1}{2}}, \tag{8.1}$$

where L is the radius of curvature of the mirrors (Fig. 3.13), is a measure of the spread of the radiation at the reflectors. In the space between the reflectors near the symmetry axis (z-axis) the transverse distribution of the amplitude follows the Gaussian curve with a spread that varies with the axial position. The variation of amplitude is approximately proportional to e^{-u}, where $u = (x^2 + y^2)/w^2$, and

$$w^2 = \frac{L\lambda}{2\pi}(1 + \xi^2). \tag{8.2}$$

Here $\xi = 2z/L$ is the displacement from the focus measured in units of the focal length. At the reflector we have $\xi = 1$ and $w = w_s$; at the focus we have $\xi = 0$ and $w_0 = w_s/\sqrt{2}$. Thus the beam at the focus narrows down to one half its cross section at the reflectors. The varia-

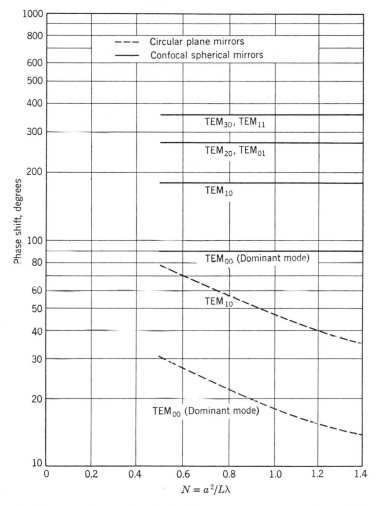

FIG. 3.14 Phase shift per transit (relative to geometrical phase shift) versus $N = a^2/L\lambda$ for confocal spherical mirrors. Dashed curves for circular plane mirrors are shown for comparison.

tion of the beamwaist* w with position is shown in Fig. 3.15. The waist contour $w(z)$ is a hyperbola whose asymptotes have the slope $\pm 2w_0/L$.

The surfaces of constant phase are nearly spherical with a radius of curvature

$$R(\xi) = \frac{1 + \xi^2}{2\xi} L. \tag{8.3}$$

* It is also called "spot size."

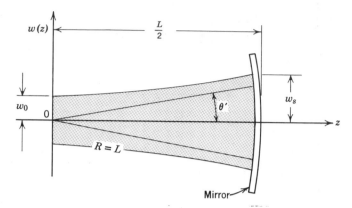

FIG. 3.15 Beam contour in a confocal spherical interferometer. One-half of the laser is shown with the heavy lines indicating the distance w from the axis where the amplitude is $1/e$ times its value on the axis; $\theta' = \tan^{-1} 2w_0/L$.

At $\xi = \pm 1$ this surface coincides with the reflector surface, as expected.

The mathematical form of the amplitude distribution remains valid beyond the output mirror and may be used to calculate the spread of the beam with distance. To obtain the angular beamwidth of the radiation pattern, we take the ratio of the spot diameter obtained from (8.2), as ξ tends to infinity, to the distance from the center of the resonator. The beamwidth between half-power points is given by

$$\theta = 2 \left(\frac{\log 2}{\pi}\right)^{1/2} \left(\frac{\lambda}{L}\right)^{1/2} = 0.939 \left(\frac{\lambda}{L}\right)^{1/2} \text{ rad.} \tag{8.4}$$

The numerical factor arises from the half-power width of the Gaussian curve. It is interesting to note that a, the radius of the reflectors, does not enter into these formulas which govern the spread of the beam; (8.1), (8.2), and (8.4) are applicable, however, only when $a > 3w_s$. It must also be emphasized that the Gaussian approximation is not valid near the edges of the reflector, and that the diameter of the reflector is an important parameter for diffraction losses.

Confocal resonators with mirrors of equal radius represent a special case of two spherical reflectors facing each other. Some other possibilities are illustrated in Fig. 3.16. Not all combinations are practical; most of these mirror combinations are such that any paraxial ray originating

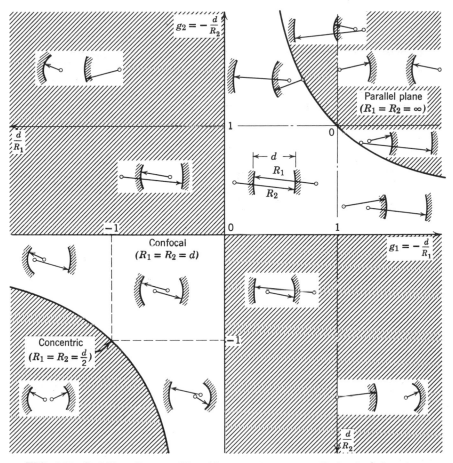

FIG. 3.16 Stability diagram. Unstable resonator systems lie in shaded regions.

within the resonator will leave the resonator after a number of reflections. This means that the structure loses its radiation even in the absence of diffraction. Such structures are called *unstable* or lossy, even in the approximative geometrical sense. *Stable structures* are those which contain rays that are periodically refocused or close upon themselves, as shown in a few examples in Fig. 3.17. Kogelnik and Li [10] have shown that the condition for stability of the resonator is

$$0 < \left(1 - \frac{d}{R_1}\right)\left(1 - \frac{d}{R_2}\right) < 1, \tag{8.5}$$

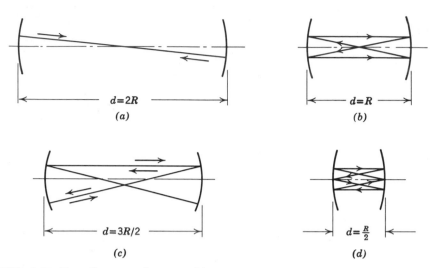

FIG. 3.17 Ray diagrams of some stable symmetric resonator structures: (a) spherical, (b) confocal, (c) and (d) structures within the stable region.

where R_1 and R_2 are the radii of the spherical resonators and d is the distance between their optical centers. It is assumed that the mirrors are so aligned that their optical axes coincide and that convex mirrors are counted with negative radii. When the variables $g_1 = 1 - d/R_1$ and $g_2 = 1 - d/R_2$ are introduced, the regions of stability and instability are separated by the hyperbola $g_1 g_2 = 1$ and by the lines $g_1 = 0$, and $g_2 = 0$. The resulting regions are shown in the stability diagram in Fig. 3.16. It is seen in this diagram that the confocal system with equal mirrors occupies a singular position on the boundary of the stable region. Although truly equal confocally placed mirrors are optimal with respect to diffraction losses, when deviations occur in the curvature of the mirrors, or when the distance between them is altered, the resulting structures have rather complicated properties.

A small deviation from equal curvature may produce a disproportionate increase in the loss of a symmetrical confocal resonator. Therefore, in order to allow for manufacturing tolerances, it is advisable to deviate deliberately from the symmetrical confocal structure or to imitate this structure by means of a spherical and plane mirror combination. Such imitated symmetric structures are shown in Fig. 3.18. They always lie on the $g_1 = g_2$ line of the stability diagram.

When a resonator structure is such that its representative point lies

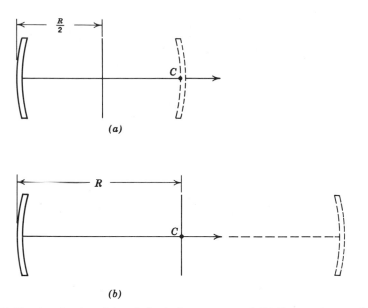

(a)

(b)

FIG. 3.18 Hemiconfocal and hemispherical resonators. *Solid lines:* mirrors: *dotted lines:* their images. C is the center of the spherical reflecting surface.

in the interior of the stability region, the field of its fundamental modes is limited by focusing rather than by the size of the mirrors. This is true, of course, only if the mirrors are sufficiently large so that they do not limit the radiation more than the focusing. The mirror, and every other aperture, has to exceed the spot size approximately by a factor of three. When this is the case, the diffraction losses of the system will be very small. In contrast to this, when the resonator structure corresponds to a point on the boundary of the stability region on the curve $g_1 g_2 = 1$, the spatial extent of the mode is limited by the mirrors. This is the case for plane-parallel–and for spherical resonators. The distribution of radiation density in various practical resonator types shown in Fig. 3.19 illustrates the situation both for focusing-limited and for aperture-limited configuration.

As already noted, the diffraction losses of confocal systems are small. According to Boyd and Gordon [15], the diffraction loss per transit of the fundamental mode of the confocal interferometer is $10.9 \times 10^{-4.94N}$, where $N = a^2/\lambda L$, as in the case of the plane interferometer. For the sake of simplicity we shall use 11×10^{-5N}.

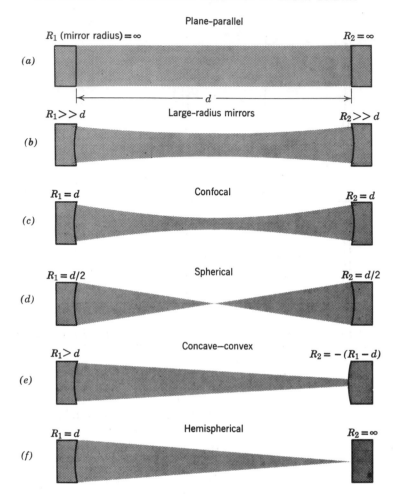

FIG. 3.19 Distribution of radiation in a few common resonator types: (a), (d), and (f) are aperture-limited; (b), (c), and (e) are focusing-limited configurations.

The following comparison may then be made between the diffraction loss of the plane and spherical interferometers of similar size:

$N = \dfrac{a^2}{\lambda L}$	1	2	4
Loss: plane	0.18	0.08	0.03
Loss: spherical	11×10^{-5}	11×10^{-10}	11×10^{-20}

All losses refer to the fundamental modes; the values for the plane interferometer are taken from the curves of Fox and Li reproduced in Fig. 3.12. Clearly, the diffraction losses of the spherical system are orders of magnitude lower than those of the plane system. The confocal spherical system possesses several other significant advantages over the plane one. The noncritical nature of the adjustment of the reflectors is an important one for the experimentalist.

One gains some insight into the structure of resonators constructed by means of various spherical mirrors by studying the propagation of a nonuniform wave field without reference to mirrors. The diffraction formulas permit the calculation of the phase and amplitude in one cross section from the same data in another cross section. It has been shown that an electromagnetic field of the type Boyd and Gordon obtained as a fundamental solution of the confocal cavity problem propagates in such a manner that it retains its essential Gaussian amplitude distribution while changing its spread and curvature in a remarkably simple manner. The amplitude and phase distribution has cylindrical symmetry, and the constant phase surfaces are nearly spherical in the neighborhood of the cylinder axis. If mirrors are placed coincident with any pair of equiphase surfaces, an open resonator with a self-reproducing field distribution is constructed. Thus the same field configuration can be generated by an infinite variety of mirror combinations. In fact, all stable configurations can be generated by a pair of confocal spherical mirrors. Mathematical analysis of the situation leads to the following result: There is an electromagnetic wave whose electric field amplitude has the following spatial variation (for small ρ):

$$E_x(\rho, z) = E_0 \exp\left[-\frac{\rho^2}{w^2} - ik\left(z + \frac{\rho^2}{2R}\right)\right], \tag{8.6}$$

where $\rho = (x^2 + y^2)^{1/2}$, $k = 2\pi/\lambda$, and the quantities w and R are essentially those introduced in (8.2) and (8.3). We write them here in the form

$$w^2 = \frac{f\lambda}{\pi}\left[1 + \left(\frac{z}{f}\right)^2\right], \tag{8.7}$$

$$R = \frac{f^2 + z^2}{z}. \tag{8.8}$$

The number f is a parameter which characterizes the field distribution. The physical meaning of the variables w and R is the same as it was in (8.2) and (8.3). The first one defines the width of the beam, the second one the radius of curvature of the phase front. These parameters vary with z. The beam has its waist at $z = 0$, where $w(0) = \sqrt{f/\lambda\pi}$. The phase front is a plane at this point: $R = \infty$. For $z = \pm f$ the radius of curvature is $R(\pm f) = \pm 2f$. With the mirrors at these locations the resonator is of the confocal spherical type $L = 2f$. As already noted, the same field configuration may be obtained by placing spherical mirrors in other positions, provided their radii have the value given by (8.8).

For the details of the studies of Gaussian beams and spherical resonators we refer to articles already cited [9, 10] and to the book by Sinclair and Bell [16] which contains an excellent summary of the mathematical aspects of this problem, together with the exposition of a practical geometrical construction for the determination of the phase front at any point on the axis. Further references to the periodical literature are also given there.

3.9   DESIGN CONSIDERATIONS FOR GAS LASERS

The mode structure discussed in Section 3.7 and 3.8 guides the choice of the parameters of gas lasers. Design considerations must include such factors as the ease of fabrication, alignment, and maintaince as well as the desired performance and efficiency of the laser.

Plane-parallel resonators are seldom used because they require extreme precision in alignment and because they operate with a large loss factor, largely due to the fact that the radiation in them is aperture-limited and not confined by focusing.

Fully spherical resonators have essentially identical disadvantages. Hemispherical resonators are somewhat easier to adjust and maintain because, once the plane mirror is aligned with the axis of the discharge tube, some angular displacement of the spherical mirror may be tolerated. These resonators, as well as the plane ones, are on the boundary of the stability region and small errors can make them unstable.

Confocal and hemiconfocal resonators have very low losses and are easy to adjust. By choosing the mirror distance d slightly less than required by the confocal geometry, stability can be assured. Once the length of the laser is chosen, the beamspread within the laser is determined by (8.1). All apertures within the laser and the mirror radii must be chosen to exceed this minimum spread by a factor of 3. Since, for a laser 1 m long, operating at a wavelength of 1 μm, w_s is only $\frac{1}{2}$mm, this is not a serious restriction. On the contrary, the confinement of the beam to a narrow region of the tube causes inefficient operation in a discharge tube of much larger diameter.

In order to utilize effectively the excitation over a wider cross section and thus to obtain a greater power output, so called large-radius resonators are often employed; they are intermediate between the confocal and the plane-parallel types. The alignment of such lasers is quite difficult and maximum power output may not be reached by manipulating each mirror separately to obtain maximum power for a fixed setting of the other mirror. For a discussion of mirror alignment parameters and for the alignment procedure applicable to large radius resonators, the reader is referred to Bloom's monograph [17].

Selection of the transverse mode configuration in which a laser operates is usually a result of the variation of losses from one transverse mode to another. When the resonator is constricted by apertures, or when a dominant focused mode is possible, only the dominant TEM_{00} modes appear. In plane-parallel resonators with large apertures, off-axis modes may be preferred because of their longer path of amplification. Often several modes of different types appear and the distribution of the energy among different mode types is a function of the power level, indicating that saturation is taking place in certain regions.

When one wishes to restrict the laser to a single-frequency operation, it is necessary to eliminate operation in any but the dominant TEM_{00}-type mode. It is further necessary to insure that only one TEM_{00n} mode lies in the range where the threshold is exceeded. Since the separation of similar modes in frequency is inversely proportional to the length of the laser [see (4.5)], lasers designed for single-mode operation are made as short as is consistent with other requirements.

REFERENCES

1. T. H. Maiman, Optical and microwave-optical experiments in ruby, *Phys. Rev. Letters,* **4,** 564–566 (1960).
2. T. H. Maiman, Stimulated optical emission in fluorescent solids I, *Phys. Rev.,* **123,** 1145–1150 (1961).
3. T. H. Maiman, R. H. Hoskins, I. J. D'Haenens, C. K. Asawa, and V. Evtuhov, Stimulated optical emission in fluorescent solids II, *Phys. Rev.,* **123,** 1151–1157 (1961).
4. E. L. Steele, *Optical Lasers in Electronics,* Wiley, New York, 1968.
5. S. Ramo, J. R. Whinnery, and T. Van Duzer, *Fields and Waves in Communication Electronics,* Wiley, New York, 1965.
6. A. Yariv, *Quantum Electronics,* Wiley, New York, 1967, Chapter 13.
7. W. R. Bennett Jr., Hole-burning effects in a He-Ne optical maser, *Phys. Rev.,* **126,** 580–593 (1962).
8. A. G. Fox and T. Li, Resonant modes in a maser interferometer, *Bell System Tech. J.,* **40,** 453–488 (1961).

9. H. Kogelnik, Modes in Optical Resonator, *Lasers,* A. K. Levine, Ed., Dekker, New York, 1966, Chapter 5.
10. H. Kogelnick and T. Li, Laser beams and resonators, *Appl. Opt.,* **5,** 1550–1567 (1966).
11. G. Toraldo di Francia, Optical resonators, *Opt. Acta,* **13,** 323–342 (1966).
12. H. K. V. Lotsch, The scalar mathematical description of the Fabry-Perot resonator, *J. Appl. Math. and Phys.* (ZAMP), **18,** 260–272 (1967).
13. A. Sommerfeld, *Optics,* Academic Press, New York, 1954, Section 35.
14. J. Kotik and M. C. Newstein, Theory of laser oscillations in a Fabry-Perot resonators, *J. Appl. Phys.,* **32,** 178–186 (1961).
15. G. D. Boyd and J. P. Gordon, Confocal multimode resonator for millimeter through optical wavelength masers, *Bell System Tech. J.,* **40,** 489–508 (1961).
16. D. C. Sinclair and W. E. Bell, *Gas Laser Technology,* Holt, Rinehart and Winston, New York, 1969.
17. A. L. Bloom, *Gas Lasers,* Wiley, New York, 1968.

Chapter 4   The ruby laser

According to its definition, the laser is a device that amplifies light by means of stimulated emission of radiation. As a matter of practice, a laser is usually a source, or generator, of radiation. The generator is constructed from the amplifier by the addition of a feedback mechanism in the form of mirrors. Having discussed the physics of the amplification process and the basic properties of the electromagnetic cavity formed by the mirrors, we are prepared for a detailed study of the ruby laser as a device. The purpose of the device described in this chapter is to generate light. Ruby lasers built specifically and solely for the amplification of light will be discussed in Chapter 6. The basic operating principles of the ruby laser were already sketched in Section 2.3. Here we shall explain the principles of its construction and then describe its components and performance.

4.1   CONSTRUCTION OF A RUBY LASER

The working element of the ruby laser is a cylinder of pink ruby, usually between $\frac{1}{2}$ and 1 cm in diameter and 2 to 10 cm long. The end faces are ground and polished so that they are plane and parallel to a high degree of accuracy. One of the end faces is provided with a completely reflecting surface, the other is partially reflecting.

The ruby is made to amplify by optical pumping. It is irradiated on its side by light from a flashlamp operated usually for a few milliseconds at a time. In order to utilize most of the light available from the lamp, close optical coupling is provided between the flashlamp and the ruby. An arrangement using a helical flashlamp is shown in Fig. 4.1; arrangements using straight flashlamps are shown in Figs. 4.2 and 4.3. When the lamp is flashed, most of the stored electrical energy is converted to heat. A fraction of the energy, however, is emitted by the flashlamp as blue and green light which is absorbed by the ruby.

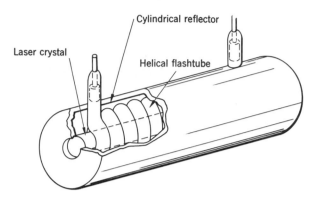

FIG. 4.1 Ruby laser with helical flashlamp.

This energy provides the excitation. The ruby funnels the energy it absorbs over a broad spectral region into a narrow energy level of the trivalent chromium ion. When the rate of excitation is sufficiently high, population inversion results in the chromium ions and the ruby eventually becomes amplifying in a narrow region around 6943 Å. When the amplification is sufficient to overcome the losses in the system, a bright coherent beam emerges through the partially reflecting end of the ruby.

The duration of the flash is usually 1 to 5 msec, that of the coherent output is shorter. A great deal of heat develops when a laser is flashed; this must be removed before it can be flashed again.

The design of a ruby laser requires a quantitative understanding of

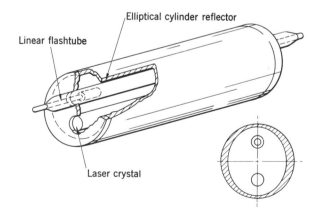

FIG. 4.2 Ruby laser with straight flashlamp in elliptical cylinder configuration.

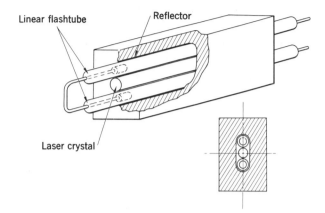

FIG. 4.3 Ruby laser with two flashlamps in close coupling configuration.

many physical and technical properties of its components. It is necessary to know the detailed structure of the energy levels of ruby that are involved in optical excitation and stimulated emission. It is also necessary to know some of the spontaneous transition rates between these levels. In addition, one should know as much as possible about the effects of temperature on the optical properties of ruby. Other physical properties, such as specific heat and heat conductivity, also enter among the design parameters because a large amount of heat is dumped into the ruby in every pulse. The electrical and optical parameters of the exciting lamp are relevant to the design of the energizing circuits and to the calculation of the excitation rate that may be achieved with a system. The latter also depends on the concentration of the available radiation on the ruby and the distribution of the exciting radiation within the ruby. The properties of the emitted laser radiation depend on the modes excited within the laser, and these in turn depend on design factors such as the reflectors and the distribution of the exciting radiation as well as the orientation of the crystallographic axis of the ruby.

All these matters are now discussed in detail.

4.2 THE RUBY

Description of the Crystals and Their Preparation. The heart of the ruby laser is the pink ruby crystal whose chemical composition is Al_2O_3 with 0.05% (by weight) Cr_2O_3. Without the chromium the crystal is known as sapphire. The ordinary physical properties of sapphire and ruby are identical with the exception of the spectroscopic properties.

Ruby has almost cubic symmetry with a small distortion along one of the body diagonals. As a result of this distortion the true symmetry of the crystal is rhombohedral. Its symmetry element is a threefold axis of rotation. As a natural consequence of the rhombohedral symmetry, the crystal is uniaxial; the optic axis coincides with the threefold axis.

Ruby with 0.05% chromium is pale pink. It contains 1.58×10^{19} Cr^{3+} ions per cubic centimeter. At this low concentration the chromium ions are so far apart from one another that their mutual interaction may be neglected. When the concentration is increased to 0.5%, the ruby is red. At this concentration the interaction of adjacent chromium ions is no longer negligible. The spectrum of the red ruby contains new lines which reflect energy levels resulting from the coupling of the paramagnetic chromium ions. This red ruby is used only in special lasers which emit light with a wavelength of 7009 and 7041 Å. In the following, unless specifically stated otherwise, we confine our attention to the pink ruby, the basic material for the common ruby laser whose radiation has a wavelength of 6943 Å. The ordinary and extraordinary indices of refraction of this ruby for red light are $\eta_o = 1.764$ and $\eta_e = 1.756$. Detailed data concerning the variation of the refractive indices with wavelength are available in the literature [1].

Some mechanical and thermal properties of ruby (and sapphire) are listed in Table 4.1. It should be noted that the thermal properties of ruby vary greatly with temperature. As the temperature increases from 60 to 300°K, the thermal conductivity drops almost two orders of magnitude. Since the variation is not linear, interpolation between the points in our table is inadvisable, and the results of experiments published in the literature [1] should be used.

TABLE 4.1

SOME PHYSICAL CONSTANTS OF SAPPHIRE
AND RUBY

Density	3.98 g/cm³
Melting point	2040°C
Specific heat	
at 20°C	0.18 cal/g-°C
at 77°K	0.025 cal/g-°C
Thermal conductivity	
at 20°C	0.092 cal/cm-°C-sec
at 77°K	2.3 cal/cm-°C-sec

Ruby cylinders used in lasers are cut from large crystals of synthetic ruby grown specifically for this purpose. Originally, most ruby crystals intended for use in lasers were grown by the Verneuil flame-fusion technique. The starting material in this technique is a fine powder of alumina (Al_2O_3) mixed with the proper amount of chromium oxide (Cr_2O_3). The powder is contained in a hopper, which is tapped periodically, releasing a small amount of powder which falls through an oxyhydrogen flame. In the flame the powder is heated to about 2050°C and deposited in a molten state on a seed crystal, which is slowly withdrawn from the furnace. With the expansion of laser technology in the early sixties, other methods of growth were perfected for ruby. It was found that better quality rubies could be grown by the Czochralski process. This process requires the production of a melt in a crucible and the careful regulation of the temperature so that the surface of the melt (which is the coolest portion) is maintained around the point of solidification. A small single crystal is then immersed into the surface and is slowly withdrawn from the melt. After 1965 the best crystals for ruby lasers were produced by this process.

The optical quality of the ruby is a critical factor in laser operation. Not only are easily detectable scattering centers detrimental, but so are all variations of the optical path from one end to the other. The mode structure and the radiation pattern of the radiation generated by the laser are largely determined by optical path variations [2]. The optical quality of the crystals and the parallelism of the end faces may be examined by a Twyman-Green interferometer.

Since the growth and preparation of laser rubies is a very specialized art, it is customary to purchase the cylinders necessary for laser construction from suppliers who specialize in crystals for lasers. Most rubies used for lasers in the United States are obtained from the Union Carbide Corp. (Linde Air Products), which grows the crystals, or from the Adolf Meller Co., an importer.

Rubies cut and ground in shapes suitable for use in lasers are available in stock, coated with reflective layers. An investigator who does not plan to fabricate lasers on a large scale will find it to his advantage to purchase rubies cut and coated according to his specifications. The precision necessary for the fabrication of good laser rubies is exacting. Flatness and parallelism of the end faces should be accurate to about one fringe over the entire face. The orientation of the end faces with respect to the crystallographic axis is also of consequence because stimulated emission takes place preferentially polarized in a plane perpendicular to the optical axis. It is customary to cut the rubies so that the cylinder axis is at 60 or 90° to the optical axis.

Defects in the ruby crystal and errors in its preparation result in an increase of the threshold energy required and in deterioration of the far field radiation pattern [2].

The Spectroscopy of Ruby. Fluorescence and stimulated emission of radiation in ruby have already been described in terms of a three-level scheme, illustrated in Fig. 2.2. Although such a description serves a didactic purpose, it represents an oversimplification. The finer details of the optical processes that take place in ruby must be comprehended in the light of a more accurate picture of the relevant energy levels.

It has already been stated that the active material in ruby is the Cr^{3+} ion. This ion has three d electrons in its unfilled shell; the ground state of the free ion is described by the spectroscopic symbol ^4F which indicates that $L = 3$ and $S = \frac{3}{2}$. The ^4F term is a quartet, as indicated by the superscript 4 preceding the letter symbol; its total multiplicity is $(2S + 1)(2L + 1) = 28$. The next group of states of the free Cr^{3+} ion is characterized by the quantum numbers $L = 4$, $S = \frac{1}{2}$; their term symbol is ^2G. The multiplicity of this group is 18.

In ruby the chromium ion is surrounded by a crystal field of nearly octahedral symmetry. This field causes a splitting of the degenerate levels of the free ion. The number of ways in which the levels are split and the remaining degeneracy of the split levels are determined by the symmetry of the crystal field; the extent of the shift of the split levels with respect to the levels of the free ion is determined by the strength of the crystal field, which, in turn, depends on the lattice constant, that is, on the separation of the adjacent like atoms in the crystal.

By applying group theory, it is shown that in an octahedral field the ground level of the free Cr^{3+} ion splits into three levels. It is customary to denote these levels by the symbols $^4F_1$, $^4F_2$, and $^4A_2$; their multiplicities are 12, 12, and 4, respectively. The symbols F_1, F_2, and A_2 arise from group theory; they refer to matrix representations of the octahedral group and are not to be related to values of the orbital angular momentum. In this shorthand notation, the upper index 4 is the only reminder that the three levels in question arise from the ground level of the free Cr^{3+} ion, namely, ^4F. The next level, ^2G, splits into four sublevels, which are designated by the symbols $^2A_1$, $^2F_1$, $^2F_2$, and ^2E, with multiplicities 2, 6, 6, and 4, respectively. Here again the symbols A_1, F_1, F_2, and E are borrowed from group theory, and the only symbol retained from the spectroscopy of the free ion is the prefix 2, indicating the doublet (G) level.

Some of the levels described are outside the range of interest for stimulated emission in ruby. The relevant levels are shown in Fig. 4.4. It

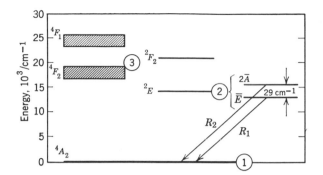

FIG. 4.4   Energy-level diagram for chromium ion in ruby.

is to be noted that the ^2E level is shown not as a single level but as a pair of nearby levels. The splitting is caused by the fact that the symmetry of the crystal is not completely octahedral but rhombohedric. As a result of the lower symmetry, the ^2E level splits into two levels of multiplicities 2 each. These are only 29 cm^{-1} apart; therefore, in equilibrium at room temperature, their populations are nearly equal.

It is of some interest to note that the $^4F_1$ and $^4F_2$ levels are broad bands, while the ^2E levels are sharp. The reason for this was discovered by Sugano and Tanabe [3], who showed that the splitting of the ^2G levels is nearly independent of the crystal field for moderate fields, but the splitting of the ^4F level depends strongly on the crystal field. As a consequence of thermal vibrations of the lattice, the crystal field constantly undergoes rapid fluctuations. These fluctuations are reflected in the splitting of the ^4F level. They cause a rapid variation in the energy of excitation of the $^4F_1$ and $^4F_2$ levels with respect to the ground level $^4A_2$, but they do not significantly affect the doublet levels ^2E and $^2F_2$. The difference between the widths of these levels is of great importance for laser operation because a broad level is needed for absorption and a narrow one for emission.

The absorptive properties of pink ruby, shown in Fig. 4.5, are dependent on the direction of the light propagation with respect to the optic axis. The peaks of the absorption curves clearly correspond to centers of the bands shown in Fig. 4.4. Fluorescence of ruby consists of the R_1 and R_2 lines (Fig. 4.5), but ordinarily laser action will take place only at the R_1 line (6943 Å) because the transition probability for this line is greater than for the R_2 line. Once the level $^2E(\bar{E})$ is depleted by stimulated emission, a rapid transfer of electrons takes place from the level $^2E(\bar{A})$ to the level $^2E(\bar{E})$, and in this manner the population of the upper level

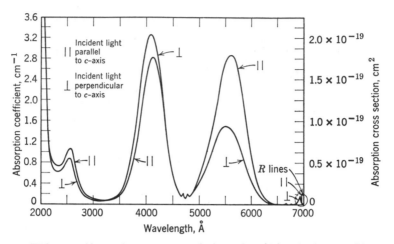

FIG. 4.5 Absorption spectrum of pink ruby. (After Maiman [4].)

never quite reaches the threshold value for laser operation. The ratio of transition probabilities corresponding to the R_1 and R_2 lines is 7 to 5 [5]. The relaxation between the components of the ^2E level occurs with the assistance of lattice vibrations, which are plentiful in the required energy region; therefore the relaxation time will be short ($\approx 10^{-9}$ sec). The levels $^4F_1$ and $^4F_2$ jointly constitute level 3 of the simplified scheme used in Fig. 2.2, whereas the ^2E levels constitute level 2.

It is apparent from Figs. 4.4 and 4.5 that excitation of ruby is best accomplished by irradiation with light in the regions between 3600 and 4500 Å, and also between 5100 and 6000 Å. Laser action is determined by the characteristics of the fluorescence spectrum; that is, the lines R_1 and R_2. These depend in a rather complicated manner on the temperature. At room temperature and above, the shape of the fluorescence lines is that shown in Fig. 4.6. In the temperature range from 20 to 80°C, the position of the peak of the R_1 line is described by the empirical equation

$$\lambda = 6943.25 + 0.068\,(T - 20), \qquad (2.1)$$

where T is the temperature in Celsius degrees. Figure 4.7 shows the temperature dependence of the wavelengths of these lines over a wider range [6]. The width of the R_1 line, as shown in Fig. 4.6, is about 7 Å, or, in terms of wave numbers, $\Delta(1/\lambda)$ is about 15 cm^{-1}. The figure represents the results of Schawlow's measurements at 60°C [7]. At room temperature the width of the R_1 line is about one-half this value. The widths of the R lines vary rapidly with temperature in the range between

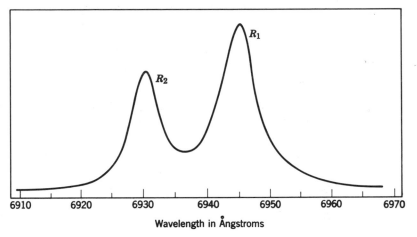

FIG. 4.6   The R lines of ruby in fluorescence at 60°C.

90 and 350°K. This variation is shown on the logarithmic scale in Fig. 4.8, which represents the results of McCumber and Sturge [8]. Below 77°K the linewidths are essentially constant, about 0.1 cm⁻¹. Near liquid helium temperature, however, the fluorescence of ruby changes its character, and further splitting of the lines is observed. The spectroscopy of

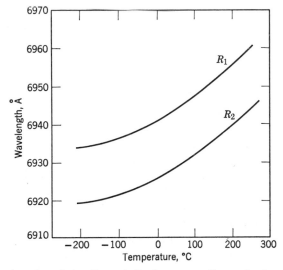

FIG. 4.7   Wavelengths of the R_1 and R_2 fluorescent lines of ruby as functions of temperature. (After Wittke [6].)

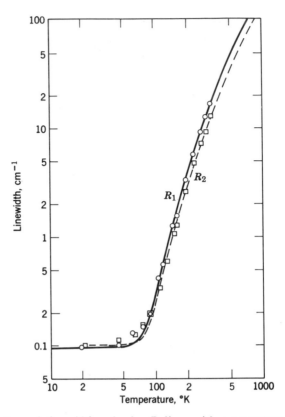

FIG. 4.8 Variation of the widths of ruby R lines with temperature. (The curves represent the values calculated from the theory of McCumber and Sturge [8].)

ruby at very low temperatures (4 to 20°K) is of importance in maser technology and has been studied extensively [2, 7, 8, 9, 10]. Since it is less directly relevant to lasers, we shall not go into its details.

The lifetime τ of the fluorescence radiation of ruby is an important variable because it is one of the parameters which determines the minimum rate of excitation that must be provided to attain the threshold of laser operation. It is related to the spontaneous emission rate A_{21}, but not as directly as one would surmise at first sight. If all electrons excited to the $\bar{E}$ level would return to the ground level by emission of radiation in the R_1 line, then the reciprocal of the lifetime of the $\bar{E}$ level, τ^{-1}, would be equal to A_{21}. Actually, the electrons have a greater freedom in the choice of their return path. First, there are two nearby levels, $\bar{E}$ and $2\bar{A}$, among which interchanges take place at a very fast rate. Decay to the

ground level may take place from either one of these nearby levels, and
the rates of transition from these levels to ground level are unequal.
The experimenter measures the weighted averages of these two return
rates. The weight factors are the populations of these levels. The distribution of the population among the $\bar{E}$ and $2\bar{A}$ levels depends on the
temperature when kT is not very large compared to their energy difference (29 cm^{-1}). The result is a temperature dependence of τ, which is
observed at low temperatures. Another complicating effect, which is
dominant at temperatures over 100°K, arises from the fact that the red
fluorescence of ruby is not confined to the R_1 and R_2 lines, but is spread
over a range of about 800 Å as a result of phonon-assisted transitions.
In these transitions a part of the excitation energy is conveyed to the
crystal lattice, and the remainder is radiated at a wavelength other than
that of the R lines. The total red fluorescent efficiency is independent
of temperature, but, because of the temperature dependence of the
phonon-assisted transitions, the R-line quantum efficiency decreases
from about 0.55 at 0°C to about 0.10 at 300°C. Thus the transitions
which do not result in emission in the R-lines, increase in importance
with increasing temperature [10].

At room temperature the lifetime of the R_1 fluorescence is 3 msec. On
decreasing the temperature, the lifetime increases to a maximum of 4.3
msec, which is reached around 100°K. Below 100°K the lifetime again
decreases, approaching about 3.7 msec for very low temperatures [9].

4.3 EXCITATION OF RUBY LASERS

General Requirements of Optical Excitation. The process of optical
pumping consists of changing the atoms of the active material from
their ground state to an excited state by means of light generated in
the pump and absorbed in the active material. It is necessary to generate
light suitable for this excitation and to convey it to the active material.
The generation and transfer of light must be accomplished with a high
degree of efficiency because inefficiency contributes to unnecessary heating and inhibits the attainment of the threshold excitation necessary
for laser action.

To obtain an estimate of the orders of magnitude involved, we note
that density of chromium ions in ruby is 1.6×10^{19} cm^{-3}, and that the
energy necessary to raise a chromium atom from the ground state into
the $^4F_1$ band is about 4.8×10^{-12} erg. Because the laser transition in
ruby terminates on the ground level, amplification takes place only if
at least one half of the chromium ions are removed from the ground

state. Therefore, the optical energy absorbed by the ruby in a short exciting pulse must be at least

$$q = \tfrac{1}{2} \times 1.6 \times 10^{19} \text{ cm}^{-3} \times 4.8 \times 10^{-12} \text{ ergs} = 3.8 \text{ J/cm}^3.$$

Because only a small fraction of the electrical input of the exciting lamp is converted into useful radiation, and because much of this radiation does not reach the ruby, or is not absorbed in it, a practical ruby laser requires at least 100 J of electrical input per cubic centimeter of ruby to operate. The time during which the excitation energy is delivered is also significant, because the excitation in ruby at room temperature decays with a time constant of 3 msec. Consequently the rate at which power must be absorbed just to maintain a ruby at the threshold of amplification is

$$P = \frac{3.8 \text{ J/cm}^3}{3.0 \times 10^{-3} \text{ sec}} = 1.27 \text{ kW/cm}^3.$$

Since these energy and power levels are quite large, the optical pump system must be so planned that the heat developed in the exciter-laser system can be removed fast enough to prevent an intolerable rise of temperature in all components.

Ideally, one would wish to have a lamp that produces an output only in the spectral region most suitable for the excitation of the active material (ruby). This lamp would then be combined with an optical system that concentrates all radiation emitted from the lamp in the active crystal in such a manner that all parts of the active crystal are equally irradiated. Such uniform irradiation would produce a laser in which the threshold is reached simultaneously throughout the entire active material, and, therefore, the laser pulse could develop uniformly over the entire end face of the laser. An ideal lamp used for continuous operation would have a steady output; one used for pulse excitation would emit a square pulse. Naturally none of these ideals are attainable and a number of compromises have to be made.

Production of the Exciting Radiation. Optical pumping of ruby is possible by means of blue and green light because of the presence of two absorption bands in ruby (see Figs. 4.4 and 4.5). Electrons excited into the $^4F_1$ and $^4F_2$ levels are transferred without emission of radiation to the ^2E levels with their energy difference going into heat. Since less heat develops when the optical excitation takes place by means of green light into the lower lying $^4F_2$ level, an optical pump, emitting only in the 5100- to 6000-Å interval, would be preferred. Such an ideal source does not exist. Fortunately, xenon-filled photographic flashlamps, when

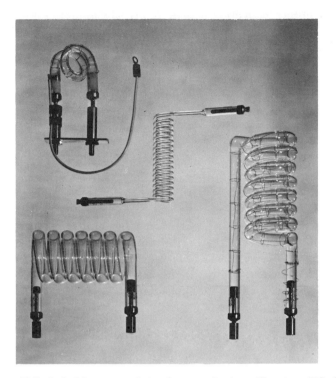

FIG. 4.9 Helical flashlamps used for laser excitation. (Courtesy EG & G.)

properly operated, emit a significant fraction of their optical output
in the favorable spectral region. These flashlamps are used either in
the form of helical coils surrounding the ruby crystal, or in the form
of straight rods placed alongside the ruby rod, or in the focal line of
a reflector. Some helical flashlamp types used for laser excitation are
shown in Fig. 4.9. A straight flashlamp with a jacket for the circulation
of a cooling liquid is shown in Fig. 4.10.

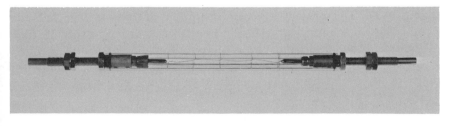

FIG. 4.10 Straight flashlamp with jacket for coolant. (Courtesy EG & G.)

The flashtubes convert between 5 and 10% of the electrical input energy into radiation in–and near the visible range. The total brightness and the spectral distribution of the radiation depend on the electrical input. At moderate current densities (below 1000 A/cm^2) a xenon lamp operates mostly as a spectral lamp; its radiation arises largely from the electronic spectrum of xenon. At very high current densities (over 20,000 A/cm^2) the radiation of the lamp approximates that of a blackbody. The optical and electrical properties of the flashtubes depend in a complicated manner on gas pressure, current density, and tube geometry. Although considerable general information is available in the pertinent scientific literature [11, 12, 13, 14], the manufacturers' catalogs are probably the best sources of practical information for the user of these tubes.*

Commercial flashtubes suitable for laser excitation usually contain xenon at a pressure of 300 to 600 torr. Their spectral output in all spectral regions increases with increasing current density and the center of the spectrum shifts toward the higher frequencies, in accordance with the known properties of blackbody radiators.

The amount of the electrical energy dissipated in a flash and the evolution of the electrical pulse in time affect the lifetime of the flashtube. A uniformly high current density over the entire cross section of the tube is desirable for efficient light generation. Experience shows that when a high voltage is placed across the tube, and the tube is triggered in the usual manner, a stable arc does not develop over the entire cross section of the tube. The discharge begins in a filament with extremely high current density, and a shock wave is generated which spreads from the filament to the walls of the tube. Unless the rise of this initial current is limited, the shock wave may reach destructive proportions. It is therefore advisable to provide in series with the tube an inductance of the order of a few hundred microhenries. The rate of current rise and fall is determined by this inductance and by the other circuit parameters, including the impedance of the tube in operating condition and the capacitance of the capacitor used for energy storage.

A diagram of a circuit used for the operation of a flashtube is shown in Fig. 4.11. The resistor R_1 is of the order of 1000 Ω. The inductor L and the capacitor C_1 are chosen to conform with the specifications of the tube. The initiation of the discharge through the flashtube requires a voltage of 10 to 15 kV. It may be accomplished, as shown in the diagram, by obtaining a spark from a spark coil connected to a wire

---

* The manufacturers of flashtubes in the United States include the firms: E G & G, Boston, Mass.; General Electric Co., Photo Lamp Dept., Cleveland, Ohio; P E K, Inc., Sunnyvale, Calif.; Kemlite Laboratories, Chicago, Ill.

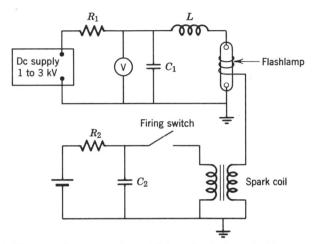

FIG. 4.11 Power supply and firing circuit for a flashlamp.

wound around the flashtube. Alternatively, the tube may be fired by coupling the spark in series with the tube with a transformer. Since it is often inconvenient to pass high current through the firing switch, the simple firing circuit in the lower half of Fig. 4.11 is usually replaced by one involving a thyratron tube, which discharges the capacitor C_2 when its grid voltage is switched.

In a well-designed system most of the electrical energy stored in the capacitor is dissipated in the flashtube. It is possible to deliver a given amount of energy to the tube by using different capacitor and voltage combinations. The evolution of the pulse in the flash tube depends on the combination used. A higher voltage produces a faster development of the arc; the use of a larger capacitor results in a longer-lasting discharge.

The studies of Marshak [13] show that in order to prevent the premature destruction of the flashtube, the following inequality should not be violated

$$CV^4 \leq Kl^3. \tag{3.1}$$

Here l is the length of the tube and K is a constant which depends on the tube material and the tube diameter. Thus the total pulse energy times the square of the voltage must remain below a fixed parameter characteristic of the tube.

Manufacturers of flashtubes provide limiting parameters, usually in the form of curves or formulas specifying the maximum energy dissipation in the tube as a function of flash duration. The circuit must be so designed that a pulse with a given energy cannot be of less duration than the tube data specify. Although tubes can be overloaded, such overloading greatly shortens their lifetime. On the other hand, a tube operated below 70 % of its rated capacity has a long life expectancy. Straight flashtubes have generally low resistance (0.2 to 1.0 Ω) and are designed for short pulses (0.2 to 1.0 msec). Helical flashtubes have higher resistance (1 to 5 Ω) and give longer pulses (1 to 3 msec). As already noted, the destruction of a flashtube is caused by the rapid rise of current during the initiation of the discharge when the discharge has not yet formed over the entire cross section of the tube. It is possible to drive flashtubes well beyond their rated capabilities if the peak current is delayed until after a stable arc has developed across the tube. This can be accomplished by first establishing a discharge by the use of a capacitor charged to a lower voltage and then switching in a second, more highly charged, capacitor [14].

Concentration and Confinement of the Exciting Radiation. We shall now be concerned with the physical arrangement of the flashlamps and the ruby. The guiding design principles are that the arrangement should result in uniform irradiation of the ruby, that it should be as efficient as possible, and that it should permit cooling of the ruby and the flashlamp.

The original arrangement with the ruby in the center of a helical flashtube, as shown in Figs. 2.1 and 4.1, is still widely used. There are, however, several other configurations available for use with straight flashtubes. Two examples were shown in Figs. 4.2 and 4.3. We shall now examine the properties of such design configurations.

The elliptical configuration shown in Fig. 4.2 consists of a straight flashlamp and a ruby cylinder of nearly equal size placed in a highly reflecting cylindrical cavity of elliptic cross section. The cavity is usually machined-out of a block of aluminum and is highly polished. It is provided with reflecting plane end plates, which hold the ruby crystal and the flash tube in position so that they are centered around the focal lines of the elliptic cylinder. The first such laser was described by Ciftan and co-workers [15], who operated a ruby laser 6 cm long and 0.6 cm in diameter, with an input energy of only 150 J. This is only a small fraction of the energy required for the same ruby in the helical configuration. An objective scale is not available for direct comparison of laser performance because the excitation energy required for reaching threshold varies with the efficiency of the lamp, the quality of the ruby, and the reflective coating as well as with the size of the ruby.

The elliptic configuration is based on the geometrical theorem which assures that rays originating from one focus of an ellipse are reflected into the other focus. The application of this simple theorem leads to a number of complications when the objects to which it is applied are no longer mathematical points and lines, but physical bodies of finite dimensions.

In order to concentrate a large amount of power in a small volume, we would like to make the laser rod thin and the source lamp thick because its power-handling capability increases with its diameter. We shall see, however, that the efficiency of radiative power transfer is greater in the reverse case, namely, when the laser is thick and the source is thin. Moreover, calculations show that this efficiency depends on the eccentricity of the ellipse.

There are two principal reasons why not all radiation emanating from a source of finite size will reach the laser. First, rays that leave the source at an angle to the surface normal will be imperfectly focused. Second, some rays will be blocked by the source itself, that is, they will be intercepted after reflection before they can reach the laser. It is sufficient to examine this problem in two dimensions because all rays may be replaced by their projections in the plane perpendicular to the cylinder axis. The complication caused by defocusing is illustrated in Fig. 4.12. A cylindrical source of radius R_1 is placed at the focal line F_1, whereas the laser, whose radius is R_2, is at the focal line F_2. We contemplate the rays reflected at a fixed point P on the ellipse; this point may be characterized by the angle θ included between F_1P and

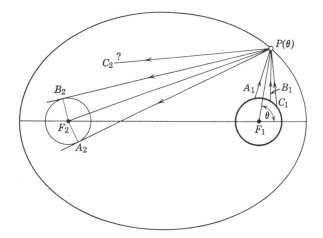

FIG. 4.12 Defocusing as a source of loss in an elliptic configuration.

the major axis of the ellipse. The lines PA_2 and PB_2 are drawn tangent to the laser and the corresponding rays are extended back to the source to points A_1 and B_1. It is clear that A_1PA_2 and B_1PB_2 represent limiting rays for the point P. Any ray arriving at P which originates outside of the arc A_1B_1 will miss the laser. Such a ray is illustrated by C_1P. One could argue that such a ray would eventually reach the laser after many reflections or would be returned to the source. The many reflections, however, are not immaterial because the reflection coefficient of the best reflector usable in this case is considerably less than 1. The loss of energy transfer by source-blocking may be demonstrated by following the path of the ray emanating from C_1 and directed toward the nearest vertex of the ellipse. A ray returned to the source does not represent energy completely lost; nevertheless, it is not communicated to the laser on the first leg of its journey and it suffers a certain loss by reflection. Similarly, it should not be assumed that any ray reaching the laser thereby delivers the full energy associated with it to the laser. Nevertheless, there is merit in defining the efficiency of the system in terms of the relative percentage of rays which travel from the source to the laser and reach it, suffering no more than one reflection on the elliptic cylinder. In counting the rays it is satisfactory to assume that the source radiates according to Lambert's law. Such efficiencies were calculated by Schuldt and Aagard [16] who obtained the curves of Fig. 4.13 for ellipses of varying eccentricities. The values of the efficiency η shown were obtained neglecting source-blocking. Large efficiencies are obtained for nearly circular ellipses which require close arrangement

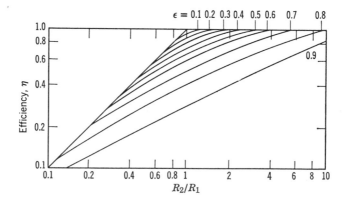

FIG. 4.13 Efficiency of an elliptical cylinder configuration. Calculated for a Lambertian source and negligible source-blocking. Notation: R_1 source radius, R_2 crystal radius, ϵ eccentricity.

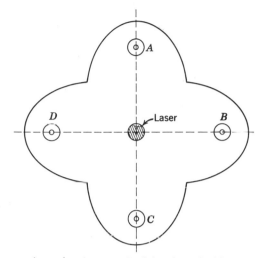

FIG. 4.14 Cross section of a laser excited by four flashlamps (*A*, *B*, *C*, and *D*.)

of the source and the laser. Source-blocking is relatively unimportant when the eccentricity of the ellipse does not exceed 0.4.

Lasers may be constructed with multiply elliptical cavities. These consist of several elliptical cylinders with a common focal line, which is the location of the ruby crystal. Flashtubes occupy the other focal lines. The cross section of such a laser utilizing four flashtubes is shown in Fig. 4.14. It has been demonstrated that the efficiency of these multiple elliptical structures is low relative to that of the single elliptical structure [17]. In fact, the efficiency drops so fast with the number of ellipses that it is barely possible to provide more excitation with two flashlamps in two elliptic cylinders than with one flashlamp in one cylinder. Nevertheless, the use of multiple ellipses has a significant advantage. When one ellipse is used, the illumination of the laser rod is far from uniform. The nonuniformity has two adverse effects. First, it produces thermal gradients and stresses in the laser rod, which tend to distort the optical path within the laser. Second, the nonuniform excitation tends to favor the formation of asymmetric modes. A laser excited by four flashlamps arranged as shown in Fig. 4.14 is for all practical purposes uniformly irradiated and tends to be free of the distortions that would result from asymmetry. An illuminating configuration that produces cylindrically symmetrical irradiation of a ruby rod by means of a single flashlamp was invented by Röss [18]. The laser and the flashlamp are aligned along the axis of rotation of an ellipsoid of revolu-

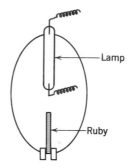

FIG. 4.15 Ellipsoidal pump configuration.

tion (prolate spheroid), as shown schematically in Fig. 4.15. This configuration is particularly suitable for the excitation of small rubies operated at a high repetition rate or even continuously. These and several other excitation geometries are described in Röss' book [19], which contains many drawings and performance data of interest to the designer.

Distribution of the Exciting Radiation in the Ruby. Uniform and powerful irradiation of the laser does not in itself ensure uniform excitation or efficient distribution and utilization of the incident radiative energy in the laser. When a dielectric rod is irradiated by isotropic radiation from the outside, not every portion of the rod is subjected to illumination of the same intensity. The reason for this variation is the refraction of the rays at the surface, which is illustrated in Fig. 4.16. Light, incident at any point, will be confined to a cone within

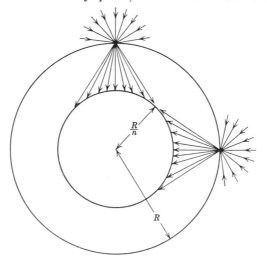

FIG. 4.16 Rays of light incident on the side walls of a dielectric cylinder.

the dielectric, whose aperture is determined by the angle of total reflection $\theta_{\max} = \sin^{-1} 1/n$, n being the refractive index. For sapphire and ruby n is equal to 1.76, so that $\theta_{\max} = 35°$. A ray refracted at this angle passes the axis of the cylinder with a minimum distance $r = R/n$, where R is the radius of the cylinder; that is, all the light that penetrates the surface of the cylinder of radius R eventually passes through a smaller internal cylinder of radius R/n. The calculation of the energy density u in a dielectric cylinder is complicated by the fact that the light entering the cylinder is partially reflected internally every time it strikes the cylinder surface; the reflection coefficient depends on the angle of incidence as well as on the direction of polarization.

The calculation of the density of the exciting radiation in a ruby cylinder is an extremely complicated task, if one takes into account the absorption in ruby and the fact that rays may enter the ruby from any direction. Considerable simplification can be achieved by replacing the three-dimensional problem by a two-dimensional one and by neglecting the absorption in the ruby in the first approximation. The results of such a calculation provide adequate orientation concerning the main features of the general problem.

We shall derive the distribution of light in a transparent (nonabsorbent) circular cylinder of refractive index n, illuminated by radiation incident in the plane perpendicular to the cylinder axis and distributed uniformly about this axis. The objective of the calculation is to find the energy density u as a function of ρ, the distance from the cylinder axis for irradiation having cylindrical symmetry.

Let us first assume a *single* plane wave incident on the dielectric cylinder of radius R and refractive index n. The radiation density within the cylinder created by this plane wave will vary not only with the coordinate ρ, but also with the angle θ included between the normal of the plane wave and the position vector. The *average* radiation density over a cylindrical ring which corresponds to an increment $d\rho$ of ρ is, however, *independent* of θ; hence it is independent of the orientation of the incident plane wave. It is therefore permissible to arrange the calculation so as to consider the effects of a single plane wave.

The plane wave on the outside of the dielectric cylinder is characterized by its intensity S_o, its propagation vector $\mathbf{k}$, and its velocity c. The energy density associated with this radiation field is denoted by

$$u_o = \frac{S_o}{c}. \tag{3.2}$$

When a plane wave represented by a bundle of rays arrives at an air dielectric interface, it is partly reflected and partly refracted according to Snell's law:

$$\sin \alpha = n \sin \beta. \tag{3.3}$$

In lieu of the customary Fresnel coefficients, which relate the electric amplitudes of the refracted and reflected waves to the amplitude of the incident wave, we introduce the energy-flow coefficients defined as follows: The reflection coefficient r_{11} is the power reflected into medium 1 from a given area of the interface divided by the power incident on the same area. The transmission coefficient t_{12} from medium 1 to medium 2 is the power transmitted through a surface element divided by the power incident on the same surface element from medium 1. Clearly, the cross section of the bundle changes on transmission. The coefficients t_{21} and r_{22} are defined similarly, and it is known [20] that $t_{12} = t_{21}$ and $r_{22} = r_{11}$. Therefore the subscripts may be omitted, provided that it is understood that these coefficients depend on the angle of incidence and that $r_{11}(\alpha) = r_{22}(\beta)$ where α and β are related by Snell's law.

Conservation of energy at the interface requires that

$$r + t = 1. \tag{3.4}$$

Figure 4.17 illustrates the path of a representative ray incident at the point P_0 on the dielectric cylinder. The minimum distance of this ray from the cylinder axis is

$$\rho_0(\alpha) = R \sin \beta = \frac{R \sin \alpha}{n}. \tag{3.5}$$

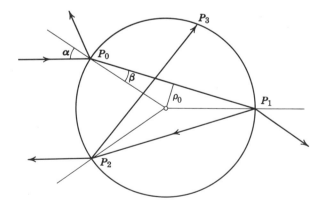

FIG. 4.17 Refraction and reflection in a dielectric cylinder.

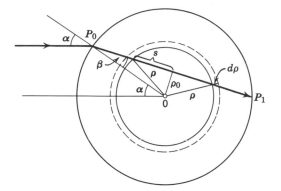

FIG. 4.18   Geometry of rays in a dielectric cylinder.

The ray strikes the cylinder surface again at P_1. It is reflected there, with a reflection coefficient r and passes through the cylinder again on a path that is geometrically similar to the segment P_0P_1.

Let us consider a section of the dielectric cylinder having unit length, and let us denote the surface element of the cylinder by $dA = R\, d\alpha$. A bundle of rays incident at P_0 on a surface element dA has a cross-sectional area $\cos \alpha\, dA$, therefore energy enters the cylinder through this surface element at the rate of $S_o t \cos \alpha\, dA$.

We now calculate the radiative energy introduced into the ring of thickness $d\rho$ through the surface element dA at P_0. Such a ring is shown in Fig. 4.18, where the radius ρ is so chosen that the ray through P_0 intersects the ring; that is, $\rho > \rho_0(\alpha)$. Denoting the length of the ray segment within the cylinder of radius ρ by $2s$, we have

$$s^2 = \rho^2 - \rho_0^2, \tag{3.6}$$

for $\rho > \rho_0$, and $s = 0$ for $\rho < \rho_0$. Two segments of the ray, each of length ds, will lie in the ring bounded by ρ and $\rho + d\rho$. Here

$$s\, ds = \rho\, d\rho, \tag{3.7}$$

for $s \neq 0$, and $ds = 0$ for $s = 0$. The bundle represented by the ray P_0P_1 produces, within the ring of thickness $d\rho$, radiation whose intensity is $S_o t \cos \alpha$, t being the coefficient of transmission. This radiation progresses through the dielectric with the speed c/n, therefore the segment of the bundle whose length is $2ds$ contains the energy

$$dQ_1 = 2nt S_0 \cos \frac{\alpha\, dA\, ds}{c}. \tag{3.8}$$

After reflection at P_1, the ray again passes through the same ring and delivers the energy $r dQ_1$ and so on. The total energy delivered into the ring on all passages is then

$$dQ = dQ_1(1 + r + r^2 + \cdot \cdot \cdot) = \frac{dQ_1}{1 - r}, \qquad (3.9)$$

Since $t/(1 - r) = 1$, and, since $S_o = u_o c$, we have

$$dQ = 2n u_o \cos \alpha \, dA \, ds. \qquad (3.10)$$

On substituting ds from (3.7) and on introducing $dA = R \, d\alpha$, we get

$$dQ = \frac{2n u_o \cos \alpha R \, d\alpha \, \rho \, d\rho}{s}; \qquad (3.11)$$

when $s > 0$, and $dQ = 0$ when $s = 0$. This energy is distributed over a volume $2\pi\rho \, d\rho$. The *average* energy produced by the *entire* plane wave in this ring is obtained by dividing (3.11) by $2\pi\rho \, d\rho$ and integrating over all such α which give a positive s. Thus

$$u(\rho) = \frac{u_o n}{\pi} \int \frac{R \cos \alpha \, d\alpha}{s}. \qquad (3.12)$$

Since

$$s = \left(\rho^2 - \frac{R^2 \sin^2 \alpha}{n^2} \right)^{\frac{1}{2}}, \qquad (3.13)$$

two cases have to be distinguished.

1. When $\rho > R/n$, the angle α may assume any value between $-\pi/2$ and $+\pi/2$. With the introduction of the variable $\rho_0 = R \sin \alpha / n$, we have

$$u(\rho) = \frac{u_o n^2}{\pi} \int_{-R/n}^{R/n} \frac{d\rho_0}{\sqrt{\rho^2 - \rho_0^2}} = \frac{2u_o n^2}{\pi} \sin^{-1} \frac{R}{n\rho}. \qquad (3.14)$$

2. When $\rho \leq R/n$, the integration according to ρ_0 goes from $-\rho$ to $+\rho$; therefore

$$u(\rho) = u_o n^2 \qquad (3.15)$$

The analytic forms of $u(\rho)$ given in (3.14) and (3.15) were derived by Devlin, McKenna, May, and Schawlow [21].

For ruby $n = 1.76$, and the above formulas lead to the top curves in Figs. 4.19 and 4.20. When absorption of radiation within the cylinder is taken into account, the calculation leads to a set of curves with the parameter $d = \alpha R$ where α is the intensity absorption coefficient. Figure 4.19 shows four such curves calculated by McKenna [22]. It is to be noted that the absorption coefficient of ruby varies greatly with wave-

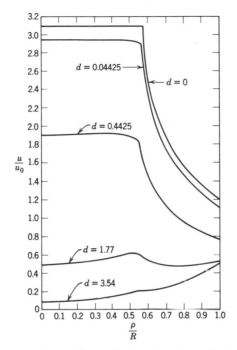

FIG. 4.19 Distribution of pumping radiation in ruby cylinders. Two-dimensional model.

length. Using Maiman's data of absorption (Fig. 4.5), we observe that the absorption coefficient of the unexcited ruby at 4300 Å is about 2.2 cm^{-1}. Therefore, the curves shown in Fig. 4.19 with parameter values $\alpha R = 0.44$, 1.77, and 3.54 correspond to cylinders with radii of 0.2, 0.8, and 1.6 cm, respectively.

The two-dimensional irradiation problem, from which the curves of Fig. 4.19 derive, is an artificial one. More realistic is a three-dimensional problem based on the assumption of an isotopic radiation field outside the ruby, with the polarization of the radiation distributed uniformly among all possible orientations. This problem is mathematically more difficult than the plane problem. Its solution for the loss-free case leads to expressions involving complete elliptic integrals. Figure 4.20 permits a comparison in the loss-free case between the results of the two-dimension calculation [(3.14) and (3.15)] on one hand, and the three-dimensional calculation on the other. The mathematical analysis of the lossy case in three dimensions is very involved. Calculated curves may be found in the papers of McKenna and associates [22, 23]. It is seen from such curves that for small values of the absorption parameter

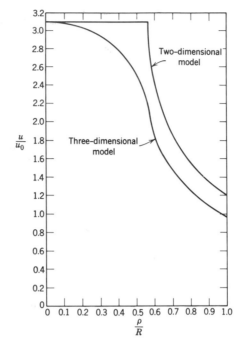

FIG. 4.20 The pumping radiation in a nonabsorbent ruby cylinder. Comparison of the density distributions calculated from a two- and three-dimensional model.

αR, say $\alpha R < 0.5$, the focusing effect predominates. For large values of αR, the radiation density is largest at the periphery because much of the radiation is absorbed before it reaches the center.

The curves in Figs. 4.19 and 4.20, as well as those in the papers quoted, do not give an adequate picture of the rate at which excitation takes place in the cross section of a ruby rod. They represent radiation densities produced within the ruby by means of a *monochromatic* radiation of appropriate wavelength, whose distribution on the outside approximates the mathematical model on which the calculation was based. Actually, the ruby is excited by means of the *total* output of the xenon lamp, therefore the absorption coefficient α of the radiation is variable over a large range. The correct total energy density distribution within the ruby must be calculated by using the spectral distribution $f(\nu)$ of the xenon lamp and the spectral variation of the absorption coefficient $\alpha(\nu)$. For a given ν, $\alpha(\nu)$ must be found first by using the absorption spectrum of ruby. Then the distribution of light of frequency ν is calculated in the ruby for the applicable value of αR, using the analysis which led to the curves referred to earlier. This distribution is $u(\nu, \rho)$. The total energy density is obtained from the integral

$$U(\rho) = \int f(\nu) \, u(\nu, \rho) \, d\nu. \tag{3.16}$$

The energy density so calculated is not the rate at which chromium ions are excited from the ground state. If irradiation was accomplished by monochromatic light of frequency ν, the rate of excitation would be proportional to $\alpha(\nu) \, u(\nu, \rho)$. Therefore the excitation rate for a realistic lamp will be proportional to

$$N(\rho) = \int \alpha(\nu) f(\nu) \, u(\nu, \rho) \, d\nu. \tag{3.17}$$

Curves of such excitation rates were calculated [23] for various values of the relevant parameters, and the general predictions from these calculations were verified experimentally by observation of the distribution of fluorescence over the cross section of the ruby [24].

The detailed conclusions drawn from these elaborate calculations are not very relevant to the excitation of a large ruby cylinder into laser action. In order to obtain the necessary population inversion in ruby, more than half of the chromium ions must be removed from the ground state. When this is done the ruby is bleached; that is, its absorption coefficient is reduced in proportion to its ground-state population. Therefore in the process of the intense excitation of ruby those parts of the crystal that are already highly excited becomes less absorbent, and the local decrease of absorption results in a redistribution of the radiation density over the cross section. The overall effect of this bleaching is a more uniform excitation than would otherwise result.

The general conclusion that follows even from the simplest model is still valid: the concentration of pumping radiation in the peripheral region of a ruby will be less than in the central region, and therefore, in a solid rod of ruby, the threshold of oscillations will be reached in the central region before it can be reached near the periphery. Light absorbed near the periphery may be completely wasted because the threshold level may not be reached before the center part is discharged. On recognizing this lack of homogeneity in the distribution of the pumping radiation, Devlin and co-workers [21] invented a composite rod laser, shown in Fig. 4.21. The rod is constructed of sapphire (pure Al_2O_3), with chromium ions added to the central region, so that the refractive index n is uniform throughout the rod. The radius of the core and the radius of the entire rod are in the proportion $1:n$. A composite rod can be made by first preparing a core rod of a single crystal ruby and then using it as a seed onto which sapphire is deposited. With a composite

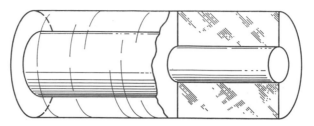

FIG. 4.21 Composite laser. The core section contains active material, the outer shell is clear sapphire.

rod of this sort the light intensity at the ruby core is greater than it is without the sapphire sheath.

Removal of heat is also facilitated by the composite rod structure. Sapphire is a good conductor of heat, especially at low temperatures. The worst barrier to heat flow is at the surface of the crystal, and the surface area of the composite rod is considerably larger than that of the core.

Composite rods of 2-mm inner–and 5-mm outer diameter and of 2.5-cm length were excited to laser action below 100°K by input energies that ranged from 460 to 490 J, whereas ruby rods of the size of the core required 750 J or more when no sheath was provided. These data refer to excitation by helical flashlamps. With a straight lamp and rod at the two foci of an elliptic cylinder, laser action was achieved with input energies as low as 44 J per flash at 77°K. In order to obtain a true measure of efficiency of a composite rod laser, one must take into account the effect of sheathing on the entire geometry, including that of the exciting source. Assuming that the active core of the laser is unchanged, the introduction of the sheathing requires that the inner radius of the helical flashtube be enlarged. Alternatively, if an elliptic configuration is used, the sheathed rod will occupy a larger volume surrounding the focal line of the elliptic cylinder and will by this fact affect the efficiency of the elliptic reflector system.

4.4 RUBY LASER OUTPUT CHARACTERISTICS

General Observations. The bright flash of light that appears on the screen for about a millisecond is the product of several complex processes that operate simultaneously. It is not surprising therefore that, when closely analysed, the output of the ruby laser reveals a complex structure that is highly variable depending on many parameters, some of which are not readily controllable.

The most spectacular observation about the complexity of the situation is obtained by recording the intensity of the output of the ruby laser as it appears on an oscilloscope. The resulting patterns, of which we have seen samples in Figs. 2.4 and 2.5, never quite repeat themselves. Their complexity and lack of reproducibility was a source of frustration to early experimenters, who would have liked a nice single peak, a constant output, or, at the very least, regular sinusoidal pulses on which to make measurements. We will delay the generation of single or repeated pulses by the so-called Q-switching techniques until Chapter 6 and will concern ourselves here with the properties of the radiation from ruby lasers generated by the techniques already described.

The distribution, in space and time, of the radiation emanating from the laser is determined by the distribution of the electromagnetic field within the laser, that is, by the excitation of the different modes of oscillation, and by the coupling of the oscillations from the laser to the outside. If a single one of the modes described in Chapter 3 was excited, the observed intensity pattern would not result. Usually many modes are excited in competition with one another and the excitation shifts rapidly among different modes.

Another disturbing observation had been made by early experimenters. A completely homogeneous ruby cylinder with parallel, uniform end surfaces does not emit coherent radiation uniformly, or even nearly uniformly, over its end surface. Small, very bright spots appear at the end faces which vary in size with the excitation of the ruby. The coherent radiation originates primarily from these bright spots. When only one such spot is present, the angular extent of the far-field pattern corresponds to that proper for a radiator of the size of the spot, not the entire ruby end surface. From the theory of the plane Fabry-Perot resonators one would not expect the formation of a spot, unless the excitation was confined to a center filament of the ruby, which is ordinarily not the case. When several spots are present, the situation is even more complicated.

A study of the laser output must include an effort to understand these highly irregular charateristics of the laser which are not consistent with assumptions of homogeneity and stationary operation. Naturally, the analysis is carried out in terms of such variables as polarization, spectral distribution of energy, coherence, and intensity distribution of the radiation. These variables are related to the geometry of the configuration, the variations of the exciting radiation, and the temperature. They are not static parameters; some vary considerably during the time laser emission takes place. Therefore they must be studied by techniques which permit measurements over very short time intervals.

Polarization. The polarization of the laser output depends on the orientation of the optic axis of the ruby with respect to the cylinder axis. According to the measurements of Nelson and Collins [25], the output is completely unpolarized when these axes are parallel. Ruby rods with their optic axes at a 60- or 90° angle with the cylinder axis give a 100% linearly polarized output with the electric vector perpendicular to the plane containing the optic axis.

Spectral Characteristics. The gross spectral characteristics of the ruby laser are determined by the peak of the fluorescent emission of ruby. At room temperature, the peak of the R_1 line measured in air is 6943 Å. The variation of this wavelength with temperature is described in Section 4.2. The laser emission takes place at the peak of the ruby fluoresence line with an overall linewidth that depends on the degree of excitation. This overall linewidth is frequently between 0.1 and 1.0 Å. It is without genuine physical significance because it is only an indication of the width of the spectral region within which several distinct oscillations take place, each with its own frequency. These individual modes of oscillation have a much narrower linewidth, but they shift around in frequency during a single pulse due to the temperature variation of the ruby. Since a 10°C increase in temperature—which may easily occur—increases the fluorescent wavelength by almost 0.7 Å, the warming of the ruby during the emission of the pulse may shift the wavelength by a significant fraction of one ångström.

In a normal operation many of the oscillation modes are excited and the laser pulse contains a large array of different frequencies, even when measured over a period of time so short that frequency changes due to thermal tuning are negligible. These frequencies are generally too closely spaced for separation by ordinary spectroscopic techniques. Their presence can be detected by mixing the frequencies in a nonlinear detector.

It was shown in Chapter 3 that the frequency difference of adjacent axial modes of a plane interferometer is $\Delta\nu = c/2\eta L$, where η is the refractive index and L the distance of the reflectors. The index of refraction of ruby for red light is 1.76 so that, for a ruby rod 5 cm long, the axial modes are spaced 1700 MHz apart. Beat frequencies between such modes were found by McMurtry [26] and Stickley [27] when the rubies were excited at room temperature. As the rubies were cooled to liquid nitrogen temperature, the beats disappeared, indicating that the fluorescent linewidth of ruby narrowed so that only one of the axial modes was excited.

The excitation of nonaxial, that is, transverse modes is also detectable in the frequency spectrum of the ruby laser. The beat frequencies of modes which differ by one unit in one of the transverse-mode indices are almost two orders of magnitude smaller than the frequency differences between two consecutive axial modes. Stickley [27] detected and identified such beat frequencies in the 20-to-70-MHz range.

The time variation of the spectral distribution of the output of the ruby laser has also been studied using a Fabry-Perot interferometer and a streak camera. These studies reveal the details of the mode structure and the variation of the excitation of these modes during the laser pulse [2].

In summary, the spectral content of the output of a ruby laser over a short period, say, 1 μsec, resembles a comblike structure with the teeth of the comb representing the frequencies of the different modes excited. This comb then shifts around in time and the time variation is due to two principal causes: The shift of the center frequency, as the temperature of the ruby rises, and the shift of the interferometer resonances, as the distance between the mirrors expands. All this variation is within the 0.1-to-1.0-Å width of the laser output that is measured by ordinary spectroscopy.

Mode Structure. In Chapter 3 we described modes of electromagnetic oscillation, characteristic of laserlike idealized structures. Each of these modes is characterized by a frequency, an electromagnetic field distribution pattern, and a quality factor Q. In the idealized structure every one of these modes of oscillation can be excited by itself. It has been noted that only those modes need to be considered whose characteristic frequencies lie within the amplifying range of the laser material and which satisfy the threshold condition. Radiation will not build up in other modes.

Although oscillations in different modes may coexist without disturbing each other, all oscillations in a laser derive their energies from the same supply, the excited atoms of the active material. They are therefore in competition with one another. Moreover, as we have seen, stimulated emission is a nonlinear process. The rate of growth of the intensity of the radiation is proportional to the existing intensity, or, expressed in terms of quantum theory, the number of photons created in a mode is proportional to the number of photons already in that mode (plus one).

The complexity of the oscillation problem in a laser is largely due to this competition of the different modes of oscillation for the limited supply of excited atoms whose excitation energy can be converted to radiative energy. The different modes of oscillation may be regarded as resonant circuits, whose driving force varies with frequency because the stimulated emission rate is the largest near the center of the atomic line. But the rate at which energy builds up in these modes varies not only because of this dependence on the driving force (determined by the relation of the characteristic frequency to the center of the atomic line), but also because the rate of dissipation $(1/Q)$ varies from one mode to the other. The dominant fraction of the energy goes to a small number of modes which have the highest Q's and whose frequencies are near the peak of the atomic resonance. Modes with more energy than the others tend to grow faster, and their growth rate increases with increasing Q. Because of this competition for the limited amount of energy available, the rich modes get richer, the poor ones poorer until almost all energy is concentrated "in the hands of a privileged few."

On the basis of the above description one might expect that a ruby laser will naturally tend to oscillate in a single mode, or in a very few modes only, and that these modes are the simple axial types which exhibit complete symmetry about the laser axis. Practically, this is not the case. Single mode oscillations can be attained only under rare, ideal circumstances. The situation in a ruby laser is not so simple nor so symmetric as the mathematical model.

We have already referred to the formation of bright spots on the end surfaces of the laser and the evidence that coherent radiation seems to originate primarily from these spots and not the entire end surface of the ruby. Assuming uniform excitation, we would expect that the radiation at the ends of a plane parallel ruby cylinder follows the curve a on Fig. 4.22, that is, the radiation drops off near the edges but is otherwise nearly uniform. Experimentally, a distribution, represented by curve b is observed under the most symmetric circumstances. Such confinement of stimulated emission to a narrow central region is not consistent with the model of the plane Fabry-Perot interferometer. It follows, however, from the theory of Fabry-Perot interferometers with curved mirrors, when the curvatures are such as to yield a stable structure (see section 3.8). Confinement of the field in the manner observed thus indicates that there is a refractive index variation within the ruby which makes the plane structure optically equivalent to a structure with mirrors curved concave from the inside [28].

Generally, not only one but an entire collection of bright spots appears, indicating the presence of several modes of oscillation. The minimal number of spots appears at the threshold excitation. Their size and number increases as the excitation is increased, indicating the increased number of modes participating in laser action. The increase in the number of modes is confirmed by the observation of beat frequencies between modes.

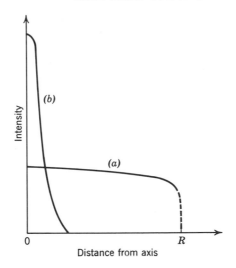

FIG. 4.22 Radial distribution of light (near field pattern) emitted from a ruby laser operating in an axially symmetric mode: (*a*) distribution expected from the plane Fabry-Perot resonator theory. (*b*) distribution actually observed in a typical ruby with plane ends.

The spot patterns seldom are entirely symmetric. Evtuhov and Neeland [29] observed that as a ruby rod excited in an elliptic configuration is rotated there is a change in the spot patterns. This is understandable, since the pump energy is not symmetrically incident upon the ruby in such a configuration. Rotation of the cylinder changes the azimuthal pump distribution. Similar changes in the spot pattern were observed when shielding-strips were placed along the side of the ruby.

Simple spot patterns are obtained only when excitation is near the threshold and when precautions are taken to eliminate all unwanted asymmetries. At higher levels of excitation, the spot pattern changes very rapidly during the laser pulse. As we have noted already, the laser pulse that has a duration of the order of 1 msec consists of many irregular bursts of about 1-μsec duration. High-speed photographs of the laser show that the spot pattern may vary from one burst to the next, indicating a rapid variation of the energy distribution between different types of laser modes. This variation is called *mode hopping*. It is normally present in all ruby lasers when special precautions are not taken to avoid it.

The primary reason for mode hopping is exhaustion of the population inversion in those regions in which the electric intensity is large, whereas the population inversion is not depleted significantly in the vicinity of the nodal surfaces where the electric field is zero. The nodal surfaces of the different modes do not coincide. Therefore, after stimulated emission in one mode depletes population inversion in certain areas, conditions become more favorable for the development of another mode. A shift to a new mode with a different frequency may take place without a change in the transverse mode pattern. In fact, the most common fluctuation observed is that between axially symmetric modes with different axial indices. The shift of excitation from one mode to another explains the observed fact that the output power fluctuates rapidly.

Numerous attempts have been made to construct a mathematical theory to predict the existence and the various parameters of laser pulsations. The common basis for all these theories is the pair of nonlinear differential equations connecting the variation of population inversion and photon density. They differ in their manner of taking into account the numerous physical variables which influence the distribution of the radiation field within the crystal. The mathematical theories invariably predict regular, periodic or damped oscillations, whereas experiments usually show the presence of highly irregular pulsations. The theories are predicated on a homogeneous and essentially isotropic medium excited uniformly, preferably not far above threshold. The predicted relaxation oscillations are very sensitive to small variations in the physical parameters, and the irregular spikes are probably formed by the random interaction of many variables which are not completely controlled. In very homogeneous crystals and under carefully controlled experimental conditions, regular (periodic) pulsations and even steady output may be observed. Regular pulsations are shown in Fig. 4.23. They were ob-

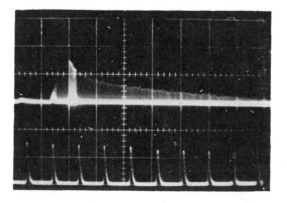

FIG. 4.23 Regular pulsations in ruby output. Time scale: *upper*—200 μsec/cm; *lower*—5 μsec/cm. Power scale—5 kW/cm. (Hughes Research Laboratories.)

tained on a ruby 6.35 cm in length and 1 cm in diameter operating at room temperature. It is found experimentally that such steady, or regular, operation is favored when the laser is forced to operate in a few selected frequencies. This may be accomplished by restricting the oscillations to a few special modes which are favored by the apertures introduced into the laser [2], or by incorporating a frequency filter into the laser cavity which excludes the excitation of all but a few modes [30]. Such a filter may be constructed by using a third partially reflecting mirror which forms a Fabry-Perot etalon with one of the other mirrors.

It should be noted that regular pulsations do not imply single-mode operation. Such pulsations were obtained in highly degenerate laser structures in which many modes of the same frequency were available [31]. They were also obtained in situations where beat frequency measurements have definitely shown the presence of more than one frequency in the laser output. Although it is not possible to state necessary and sufficient conditions for the occurrence of regular pulsations, it is clear from the experimental evidence that such pulsations are favored when the frequency range and the mode range are restricted and when all obvious inhomogeneities are reduced to a minimum. Lowering the temperature of the ruby laser from room temperature to liquid nitrogen temperature may also cause the appearance of regular pulsations. This effect is caused by the narrowing of the fluorescence at the lower temperature and the consequent reduction of the number of excitable modes.

Coherence and Radiation Pattern. Coherence of the light emerging from different areas of the partial reflector and sharp directionality of the beam are related phenomena; directional characteristics of the beam are determined in a well-known manner by the phase and amplitude distribution over the radiating aperture. In fact, the most conspicuous proof of the appearance of stimulated radiation in ruby is the sudden change in the directional distribution of light emitted when the threshold of excitation is exceeded. Collins and others [32] observed this change in directionality with a carefully prepared ruby rod, 0.5 cm in diameter, whose ends were parallel within 1 min. The fluorescent radiation was essentially nondirectional until the threshold was reached; at this point a beam appeared, which was confined to an angle of 0.3 to 1° from the axis of the sample. In general, the collimation of the beam varies greatly from one ruby laser to another; it depends on the qualities of the crystal, the geometry, the reflectivity, and the alignment of the mirrors, as well as on the degree of excitation. A 10-mrad beam is easy to obtain, and a 1-mrad beam can also be achieved, although with considerable care. These beamwidths are considerably larger than we would expect from diffraction theory, assuming radiation from a circular aperture of diameter d radiating in phase and with uniform amplitude. The theory leads to the value

$$\theta_0 = \frac{1.22\lambda}{d}$$

for the angular distance of the first zero from the center of the diffraction pattern. For $d = 1$ cm and $\lambda = 6493 \times 10^{-8}$ cm, the value of θ_0 is 0.085 mrad. The explanation for the much larger observed beamwidths lies in the fact that the entire surface area does not, as assumed, participate in radiation with uniform phase and amplitude.

We already noted that isolated luminous spots appear on the surface of the ruby and stimulated radiation emanates mainly from these. The divergences of the beams radiated from the lasers have the correct magnitude when related to the sizes of these spots and when the possibility is taken into account that different spots may radiate out of phase.

Coherence of the laser output was also studied by means of interference experiments with radiation emanating from different parts of the ruby surface. Direct examination of coherence by means of interference experiments has confirmed expectations. Immediately after the discovery of the ruby laser, investigators at Bell Telephone Laboratories [32, 33] obtained diffraction patterns from a rectangular aperture of 50 by 150 μm on one of the reflectors and interference patterns from a pair of long parallel slits, 7.5 μm wide, separated by 54 μm, also on one of the reflectors. Examination of these patterns discloses that the areas of the reflector undergoing simultaneous laser action are of the order of at least 0.005 cm in diameter. The narrowest beamwidths obtained from these lasers indicate that the area of coherence is at least an order of magnitude larger than this. Experiments of a Russian team [34] show that coherence, in fact, may extend over the entire end surface of the crystal, notwithstanding the broadness of the beam which suggests a more restricted region of coherence. The divergence of the beam results from the fact that the phase is not uniform over the crystal surface.

Power Output and Efficiency. The peak power radiated from a laser without control of the pulsations in intensity is a somewhat indefinite quantity since the peaks of intensity are irregular. Under favorable circumstance, peaks of 20 to 30 kW were obtained from lasers 1 cm in diameter and 4 cm in length. The average during a flash of a millisecond or two is considerably lower. More meaningful and easier to measure is the total energy radiated in one flash. This depends on the excitation and on the size of the ruby; values between 0.1 and 1.5 J may be typical for lasers about 1 cm in diameter and 4 cm in length. Much depends, of course, on the excitation, on the quality of the ruby, and on the quality and alignment of the reflectors. Larger lasers which deliver hundreds of joules in a single flash are commercially available, and the records of energy output per flash are superseded almost every month.

The overall efficiency of the laser expressed in coherent energy output per input electrical energy in the flashlamp is a function of many variables. Among the most significant variables are the spectral characteristics of the flashlamp, which in turn depend on the composition and pressure of the gas in the lamp. A xenon lamp at a pressure of 150 torr seems to be most efficient. With ordinary ruby lasers of 1-cm diameter and 4-cm length it is customary to use an exciting flash of 500 to 1000 J energy. The output under favorable circumstances is about 1 to 1.5 J. The overall efficiency of energy conversion is between 0.1 and 1%.

Extremely high peak values of radiated power can be obtained by preventing the irregular pulsations and causing all the stored energy to be radiated in one burst. This technique leads to the so-called "giant pulse," a phenomenon discussed in Chapter 6.

It should be kept in mind that in order to produce laser action a certain minimum energy has to be supplied per unit volume of the ruby, and that it must be supplied above a minimum rate sufficient to overcome spontaneous decay. The energy absorbed in the ruby is, of course, only a small part of the radiative output of the flashlamp since a significant part of this radiation will not reach the ruby, and only radiation in the proper absorption region of the ruby is effective in providing excitation.

The most serious engineering problems encountered in connection with laser design pertain to the concentration of the available radiation on the ruby and the efficient removal of that overwhelming portion of the input energy which is converted into heat.

4.5 RUBY LASERS OPERATING AT UNCONVENTIONAL FREQUENCIES

The energy-level diagram of the ruby (Fig. 4.4) discloses two closely spaced ^2E levels. We have been concerned so far only with laser action resulting from a transition from the lower one of these to the ground state, the R_1 transition. It has been demonstrated that the R_2 transition can be used for the construction of a laser operating at $\lambda = 6929$ A at room temperature. The two levels from which the R_1 and R_2 transitions originate are separated by an energy difference corresponding to about 29 cm⁻¹. The relaxation time between these nearby levels is short compared with the lifetimes and relaxation times involved in laser action. Under ordinary circumstances, laser action will occur only in the R_1 transition because the threshold condition for this line is met at a lower level of excitation than that for the R_2 line. Once laser action commences in the R_1 line, the level from which the R_1 line originates becomes depleted and population transfer from the other nearby level proceeds at a rate so fast that the threshold population is never reached for the R_2 line. Quantitatively, the situation is as follows: The threshold condition is given by (1.1) of Chapter 3. The coefficient α is proportional to the excitation, but it is greater for the R_1 transition than for the R_2 transition by a factor of 1.4. Thus, in the case of constant γ and increasing α's, the threshold is first reached for the R_1 line. Laser action in the R_2 line was obtained by retarding the start of oscillations in the R_1 line by artificially increasing the loss γ for that line. This was accomplished by employing highly selective interference filters as reflectors [35] and also by the use of a polarizer and birefringent combination [36] designed to discriminate against the R_1 line.

Still another type of laser action is observed in red ruby. This material has a chromium concentration of about 0.5%, about 10 times that of pink ruby. It is known that the energy-level structure of chromium in the red ruby is different from that of the pink ruby because of the exchange interaction between the strongly magnetic chromium ions. Fluorescence in red ruby is known to have peaks at 7009 and 7041 Å, and it is also known that these lines, which terminate above the ground state, must be attributed to interaction between chromium ions.

Wieder and Sarles [37] at Varian Associates and Schawlow and Devlin [38] at Bell Telephone Laboratories demonstrated simultaneously that laser action can be obtained at these wavelengths. These are called the *satellite* N_1 (7041 Å) and N_2 (7009 Å) *lines*. In contrast to the R_1 and R_2 lines, laser action in the N_1 and N_2 lines may appear simultaneously, indicating that their initial levels are not as closely coupled as those of the R_1 and R_2 lines.

Laser action in the satellite lines is observed at liquid nitrogen temperature. Cooling is desirable to depopulate the terminal levels which are about 35 cm⁻¹ above ground state. Although the red ruby laser is of the four-level type, the pump power required for its operation is not less than that required for the R_1 laser.

4.6 CONTINUOUSLY OPERATING RUBY LASERS

The early ruby lasers were so constructed that the user was forced to wait several minutes between pulses, to allow the laser and its auxiliaries to cool. That long a waiting period is clearly not necessary when the laser is so constructed that the heat developed in the system is quickly and efficiently removed. Normal ruby lasers produced after 1962 can be fired a few times per minute.

It is considerably more difficult to design a laser for continuous or quasicontinuous operation. We designate as "quasicontinuous" a laser which can be pulsed so rapidly that the eye perceives a continuous beam. This is certainly the case when the pulse repetition rate exceeds 20 Hz. The difficulty lies with the engineering of a compact source of excitation of the required power handling capability and with the efficient removal of the large amount of heat developing.

We calculated in Section 4.3 that exciting a ruby to the point of amplification requires about 3.8 J/cm^3 energy absorbed in the ruby; maintaining it at this level, that is, compensating for the spontaneous luminescent decay, requires a power of about 1270 W/cm^3. This amount of energy (or power) does not include energy converted into coherent radiation. We noted also that the energy dissipated in the exciting lamp is more than one order of magnitude greater than the energy absorbed by the ruby in the proper energy band. Since the rate of heat removal is limited by the thermal conductivity of the ruby, continuous operation is easier to achieve in thin ruby rods and at low temperatures.

The first continuously operating ruby laser was constructed at Bell Telephone Laboratories within a year after the invention of the laser. The active element of this continuous laser was a tiny trumpet-shaped combination of sapphire and ruby. Its active region was only 0.6 mm in diameter and about 1 cm in length. It was kept in a Dewar flask because it had to be operated at liquid nitrogen temperature. This trumpet-shaped laser of Nelson and Boyle [39] was a triumph of engineering ingenuity but without much scientific or practical significance.

In 1964, Evtuhov and Neeland [40] at the Hughes Research Laboratories constructed a rather simple water-cooled ruby laser capable of continuous operation at room temperature. The cooling water circulates directly through the elliptic cylinder which contains the ruby in one of its focal lines and a mercury arc in the other. Figure 4.24 is a photograph of such a laser which shows how the elliptic cavity is assembled. The following dimensions were used: ruby diameter 2 mm, ruby (and cavity) length 25 mm, eccentricity 0.4 and major axis 25 mm. With a reflector having 1% transmission, threshold was reached for 840 W input to the arc. A continuous power output of 70 mW was obtained when the input was raised to 2000 W. Some time later, an output of 2.4 W was reached with a ruby 75 mm long which was excited with an electrical input of 5080 W [41].

Almost simultaneously, two different types of continuous wave ruby lasers were developed at Siemens Central Laboratories in Munich. Gürs [42] constructed an elliptic cylinder configuration which differed from the Hughes continuous wave laser in that the cooling liquid was confined to transparent jackets surrounding the ruby and the mercury arc. Röss [43] developed a continuous wave laser utilizing an ellipsoidal configuration, illustrated schematically in Fig. 4.15. A cross-sectional drawing of this laser is shown in Fig. 4.25. A glass jacket surrounds the ruby and the mercury arc. Cooling water is circulated in the jacket. The ellipsoidal mirror is cut from a block which has fins, providing for increased heat dissipation. The dimensions of the ruby and the mercury arc are similar to those used by Evtuhov and Neeland. Somewhat higher conversion efficiency has been reported for this configuration than for the elliptic cylinder [19, 43].

The continuous wave ruby laser does not provide constant intensity output; it is subject to essentially the same high-frequency relaxation oscillations (pulsations) that are observed in pulsed ruby lasers.

In most applications the "continuous" lasers here described are not operated in a continuous wave regime. They are pulsed at a repetition

FIG. 4.24 Photograph of a continuous ruby laser. Ruby rod held in left half, mercury arc held in right half. The objects are reflected in the highly polished cylindrical surfaces. (Photo: Hughes Research Laboratories.)

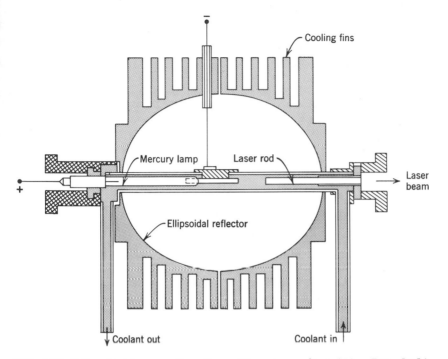

FIG. 4.25 Ellipsoidal laser configuration with water cooling. (After Röss [43].)

rate of 10 to 100 Hz by modulating the input of the mercury arc, or their output is periodically pulsed by one of the Q-switching techniques discussed in Chapter 6.

REFERENCES

1. R. D. Olt, Synthetic maser ruby, *Appl. Opt.*, **1**, 25–30 (1962).
2. V. Evtuhov and J. K. Neeland, Pulsed ruby lasers, *Lasers*, A. K. Levine, Ed., Dekker, New York, 1966, pp. 1–136.
3. S. Sugano and Y. Tanabe, On the absorption spectra of complex ions II, *J. Phys. Soc. Japan*, **9**, 766–779 (1954).
4. T. H. Maiman, R. H. Hoskins, I. J. D'Haenens, C. K. Asawa, and V. Evtuhov, Stimulated optical emission in fluorescent solids II, *Phys. Rev.*, **123**, 1151–1157 (1961).
5. F. J. McClung, S. E. Schwarz, and F. J. Meyers, R_2-line optical maser action in ruby, *J. Appl. Phys.*, **33**, 3139–3140 (1962).
6. J. P. Wittke, Effects of elevated temperatures on the fluorescence and optical maser action of ruby, *J. Appl. Phys.*, **33**, 2333–2335 (1962).
7. A. L. Schawlow, Fine structure and properties of chromium fluorescence in aluminum and magnesium oxide, *Advances in Quantum Electronics*, J. R. Singer, Ed., Columbia University Press, New York, 50–62 (1961).

8. D. E. McCumber and M. D. Sturge, Linewidth and temperature shift of the R lines in ruby, *J. Appl. Phys.*, **34**, 1682–1684 (1963).

9. D. F. Nelson and M. D. Sturge, Relation between absorption and emission in the region of the R lines of ruby, *Phys. Rev.*, **137**, A 1117–1130 (1965).

10. G. Burns and M. I. Nathan, Quantum efficiency of ruby, *J. Appl. Phys.*, **34**, 703–705 (1963).

11. I. S. Marshak and L. I. Shchoukin, Physical and technical parameters of flash-tubes, *J. Soc. Motion Picture Television Engrs.*, **70**, 169–176 (1961).

12. A. Buck, R. Erickson, and F. Barnes, Design and operation of xenon flashtubes, *J. Appl. Phys.*, **34**, 2115–2116 (1963).

13. I. S. Marshak, Limiting parameters and generalized characteristics of xenon lamps, *Appl. Opt.*, **2**, 793–799 (1963).

14. J. L. Emmett, High-intensity flash tubes, *Quantum Electronics and Coherent Light*, P. M. Miles, Ed., Academic Press, New York, 1964, pp. 339–343.

15. M. Ciftan, C. F. Luck, C. G. Shafer, and H. Statz, A ruby laser with an elliptic configuration, *Proc. IRE*, **49**, 960–961 (1961).

16. S. B. Schuldt and R. L. Aagard, An analysis of radiation transfer by means of elliptical cylinder reflectors. *Appl. Opt.*, **2**, 509–513 (1963).

17. D. L. Fried and P. Eltgroth, Efficiency of multiple ellipses confocal laser pumping configuration, *Proc. IRE*, **50**, 2489 (1962).

18. D. Röss, Exfocal pumping of optical masers in elliptic mirrors, *Appl. Opt.*, **3**, 259–265 (1964).

19. D. Röss, *Lasers, Light Amplifiers and Oscillators*, Academic Press, New York, 1968.

20. A. Sommerfeld, *Optics*, Academic Press, New York, 1954 (Section I.4).

21. G. E. Devlin, J. McKenna, A. D. May, and A. L. Schawlow, Composite rod optical masers, *Appl. Opt.*, **1**, 11–15 (1962).

22. J. McKenna, The focusing of light by a dielectric rod, *Appl. Opt.*, **2**, 303–310 (1963).

23. C. H. Cooke, J. McKenna, and J. G. Skinner, Distribution of absorbed power in a side-pumped ruby rod, *Appl. Opt.*, **3**, 957–961 (1964).

24. J. G. Skinner, Pumping energy distribution in ruby rods, *Appl. Opt.*, **3**, 963–965 (1964).

25. D. F. Nelson and R. J. Collins, The polarization of the output from a ruby optical maser, *Advances in Quantum Electronics*, J. R. Singer, Ed., Columbia University Press, New York, 1961, pp. 79–82.

26. B. J. McMurtry, Investigation of ruby optical maser characteristics using microwave phototubes, *Appl. Opt.*, **2**, 767–786 (1963).

27. C. M. Stickley, A study of transverse modes of ruby lasers using beat-frequency detection and fast photography, *Appl. Opt.*, **3**, 967–979 (1964).

28. V. Evtuhov and J. L. Neeland, Characteristics of ruby laser modes in a nominally plane parallel resonator, *Quantum Electronics III*, P. Grivet and N. Bloembergen, Ed., Columbia University Press, New York, 1965. pp. 1405–1414.

29. V. Evtuhov and J. L. Neeland, Observations relating to the transverse and longitudinal modes of a ruby laser, *Appl. Opt.*, **1**, 517–520 (1962).

30. D. Röss, Single mode operation of a room-temperature ruby laser, *Appl. Phys. Letters*, **8**, 109–111 (1966).

31. R. V. Pole and H. Wieder, Continuous operation of a ruby laser during pumping pulse, *Appl. Opt.*, **3**, 1086–1087 (1964).

32. R. J. Collins, D. F. Nelson, A. L. Schawlow et al., Coherence, narrowing direction-

ality and relaxation oscillations in the light emission from ruby, *Phys. Rev. Letters,* **5,** 303–305 (1960).

33. D. F. Nelson and R. J. Collins, Spatial coherence in the output of an optical maser, *J. Appl. Phys.,* **32,** 739–740 (1961).

34. M. D. Galanin, A. M. Leontovich, and Z. A. Chizhikova, Coherence and directionality of ruby laser radiation, *Soviet Phys. JETP,* **16,** 249–251 (1963); **43,** 347–349 (1962).

35. F. J. McClung, S. E. Schwarz, and F. J. Meyers, R_2 line optical maser action in ruby, *J. Appl. Phys.,* **33,** 3139–3140 (1962).

36. C. J. Hubbard and E. W. Fisher, Ruby action at the R_2 wavelength, *Appl. Opt.,* **3,** 1499–1500 (1964).

37. I. Wieder and L. R. Sarles, Stimulated optical emission from exchange-coupled ions of Cr^{3+} in Al_2O_3, *Phys. Rev. Letters,* **6,** 95 (1961).

38. A. L. Schawlow and G. E. Devlin, Simultaneous optical maser action in two ruby satellite lines, *Phys. Rev. Letters,* **6,** 96 (1961).

39. D. F. Nelson and W. S. Boyle, A continuous operating ruby optical maser, *App. Opt.,* **1,** 181–183 (1962). [Also *Appl. Opt. Suppl.,* **1,** 99–101 (1962).]

40. V. Evtuhov and J. K. Neeland, Continuous operation of a ruby laser at room temperature, *Appl. Phys. Letters,* **6,** 75–76 (1965).

41. V. Evtuhov and J. K. Neeland, Power output and efficiency of continuous ruby lasers, *J. Appl. Phys.,* **38,** 4051–4056 (1967).

42. K. Gürs, Ein kontinuierlicher wassergekühlter Rubinlaser, *Phys. Letters,* **16,** 125–127 (1965).

43. D. Röss, Analysis of room-temperature cw ruby lasers, *J. Quantum Electronics QE-2,* 208–214 (1966).

Chapter 5   Four-level solid lasers

5.1   PRINCIPLES OF FOUR-LEVEL OPERATION

Although the ruby laser is one of the most-used solid laser types, its operation is peculiar because its stimulated transitions terminate at the ground level. We have already noted that such operation is necessarily wasteful because the ground level of one half the active atoms must be emptied before laser action is possible, and that the energy expended on exciting these atoms is essentially wasted. Lasers whose fluorescence cycle involves four levels may operate more efficiently. Such four-level cycle was illustrated in Fig. 2.3 on p. 47. The cycle begins at the ground level, level 1, from which the atoms are raised by optical pumping to level 4. This topmost level must be broad, as in the case of three-level lasers. A rapid transition, generally without radiation, then carries the atoms to level 3, which is the starting level of the laser transition. So far the cycle of the four-level laser is identical with that of the three-level laser. The next step is different, however, because stimulated transition in the four-level laser does not take the atoms directly back to the ground level, but into an intermediate level (level 2). This level is generally empty, therefore laser action can start with a very moderate accumulation of atoms in level 3.

A good four-level laser material must possess a reasonably broad absorption band in a frequency region where adequate pump power is conveniently available. Moreover, it must have a terminal level (2), which is normally sparsely occupied and which drains rapidly to the ground state. Finally, there must be natural, fast processes which transfer the excited atoms from the broad level (4) to a narrow level (3). The calculations in Section 3.3 show that the number of atoms N_3 required to be present in level 3 for laser operation is inversely proportional to the peak value $g(0)$ of the normalized lineshape and therefore directly proportional to the linewidth. Consequently, the minimum pump power required to operate the laser is also proportional to the width of the

$3 \rightarrow 2$ transition. Although the calculations of Section 3.3 were made under highly restrictive assumptions, the conclusion that the pump power increases with increasing linewidth is quite generally true. It suggests that the search for four-level laser materials be concentrated on materials with narrow $3 \rightarrow 2$ transitions.

The active elements of a four-level solid laser are ions of a rare-earth, actinide, or transition element in a host crystal. The crystal field splits the degeneracy of multiple levels of the active ion, thus creating a relatively broad absorption band (level 4), and it facilitates the transition from level 4 to level 3, the starting level of stimulated emission.

The energy gap between level 2, the terminal level, and level 1 (the ground level) varies from ion to ion, but it is usually smaller than the energy difference of levels 2 and 3. The lifetime of level 2 must be short, or a phenomenon analogous to the stopping-up of a sink will occur. Until stimulated emission begins, the occupancy of level 2 is determined by the conditions of thermal equilibrium applicable to levels 1 and 2. In order that the terminal level 2 remain essentially unoccupied, it is necessary that $E_2 - E_1$ be greater than kT. Therefore, frequently the temperature of the crystal determines whether or not population inversion can be established at a certain pumping rate; the threshold pumping rate is a function of temperature.

Although some four-level lasers can operate continuously, most such lasers are operated intermittently and are excited by a flashlamp. When refrigeration to liquid nitrogen temperature is needed, an arrangement such as illustrated in Fig. 5.1 may be used with the helical flashcoil surrounding the narrow end of the Dewar flask.

Alternatively, the constricted end of the Dewar flask may be aligned at a focal line of an elliptic cylinder while the light source is placed in the other focal line. Since the absorption spectrum of the laser material varies from one ion to another, different sources of excitation are preferred for different solid lasers. In addition to the xenon arc, mercury arc and tungsten incandescent lamps are used.

The discovery of four-level lasers followed shortly after the discovery of ruby. In 1960 Sorokin and Stevenson [1, 2] observed stimulated emission from U^{3+} and Sm^{2+} ions incorporated in calcium fluoride crystals. Within another year many other rare-earth ions were shown to be suitable laser materials in a variety of host crystals. Practically all rare-earth ions may be used in one medium or another. The typical laser wavelengths obtained from such ions are shown in Table 5.1 with indication of the most common host materials. The wavelengths vary somewhat from one host crystal to another. Detailed description of the host mate-

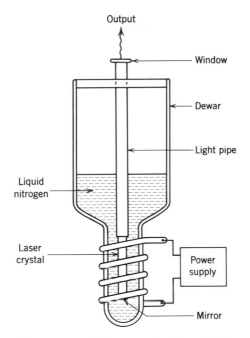

FIG. 5.1 Experimental arrangement for an optically excited laser cooled to liquid nitrogen temperature (77°K).

rials and precise listings of the laser wavelengths are found in several review articles [3, 4, 5].

Technically the most important four-level laser types employ neodymium as the active material. Neodymium ions are capable of laser action at a number of frequencies and in a variety of host crystals as well as in glass. The best results are obtained with Nd^{3+} incorporated into yttrium aluminum garnet ($Y_3Al_5O_{12}$). Very powerful lasers, operating at the wavelength $\lambda = 1.06$ μm, can be constructed from glass containing neodymium. These are competitive with ruby in most applications where the delivery of significant power is required.

Ions of some transition metals (Co, Ni), incorporated in MgF_2 or ZnF_2, can be used to produce rather peculiar lasers. In these materials stimulated emission takes place only when accompanied by the emission of a phonon. That means that the energy difference of the atomic levels is in part converted into heat and in part to stimulated optical radiation.

Systematic and exhaustive description of four-level lasers is a lengthy task because of the large number of combinations possible with a variety

TABLE 5.1
SOME SOLID ION LASERS AND THEIR TYPICAL HOST MATERIALS

Ion	Typical host	Wavelengths (μm)	Note
Cr^{3+}	Sapphire	0.6943	Pink ruby
Cr^{3+}	Sapphire	0.7009, 0.7041	Red ruby
Co^{2+}	MgF_2, ZnF_2	1.75, 1.80, 1.99, 2.05	Phonon-assisted
Ni^{2+}	MgF_2	1.62	Phonon-assisted
Pr^{3+}	$CaWO_4$	1.0468	
Nd^{3+}	$CaWO_4$, glass	0.9142	
Nd^{3+}	$CaWO_4$, CaF_2, YAG, glass	1.04–1.07	Several nearby lines around 1.06 μm.
Nd^{3+}	$CaWO_4$, glass	1.34–1.39	Several lines
Sm^{2+}	SrF_2	0.6969	20°K and below
Sm^{2+}	CaF_2	0.7083	20°K and below
Dy^{2+}	CaF_2	2.36	
Ho^{3+}	CaF_2	0.5512	
Ho^{3+}	$CaWO_4$, $CaMoO_4$, CaF_2, glass	2.05–2.07	Several nearby lines: host dependent.
Ho^{3+}	YAG	2.09–2.12	Several nearby lines
Er^{3+}	$CaWO_4$, CaF_2	1.61	
Er^{3+}	YAG	1.654–1.660	
Er^{3+}	CaF_2	2.69	
Tm^{2+}	CaF_2	1.116	Continuous at 20°K
Tm^{3+}	$CaWO_4$, $Ca(NbO_3)_2$	1.91	
Tm^{3+}	SrF_2	1.97	
Yb^{3+}	YAG	1.0296	
U^{3+}	SrF_2	2.41	
U^{3+}	CaF_2	2.24, 2.51, 2.57, 2.61	

of active ions embedded in a variety of host lattices. The wavelength of the emitted light depends to some extent on the host crystal and to some extent on the temperature. In keeping with the purpose of this book, the discussion here will be confined to the exposition of general principles and to the description of the most practical four-level lasers, namely those in which neodymium is the active element.

5.2 SPECTROSCOPY OF RARE EARTH IONS IN CRYSTALS

Why are rare-earth ions in certain crystals good laser materials? The question can be answered only on the basis of knowledge about the structure of the rare-earth ions and about the interactions that take place between the host crystal and these ions. Only a sketchy and qualitative outline of this subject can be given here. The reader interested in the quantitative aspects of rare-earth spectroscopy is referred to the works of Dieke and Crosswhite [6].

The rare-earth elements fit into the periodic table following the period ending with the fifty-fourth element, xenon. In this element the shells for which the principal quantum number n has values 1, 2, and 3 are completely filled. The shell $n = 4$ has its s, p, and d subshells filled; the $4f$ subshell capable of accommodating 14 electrons is completely empty. Nevertheless, the $n = 5$ shell has acquired its first 8 electrons which fill the $5s$ and $5p$ orbits. Thus, in the customary notation, the electron configuration of xenon is written as follows:

$$1s^2 2s^2 2p^6 3s^2 3p^6 3d^{10} 4s^2 4p^6 4d^{10} 5s^2 5p^6.$$

All elements beyond xenon have this electronic structure and, in addition, have electrons in the $4f$, $5d$, $6s$, etc., orbits. The first addition takes place not in the inner $4f$ orbits, but in the outer $6s$ orbits. Cesium and barium, the elements following xenon, have one and two $6s$ electrons, respectively. Rare-earth elements begin after barium with the building of the inner vacant $4f$ orbits. The electronic structure of the first few rare-earth atoms is illustrated in Table 5.2 with only the $5s$ and $5p$ electrons shown of the completed xenon structure.

The energy differences between the $4f$ and $5d$ orbits are generally small, and in the case of some rare-earth elements the distribution of the electrons between these orbits in the ground state of the atom is not known.

A divalent rare-earth ion is formed when the atom gives up its outermost $6s$ electrons. When a trivalent ion is formed, the atom also loses

TABLE 5.2

ELECTRONIC STRUCTURE OF ELEMENTS 57 TO 62 IN
THEIR GROUND STATE

INNER ELECTRONS OF COMPLETED SHELLS OMITTED

N	Element	$4f$	$5s$	$5p$	$5d$	$5f$	$5g$	$6s$
57	La	0	2	6	1	0	0	2
58	Ce	1	2	6	1	0	0	2
59	Pr	3	2	6	0	0	0	2
60	Nd	4	2	6	0	0	0	2
61	Pm	5	2	6	0	0	0	2
62	Sm	6	2	6	0	0	0	2

TABLE 5.3

$4f$ ELECTRONS AND LEVELS OF RARE-EARTH IONS
(AFTER DIEKE AND CROSSWHITE [6])

R^{2+}	R^{3+}	$4f$	Ground-Level Symbol
—	La	0	$^1S_0$
La	Ce	1	$^2F_{5/2}$
Ce	Pr	2	$^3H_4$
Pr	Nd	3	$^4I_{9/2}$
Nd	Pm	4	$^5I_4$
Pm	Sm	5	$^6H_{5/2}$
Sm	Eu	6	$^7F_0$
Eu	Gd	7	$^8S_{7/2}$
Gd	Tb	8	$^7F_6$
Tb	Dy	9	$^6H_{15/2}$
Dy	Ho	10	$^5I_8$
Ho	Er	11	$^4I_{15/2}$
Er	Tm	12	$^3H_6$
Tm	Yb	13	$^2F_{7/2}$
Yb	Lu	14	$^1S_0$

its $5d$ electron if it has one, otherwise, one of the $4f$ electrons is lost. Thus divalent and trivalent ions of rare earths are simpler than the corresponding atoms. In their ground state they contain only $4f$ electrons in addition to the basic common xenon shell. Table 5.3 contains the number of $4f$ electrons of these ions together with the standard spectroscopic symbol describing the ground state of the ion.

The great complexity of the rare-earth spectra arises from the large number of states that have nearly the same energy. The excited ion configurations of the lowest energy are obtained when one of the $4f$ electrons is promoted into an empty $4f$ or $5d$ orbit. Dieke and Crosswhite [6] calculated that for the Nd^{3+} ion one obtains in this manner 241 configurations with 5393 allowed transitions. For Sm^{3+} these numbers are 1994, and 306, 604, respectively.

The nomenclature of the energy levels may be illustrated by the discussion of the Nd^{3+} ion. This ion has three electrons in the $4f$ subshell. In the ground state their orbits are so aligned that the orbital angular momentum (L) adds up to $3 + 2 + 1 = 6$ atomic units. The symbol I in the last column of Table 5.2 expresses the fact that $L = 6$, since I is the sixth term of the conventional spectroscopic series: $P, D, F, G, H, I, \ldots$.

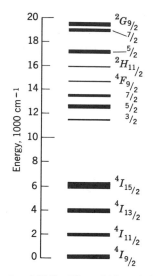

FIG. 5.2 Lowest energy levels of Nd^{3+}. The widths of the lines indicate the extent of the splitting of levels in the crystal field of $LaCl_3$. (After Dieke and Crosswhite.)

The spins of the three electrons are aligned parallel to each other, providing an additional $\frac{3}{2}$ units of angular momentum, which, when added antiparallel to the orbital angular momentum, give a total angular momentum of $6 - \frac{3}{2} = \frac{9}{2}$ units. According to the quantum rules for the addition of angular momenta, the vector sum of an orbital angular momentum of $6\hbar$ and a spin angular momentum of $\frac{3}{2}\hbar$ may result in the following four values of the total angular momentum: $\frac{9}{2}\hbar$, $\frac{11}{2}\hbar$, $\frac{13}{2}\hbar$, and $\frac{15}{2}\hbar$.* The levels corresponding to these values are $^4I_{9/2}$, $^4I_{11/2}$, $^4I_{13/2}$, and $^4I_{15/2}$. The first of these, which has the lowest energy, is the ground level; the others are among the first few excited levels of Nd^{3+}. These levels are distinguished by the orientation of the spins with respect to the resultant orbital angular momentum. Other excited levels are obtained when another combination of the orbital angular momenta is chosen. It should be noted that all these levels are multiple; they are split in electric and magnetic fields. Some of the lowest levels of Nd^{3+} are shown in Fig. 5.2.

Isoelectronic ions, such as Sm^{2+} and Eu^{3+}, appear on the same line in Table 5.3; their ground states are identical. It does not follow, however, that their excited states appear in the same order. In the case of trivalent ions (with the exception of Tb^{3+}), all levels below 50,000 cm^{-1} belong to

* When speaking of an angular momentum of $l\hbar$ we mean a vector whose magnitude is $\sqrt{l(l+1)}\,\hbar$.

$4f^n$ configurations. Transitions between these levels are forbidden by the normal selection rules. The sharp-lined visible spectra observed on trivalent and divalent rare earths in crystals are attributed to these normally forbidden transitions, which are made possible through the action of the crystal field on the ion. These $4f \rightarrow 4f$ transitions are much weaker than the allowed $5d \rightarrow 4f$ transitions. In the case of dipositive rare-earth ions the $4f^{n-1}5d$ levels lie considerably lower than in their isoelectronic tripositive counterparts. In Sm^{2+}, for example, the $4f^{n-1}5d$ levels extend down to 20,000 cm^{-1}, in Dy^{2+} to 16,000 cm^{-1}. The presence of $4f^{n-1}5d$ levels provides dipositive rare-earth ions with broad and strong absorption lines in the region where these are particularly useful for optical pumping. When the excitation is transferred to an excited $4f^n$ state, the conditions are favorable for laser action because the return to the ground state may then occur in a sequence of $4f \rightarrow 4f$ transitions, whose narrow lines facilitate the attainment of the threshold.

The aforementioned advantages of dipositive rare-earth ions are somewhat offset by the fact that the divalent form is usually not the most stable form of these ions, and therefore, without special precautions, only a small fraction of the rare-earth material incorporated in a crystal will be in the divalent form.

Some of the energy levels relevant to laser action in Sm^{2+} are shown in Fig. 5.3. The broad band above the laser levels in samarium is due to $4f^{n-1}5d$ levels. No such band is available in Nd^{3+}, whose excitation is accomplished through several $4f^n$ levels that lie above the $^4F_{3/2}$ level, shown in Fig. 5.2. These figures are not drawn to the same scale. The Nd^{3+} lines shown here in heavy lines would be resolved into groups of individual lines on a larger scale diagram.

The fluorescence spectrum of Nd incorporated in a crystal lattice consists of a multiplicity of lines corresponding to transitions between different sublevels into which these levels are resolved in the crystal field. Under suitable circumstances, a number of these transitions may be employed in a laser. The splitting of the $^4F_{3/2}$ and $^4I_{11/2}$ levels varies with the environment of the Nd ion. This environment may vary not only from one host crystal to another, but even within the same host crystal; the resulting ensemble of ionic levels presents a rather complex picture. Part of the complexity may arise from the presence of a trivalent ion in a divalent lattice. The local imbalance of charge at the site of a given Nd ion can be neutralized in a variety of ways. In a $CaWO_4$ crystal, for example, the charge balance may be restored by calcium vacancies, interstitial oxygen, or pairing of Nd ions. The charge imbalance may be relieved chemically by the addition of monovalent alkali metal ions to the crystal. It is customary to incorporate an alkali metal, usually sodium,

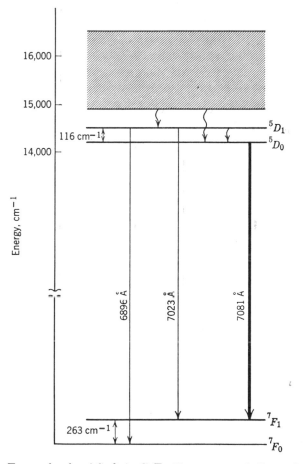

FIG. 5.3 Energy levels of Sm^{2+} in CaF_2. Heavy arrow indicates laser line.

into the host crystal to simplify the level structure of the Nd ion and thus to reduce the number of possible transitions that compete with each other. Different methods of charge compensation lead to somewhat different spectra because each Nd ion gives rise to a spectrum characteristic of the particular local crystal field in which it finds itself. Variations of this type account in part for the large number of Nd laser lines reported.

Analysis of Nd spectra in crystals requires the examination of a number of factors in addition to the charge compensation mechanism. The variation of linewidths and lineshapes with temperature and Nd^{3+} concentration affects the thresholds of oscillations of all lines. Considerable information on these matters may be found in the works of L. F. Johnson [3, 7], who studied the spectra of Nd in a number of host crystals, giving particular attention to $CaWO_4$, which was up to the discovery of yttrium aluminum garnet the preferred crystalline host for a Nd laser.

The phenomena observed with rare-earth ions other than Nd are somewhat similar. It must be borne in mind that even when the crystal is doped with only one rare-earth element, not all rare-earth ions have the same valence in the crystal lattice, and that the observed complex spectrum may be a superposition of the spectra of divalent and trivalent ions located in a variety of crystal sites.

The construction of the elements in the actinide series is similar to the rare earths; therefore the above discussion applies *mutatis mutandis* to the spectrum of uranium and related elements.

5.3 NEODYMIUM CRYSTAL LASERS

A detailed description of the Nd laser is motivated by the fact that stimulated emission has been obtained from this ion in many host lattices as well as in glass. Higher power has been obtained from Nd than from any other four-level material.

The most frequently used host crystals are $CaWO_4$, $SrWO_4$, $SrMoO_4$, $Ca(NbO_3)_2$, and $Y_3Al_5O_{12}$(YAG) with a Nd concentration of 0.5 to 2.0%.* In these crystals laser emission is observed in three groups of lines centered around 0.914, 1.06, and 1.35 μm, respectively. The transitions associated with these lines are shown in Fig. 5.4. It must be kept in mind that the levels shown in the figure are multiple, therefore the laser radiation generally contains several spectral components. The 1.06-μm group is the easiest to excite. There is some variation in wavelength from one host material to another. In CaF_2 and SrF_2, the $^4F_{3/2} \rightarrow {}^4I_{11/2}$ transition results in laser emission around 1.044 μm in contrast to the 1.06 μm in the other crystals mentioned.

The rather extensive experience with $CaWO_4$:Nd^{3+} lasers may be summarized as follows: Stimulated emission may be produced by radiation from a xenon flashtube or a mercury arc. Absorption of the pumping light takes place by excitation to several of the energy levels shown near the top of Fig. 5.2. At 77°K, the wavelength requiring lowest excitation is 1.065 μm. This line may also be excited at room temperature, but the line with the lowest threshold at room temperature has a wavelength

---

* Neodymium concentration data were unfortunately omitted in many original publications pertaining to such lasers.

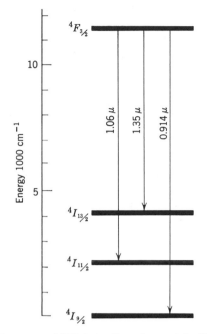

FIG. 5.4. Levels and groups of Nd laser lines in crystals. Each level shown represents a group of closely spaced levels.

of 1.058 μm. At 77°K, we find at least two more wavelengths (1.064 and 1.066 μm) arising from $^4F_{3/2} \rightarrow {}^4I_{11/2}$ transitions. Threshold energies are 1 to 7 J. The above data are applicable to Na-compensated crystals; in other crystals the wavelengths are slightly different and threshold energies are considerably higher. The transitions to the $^4I_{13/2}$ levels yield three laser lines when operated at 77°K; their wavelengths are 1.337, 1.345, and 1.387 μm. The thresholds of the first two lines are low; that of the last one is very high. The terminal levels of these transitions are approximately 4000 cm^{-1} above ground level. The $^4F_{3/2} \rightarrow {}^4I_{9/2}$ transition radiating at 9142 Å (10,935 cm^{-1}) has as its terminal level a sublevel of the ionic ground level, $^4I_{9/2}$, which is only 471 cm^{-1} above the true ground level in the crystal field. This terminal level is partially populated at room temperature; therefore it is easier to produce laser action at 9142 Å in a refrigerated crystal. The splitting of the $^4I_{9/2}$ level is in general more a help than a hindrance for Nd-laser technology because it broadens the absorption bands, so that the neodymium laser can be pumped more effectively at room temperature than at 77°K.

The charge compensation difficulty encountered in $CaWO_4$ is avoided when the Nd^{3+} ion replaces a trivalent metal ion. This takes place in crystals of the type: LaF_3, Y_2O_3, Gd_2O_3 and in yttrium garnets, the most important of which is yttrium aluminum garnet. This crystal (abbreviated YAG) has the composition $Y_3Al_5O_{12}$. It has displaced all other Nd host crystals in technical laser applications because it offers several advantages over calcium tungstate which was the favored host crystal when Nd lasers first came into use. The relevant factors are the following:

1. The fluorescence linewidths of Nd^{3+} in YAG are only one third of those in $CaWO_4$.

2. The absorption bands in the near-infrared region (levels $^4F_{3/2}$, $^4F_{5/2}$, and $^4F_{7/2}$) are stronger by a factor of three in the garnet with the result a tungsten lamp may be used as an efficient pumping source.

3. Garnets have better optical and mechanical qualities than $CaWO_4$.

As a result of these favorable properties, continuous oscillations have been obtained from a YAG–Nd^{3+} rod of 3-cm length at room temperature with only 360 W input power to an exciting tungsten lamp. Continuous operation of a $CaWO_4$–Nd^{3+} laser under similar conditions requires almost three times as much power. The YAG–Nd^{3+} laser operates on the $^4F_{3/2} \rightarrow {}^4I_{11/2}$

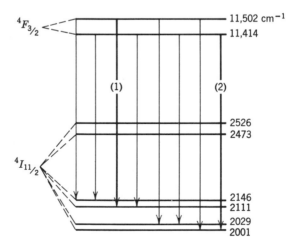

FIG. 5.5 Structure of the $^4F_{3/2}$ and $^4I_{11/2}$ levels of Nd^{3+} in YAG. After Geusic et al. [8]. (Fluorescent transitions in the 1.06-μm region are shown by vertical lines; laser lines are heavy. The energy values above ground level are valid for 300°K.)

transition with a wavelength of 1.0648 μm at room temperature and 1.0612 μm at 77°K. Several other garnets may also serve as host materials, but they do not offer the advantages of YAG [8].

The details of the relevant Nd^{3+} energy levels in YAG are shown in Fig. 5.5. Only the structure of the $^4F_{3/2}$ and $^4I_{11/2}$ levels is shown because the transitions between the sublevels of these groups are responsible for the technically important radiation whose wavelengths form a cluster around 1.06 μm. The important fluorescent transitions start at one of the $^4F_{3/2}$ levels which are 88 cm^{-1} apart at room temperature. They terminate at the lower four of the $^4I_{11/2}$ levels, which are split six ways. At room temperature, laser oscillations are readily obtained in line 1 at 1.0648 μm, whereas at 77°K line 2 at 1.0612 μm is preferred. The reason for this shift with temperature is the fact that the relaxation among the $^4F_{3/2}$ sublevels is fast. Therefore at low temperatures, relatively few ions are found at the upper sublevel. This concentration favors line 2. At room temperature, the population of the upper level is increased and line 1 with a larger transition rate has the advantage. Similar shifts had been observed earlier in LaF_3, where the separation of the upper levels is smaller than in YAG [7].

5.4 NEODYMIUM GLASS LASERS

The neodymium ion is noted for its stimulated emission in glass. It is not the only earth ion that may be used in a glass laser, but it is by far the most important one. Technically, the neodymium glass laser is the only serious competitor of the ruby laser for applications which require the delivery of a significant amount of radiative energy into a small volume during a short period of time or which require the production of a short, powerful pulse in a collimated beam.

While the neodymium crystal lasers are interesting research tools, the powerful neodymium lasers are all of the glass type. This type of laser was discovered in 1961 when Snitzer [9] obtained stimulated emission with 2 to 6% Nd^{3+} incorporated in a number of glasses, barium crown glass being the most favorable medium. The early Nd-glass lasers were thin fibers of 0.03 cm in diameter and 1 to 7.5 cm in length. They were coated with ordinary glass containing no neodymium. As technology advanced, the Nd lasers took on the shape of the ruby lasers. Ordinary Nd laser rods have a diameter of the order of one cm. It is easier to make large glass rods than ruby rods, therefore Nd glass is a preferred material for lasers whose main purpose is to deliver large bursts of energy. The construction of a Nd laser 2 m in length and 38 mm in diameter has been reported. Such large lasers can deliver 5000 J in a single pulse. The total energy available from a single pulse of a Nd-glass laser depends on the size of the laser and on the excitation. It is possible to obtain about 100 J output from a laser of 10 cm³ volume [10].

The glass lasers are excited by means of xenon flashtubes. Several specially designed tubes are used for the excitation of large lasers.

The output of a high-power glass laser differs from the output of a low-power Nd-doped crystal laser in several respects. The peak of the emitted radiation is around 1.06 μm, and the spectrum generally extends over a region much broader than that characteristic of a crystal laser. The actual width of the spectral region covered depends both on the individual laser and on the degree of excitation. Figure 5.6 shows the variation of the spectral width in a typical case. Linewidths of over 100 Å are not uncommon in highly excited Nd-glass lasers.

The power output of a Nd-glass laser may show irregular spiking as well as regular oscillations. Lasers whose output consists of random spikes produce an aggregate of sharp lines spread over the wavelength interval illustrated in Fig. 5.6. For a laser that gives damped oscillations the spectral output just above threshold consists of a single band a few ångströms wide. As the laser is driven harder, this band increases in intensity, but not in width. For a continued increase in pump power, side bands appear on each side of the central band, displaced about 8 Å. As the pump power is increased further, the side bands move farther away from the central band. For still higher pumping, additional sets of side bands occur and move out with increased power.

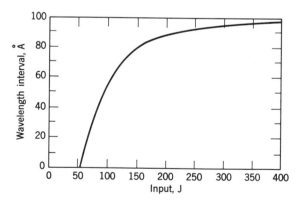

FIG. 5.6 Spectral width of a Nd-glass laser as a function of pump power. Measurements of Snitzer [10] on a barium crown glass laser with 6 wt % Nd₂O₃.

The interpretation of the spectra of power lasers is complicated by the fact that the photographic record obtained by ordinary spectroscopy registers the total energy available in different frequency regions during the entire flash; it does not record time variation during the flash. It was shown by means of time resolution spectroscopy that the spectral composition of the Nd-glass laser varies within a single flash. In the case of a laser with random spiking, the output spectrum varies from spike to spike. Whenever bands occur, as they do for a laser with damped oscillations, adjacent bands do not emit simultaneously. The time sequence indicates that a hole is burned into one part of the spectrum, and the emission shifts to an adjacent part.

Operation of the Nd-glass laser has also been achieved in the $^4F_{3/2} \rightarrow {^4I_{9/2}}$ transition, yielding radiation 9180 Å, and in the $^4F_{3/2} \rightarrow {^4I_{13/2}}$ transition, which yields radiation at 1.37 μm. As in the case for Nd-crystal lasers, laser operation in these transitions depends on the suppression of the favored $^4F_{3/2} \rightarrow {^4I_{11/2}}$ transition by proper choice of frequency-dependent mirrors. It is moreover necessary to cool the laser to 77°K before one can obtain the 9180-Å line.

Continuous operation in Nd-glass lasers is more difficult to obtain than in crystals because of the lower thermal conductivity of glass. It can be achieved in rods of small diameter even at room temperature.

The main advantages of glass as a laser host are flexibility in size and shape and excellent optical quality. There is flexibility in some of the physical properties, in particular in the refractive index, which may be varied from approximately 1.5 to 2.0 by selection of the glass. It is possible to adjust the temperature coefficient of the index of refraction so as to produce thermally stable optical cavities.

The major disadvantage of glass is its low thermal conductivity which imposes limitations on continuous operation or high repetition rate applications.

Glass hosts produce broader spectral lines than crystals. This makes it more difficult to attain threshold because a larger inversion is required to compensate for the lower peak of the lineshape. On the other hand, the general broadening of the spectrum is useful because it provides for absorption of a greater portion of the pumping radiation. The broader spectral output of the glass lasers is useful for the so-called mode-locked, giant-pulse operation which results in the production of an equally spaced series of pulses of around 10^{-11} sec duration (see Chapter 6).

Crystal and glass lasers complement each other with the glass lasers preferred for high-energy pulsed operation, crystals for continuous wave and high repetition-rate applications.

5.5 SENSITIZED LASER MATERIALS

The common disadvantage of four-level ion lasers is their inefficient utilization of the incident radiative energy. The spectrum of any one of the rare-earth ions covers only a small fraction of the visible- and near-infrared region where optical energy from pumping sources is conveniently available. Thus a glass or a crystal containing ions of only one of the rare earths is transparent to most of the pump radiation. When ions of several rare earths are contained in the same crystal, their generally nonoverlapping absorption bands will cover more of the spectrum, and the crystal will be able to retain a larger share of the radiative energy incident upon it from the pumping source. The problem then becomes this: Can the energy be concentrated in one or a few laser lines, or does one obtain a group of independently operating, weak lasers in the same crystal?

The phenomenon of *sensitized fluorescence* in crystals, containing more than one kind of impurity, has been known for some time. It has been demonstrated that excitation energy can be transferred from one impurity to another, not only by the emission of light by the first impurity and its absorption by the second, but also by means of a nonradiative process, involving coupling with the atoms of the crystal lattice. In the case of such energy transfer, moderate energy differences between those available in the two impurities can be bridged by phonons; that is, by energy of lattice vibrations. Absorption of incident optical energy by one ion, its partial transfer and reemission in the form of fluorescence by an ion of a different kind is called sensitized fluorescence. The ion which absorbs the energy and passes it to the second ion is called the *sensitizer;* the ion receiving the energy is called the *activator*.

It should be noted that when impurity ions are present in a crystal in large concentrations then cooperative phenomena may occur; the emission and absorption spectrum of the crystal is not the sum of the spectra of crystals containing only one of the impurity ions. A result of the cooperation of ion pairs is the appearance of the satellite lines N_1 (7041 Å) and N_2 (7009 Å) in red ruby (see p. 142). In the discussion that follows here we restrict ourselves to situations in which cooperative effects of ion pairs may be neglected, either because ions occur in low concentrations or because the interaction forces between nearby ions are small.

Sensitized fluorescence, or energy transfer between ions of different atoms, was first used in 1963 as a means of lowering the irradiation required for laser oscillations in thulium and holmium ions. Johnson and co-workers [11] at Bell Telephone Laboratories demonstrated that the emission of the 1.9-μm fluorescent line in Tm^{3+} incorporated in $CaMoO_4$ crystals is greatly enhanced when the crystal is prepared with the addition of a small concentration of erbium. The proportions of the impurities, Tm (0.25%), Er (0.75%), were such that the absorption spectrum of the crystals containing both kinds of ions was simply the superposition of the absorption spectra obtained on crystals containing one of these impurities. Yet the emission of fluorescent radiation of thulium increased because of the energy absorbed by erbium so that the threshold for obtaining laser oscillations at 77°K was reduced by a factor of 3. A similar transfer of excitation takes place in $CaMoO_4$ crystals doped with Er (0.75%) and Ho (0.50%). The fluorescence of holmium is greatly enhanced by the presence of erbium so that the energy required for laser oscillations at 77°K is reduced by about a factor of 2.

After the initial successes with $CaMoO_4$ as a host crystal, it was found that similar excitation energy transfers may take place in other hosts. Yttrium aluminum garnet (YAG) and glass are particularly suitable host materials for lasers utilizing the combination of several rare-earth ions [5, 10, 12]. In such materials, excitation energy is transferred from Er, Yb, and Tm to Ho, and from Nd to Yb. An efficient laser material is obtained by replacing the three atoms of yttrium in YAG by the following combination: $Y_{1.25}$ $Er_{1.5}$ $Tm_{0.2}$ $Ho_{0.05}$. Laser oscillation takes place in the holmium lines: 2.123 and 2.129 μm at 77°K. At this temperature, a continuous output of 15 W has been obtained with an excitation using 300 W input into a tungsten lamp [12]. This attained efficiency of 5% is much higher than can be obtained from a laser using a single ion.

Erbium oxide (Er_2O_3) is an interesting host crystal for lasers utilizing the fluorescence of Ho^{3+} and Tm^{3+} ions. As Soffer and Hoskins have shown, the Er^{3+} ions of the host crystal are very efficient intermediaries for the pumping of other rare-earth activators [13, 14]. Substitution of Er^{3+} by Ho^{3+} and Tm^{3+} is easily accomplished in the Er_2O_3 lattice since these ions have nearly the same ionic radius. Erbium has a number of strong absorption bands in the region where pumping sources are available. Energy transfer was demonstrated from bands located in the vicinity of 0.49, 0.53, 0.66, 0.81, 0.95, and 1.5 μm. Excellent efficiency has been obtained from Er_2O_3 crystals with 0.5 to 5% Ho concentration. Laser threshold was reached at 5 J input in a helical xenon flashlamp, used to excite a crystal 12 mm in length at 77°K. The output wavelength of the Ho laser is 2.1 μm; that of the Tm laser 1.93 μm. Among glass laser materials operating with a transfer of excitation, combinations of the ions Yb^{3+}, Nd^{3+}, and Er^{3+} were studied in some detail [10]. Energy transfer takes place from Nd to Yb, and from Yb to Er. In the first instance Yb emission at 1.015 and 1.06 μm is obtained; in the second, Er emission at 1.54 μm. A spectacular improvement of efficiency is not obtained in such mixed-ion glass lasers, but they have shown properties that might offer some advantages in amplifier applications. Glass lasers have also been constructed with uranyl $(UO_2)^{2+}$ ion as a sensitizer, and with neodymium or europium as activators [15, 16].

In conclusion it might be useful to discuss once again the mechanism of energy transfer between different ions. Such transfer may occur when the first ion emits radiation (by fluorescence) which is then absorbed by the second ion, or the energy may be transferred by means of the coupling that exists between ions and their neighbors. This latter transfer is called nonradiative; it may (and usually does) involve the conversion of some energy of excitation into energy of molecular motion (heat). Experimentally, we can determine whether the radiative or nonradiative energy exchange mechanism is dominant by the following two methods: first, we may use a suitable optical filter and excite only the sensitizer, using an optically thin sample such as a thin layer of powder. Because the fluorescence escapes in all directions, the excitation of the activator in this configuration results largely from nonradiative transfer. Second, we may measure the rate of the sensitizer's decay of fluorescence in the presence and absence of the activator. The rate of fluorescent decay is not affected by the radiative transfer of energy, but nonradiative transfer shortens the lifetime of fluorescence in the sensitizer. Although these measurements are difficult and their quantitative interpretation involves many complications, they clearly indicate that in the materials mentioned the transfer of energy is of the nonradiative type.

REFERENCES

1. P. P. Sorokin and M. J. Stevenson, Stimulated infrared emission of trivalent uranium, *Phys. Rev. Letters*, **5**, 557–559 (1960).
2. P. P. Sorokin and M. J. Stevenson, Stimulated emission from $CaF_2:U^{3+}$ and $CaF_2:Sm^{2+}$, *Advances in Quantum Electronics*, J. R. Singer, Ed., Columbia University Press, New York, 1961, pp. 65–76.
3. L. F. Johnson, Optically pumped pulsed crystal lasers other than ruby, *Lasers*, Vol. **1**, A. K. Levine, Ed., Dekker, New York, 1966, pp. 137–180.
4. Z. J. Kiss and R. J. Pressley, Crystalline solid lasers, *Appl. Opt.*, **5**, 1474–1486 (1966).
5. E. Snitzer, Glass Lasers, *Appl. Opt.*, **5**, 1487–1499 (1966).
6. G. H. Dieke and H. M. Crosswhite, The spectra of the doubly and triply ionized rare earths, *Appl. Opt.*, **2**, 675–686 (1963).
7. L. F. Johnson, Optical maser characteristics of rare-earth ions in crystals, *J. Appl. Phys.*, **34**, 897–909 (1963).

8. G. E. Geusic, H. M. Marcos, and L. G. Van Uitert, Laser oscillations in Nd-doped yttrium alumium, yttrium gallium and gadolinium garnets, *Appl. Phys. Letters,* **4,** 182–184 (1964).

9. E. Snitzer, Optical masser action of Nd^{3+} in a barium crown glass, *Phys. Rev. Letters,* **7,** 444–446 (1961).

10. E. Snitzer and C. G. Young, Glass lasers, *Lasers,* Vol. **2,** A. K. Levine, Ed., Dekker, New York, 1968.

11. L. F. Johnson, L. G. Van Uitert, J. J. Rubin, and R. A. Thomas, Energy transfer from Er^{3+} to Tm^{3+} and Ho^{3+} ions in crystals, *Phys. Rev.,* **133,** A 494–498 (1964).

12. L. F. Johnson, J. E. Geusic, and L. G. Van Uitert, Efficient, high-power coherent emission from Ho^{3+} ions in yttrium aluminum garnet, assisted by energy transfer, *Appl. Phys. Letters,* **8,** 200–202 (1966).

13. B. H. Soffer and R. H. Hoskins, Energy transfer and cw laser action in Tm^{3+}: Er_2O_3, *Appl. Phys. Letters,* **6,** 200–201 (1965).

14. R. H. Hoskins and B. H. Soffer, Energy transfer and cw laser action in Ho^{3+}: Er_2Ho_3, *IEEE J. Quant. Electrcs.,* **QE–2,** 253–255 (1966).

15. H. W. Gandy, R. J. Ginther, and J. F. Weller, Radiationless resonance energy transfer from UO_2^{2+} to Nd^{3+} in coactivated barium crown glass, *Appl. Phys. Letters,* **4,** 188–190 (1964).

16. L. G. DeShazer and A. Y. Cabezas, Large energy transfer from uranyl to europium ions in glass, *Proc. IEEE,* **52,** 1355 (1964).

Chapter 6   Giant-pulse generation and amplification

6.1   THE PRINCIPLE OF Q-SWITCHING

The irregular pulsations of the laser are greatly disturbing for almost any application, especially communications, where the timing and the control of the intensity envelope are particularly important. Fortunately, it is possible to remove these irregularities and at the same time to greatly increase the peak intensity by regulating the regeneration in the laser. This method of control was first proposed by Hellwarth [1]. It is accomplished by detaching the reflectors from the ruby and inserting a fast shutter between the ruby and one of the reflectors. With the shutter closed, excitation in the ruby can be built up far above the level of threshold with the shutter open. The shutter is kept closed until a high excitation is reached; when it is opened, the radiation builds up rapidly and all the excess excitation is discharged in an extremely short time. The intensity of the resulting short pulse exceeds that obtainable from an ordinary laser flash by several orders of magnitude. Because of its extremely high power, the flash so produced is called the *giant pulse*.

It was shown in Section 3.1 that the laser reaches the threshold of oscillations when α, the amplification per unit length reaches the value

$$\alpha = \gamma/L \tag{1.1}$$

where γ is the fractional loss of radiation in one passage through the laser and L is the length of the active material. The quantity α is proportional to the population inversion $N = N_2 g_1/g_2 - N_1$. When the laser is optically pumped, the value of N increases, but laser oscillation does not begin until the population inversion reaches the threshold value determined by (1.1). When this point is reached, radiation density builds up rapidly, and stimulated transitions proceed at an increasing rate until the inversion is exhausted and the laser material is driven below the threshold. Thus the maximum inversion reached is controlled by the loss rate γ of the laser.

The laser may be viewed as an oscillator consisting of an amplifier with a feedback device. When so regarded, the condition of oscillation is that the gain in the amplifier must be at least equal to the sum of the losses encountered in the system. The loss rate of a system is frequently described by the quality factor Q (see Section 3.6). Any disturbance that decreases the quality factor increases γ and, therefore, increases the population inversion at which oscillations begin. Because of this relationship, the techniques used to delay the onset of oscillations by temporarily increasing the losses in the laser are called *Q-switching* or *Q-spoiling* techniques.

Q-switching may be accomplished by changing the reflectivity of one of the mirrors, by inserting or removing a diaphragm, by changing the paths of the rays between the mirrors, and also by changing the transparency of the material within the laser cavity. What really matters here is a change in the ratio of gain to loss. Therefore switching may be accomplished by suddenly increasing the gain, as it was done in the original experiments of McClung and Hellwarth which will be described in the next section. For efficient production of a single large pulse it is essential that the *Q*-switching process be *fast* compared to the lifetime of a photon within the cavity and that the time of switching from low *Q* to high *Q* be chosen so as. to assure the accumulation of the greatest possible inversion in the material.

The typical result of a successful *Q*-switching experiment is the elimination of the many irregular pulsations that last, in the case of ruby, about 0.5 msec and the appearance of a single pulse of less than 0.1-μsec duration with a peak intensity at least a thousand times greater than the average power of the former pulse.

Let us assume now that it is within our power to vary rapidly the reflectivity of one of the reflectors of a laser. We make the reflectivity of the other mirror practically equal to one and use the mirror with variable reflectivity as the output port of the laser. How should we then vary the reflectivity of the output port to obtain a short pulse with the greatest available power? First, we shall select a relatively low reflectivity. This will provide a cavity with a low *Q* in which a high inversion may be built up without starting oscillations. Then we switch the reflectivity to a high value, thus increasing *Q* and causing the radiation density to build up to a high value within the laser. Although the radiation density has risen to a very high value, the lifetime of the photons within the laser will be long because of the high reflectivity, and the radiation will not be able to leave the laser at a rapid rate. Thus we get a fast rising and slowly decaying pulse. In order to get a large radiation flux outside of the laser, we have to release the radiation faster than is possible with a high reflectivity mirror. The objective of producing a short pulse of the highest possible peak intensity calls for a second change: A decrease of reflectivity programmed to occur when the radiation density within the laser reaches its peak. Such double switching was proposed by Vuylsteke [2]. The operation of a laser in this manner is called the *pulse transmission mode* (PTM). Because of the experimental difficulties of double switching, the PTM is rarely used. Most giant pulses are generated by a single switching operation.

The techniques of Q-spoiling encompass a variety of processes, most of which rely on an external signal or interference to change the regeneration of the laser cavity. These externally controlled, timed Q-switching techniques are complemented by the technique that exploits the bleaching of saturable dyes. The rising light flux within the laser is capable of decreasing the absorptivity of certain dye solutions placed in the laser cavity. The sudden decrease of absorption has the same effect as the removal of an obstacle in the path of the beam. When properly adjusted, these lasers containing saturable dyes trigger themselves to emit a giant pulse or a series of giant pulses. The various Q-spoiling techniques are the subject of the next two sections. In practice these techniques are most frequently applied to generate pulses from ruby–and neodymium lasers. They have raised the peak optical power level available from a single laser from the kilowatt region to the megawatt region. In combination with amplifiers and pulse sharpening techniques, they made optical power levels in the gigawatt range possible.

It is essential for the success of the Q-switching technique that it be possible to store a significant amount of energy in excited atomic systems for a reasonable length of time, say a millisecond. This condition is met in the common solid lasers, but not in the atomic and ionic gas lasers because the spontaneous decay times of the latter are generally too short. The situation is more favorable in molecular gas lasers, in particular, in CO_2, where the lifetime of the laser level is about 4 msec. A CO_2 laser normally designed for continuous wave operation can be Q-switched at a fast rate by using rotating reflectors. Nevertheless, the main application of Q-switching techniques is to solid lasers. Therefore our discussion of this technique will be conducted with the underlying assumption that it is to be applied to a ruby–or neodymium laser.

6.2 EXTERNAL Q-SWITCHING TECHNIQUES

Electro-Optic Switching. The first successfully Q-switched ruby laser was constructed in 1961 in the Hughes Research Laboratories [3]. In their original experiments McClung and Hellwarth employed an electro-optical shutter which, in its closed state, greatly diminished the gain of the laser. The switching scheme exploits the preferential natural polarization of the stimulated emission of ruby which is observed when the optic axis of the ruby does not coincide with the cylinder axis. The gain of the excited ruby crystal is largest for a radiation whose electric vector is perpendicular to the optic axis. Q-switching is then accomplished by means of an apparatus illustrated in Fig. 6.1. A nitrobenzene Kerr cell is placed between one of the detached reflectors and the ruby rod so that its electric field is applied at 45° to the plane of the ruby *c*-axis. When the electric field of the Kerr cell is turned on to its quarter-wave value, the polarization of the laser light which passes through the cell twice is rotated 90° from the favored plane of polarization. In this situation the gain on the reverse passage is low, and the overall regeneration is insufficient for oscillations until a much higher level of excitation is reached than is required for oscillations with the Kerr cell absent.

To produce a single giant pulse the Kerr cell is activated, the exciting flashlamp is fired, and after about 500 μsec the Kerr cell is suddenly turned off. The resulting pulseshape and pulse timing are reproducible. A typical oscilloscope trace is shown in Fig. 6.2. A total peak intensity of 600 kW was reported in the early experiments of Hellwarth and McClung with a ruby cylinder 3 cm in length and 0.9 cm in diameter. This figure contrasts with a maximum of 6 kW obtainable without the pulsed reflector on cylinders of about the same size. The name "giant pulse" is justified by the surprising increase in power obtainable by this technique. Within two years from the first experiments giant pulses exceeded the power level of 100 MW.

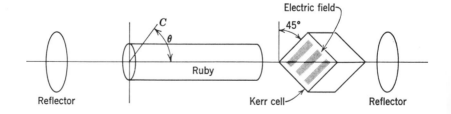

FIG. 6.1 Diagram of a giant pulse laser with a Kerr cell.

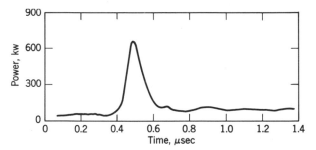

FIG. 6.2 Oscilloscope trace of a giant pulse generated with a Kerr-cell-equipped laser.

Operation of the Kerr cell requires a dc voltage supply of about 10 kV, but almost no current drain. In the interest of fast switching the shutter is kept closed with the voltage applied to the cell, and the shutter is opened by short-circuiting the plates of the cell. In this manner, switching may be accomplished in 5 nsec, a time interval much shorter than that required for the evolution of the pulse.

The shutter described here is not an absolute one which prevents laser action no matter how high the excitation. It only increases the threshold at which oscillations may set in because at sufficiently high levels of excitation amplification will take place in all polarizations. However, the onset of oscillations is delayed until a higher level of excitation is built up than is possible without the shutter, and when the shutter is opened all the excess is unloaded at once. An absolute electro-optical shutter may be constructed by incorporating a Glan-Thomson prism between the laser rod and the Kerr cell. With this addition the level of excitation can be built up higher than with a shutter based on the preferred polarization in the ruby.

The Glan-Thomson prism is a refined version of the classical Nicol polarizer. It permits light of one polarization to pass through and diverts light polarized 90° from the preferred orientation to another path. The polarizer is usually cemented together from two prisms, and the cement is frequently damaged by the high radiation density; therefore specially constructed polarizers must be used for high-power work.

Kerr cell switching is applicable to many laser types, not only to ruby. When switching a laser without preferred polarization characteristics, a polarizing prism must always be employed.

Note that if the shutter is left open and the exciting flash persists after the giant pulse for a period long enough to reach the threshold condition, the laser will break into the ordinary pulsations that take place in the absence of the shutter. Therefore if it is essential that only one pulse be emitted the shutter must be closed again. Fortunately, this need not be done very fast because the recovery time of the laser after the giant pulse is at least 10 μsec, which is ample time to close an electro-optical shutter. The time of recovery to the initial state after a pulse depends on the energy of the pulse. Low-energy pulses require proportionately less recovery time.

Mechanical Switching. The idea of employing a rotating chopperwheel to open and close the optical path between the crystal and one of the mirrors occurred almost as early as the use of the more expensive Kerr cell. Collins and Kisliuk [4] demonstrated the feasibility of such a system in 1962.

The principal disadvantage of the mechanical chopper is its slowness. With the wheel revolving at 10,000 rpm, tens of microseconds elapse from the time the slot first begins to expose the ruby until the ruby face is fully exposed. It has been suggested that faster mechanical switching can be achieved by focusing the rays of the laser and by placing the shutter in the focal plane. This arrangement is operable only for low power levels because the air will break down at the focal point.

Fast mechanical switching may also be accomplished by rotating one of the mirrors, or by replacing one of the mirrors with a rotating totally reflecting prism. The rotating prism method is the most favored mechanical method for the generation of giant pulses. The firing of the flashtube is synchronized with the rotation of the prism so that firing occurs at a predetermined time before the prism reaches the reflecting position.

Acoustic Deflection. An interesting Q-spoiling method depends on the deflection of light waves in an ultrasonic field. Such a field can be produced in a fluid cell placed between the mirrors of a laser. The ultrasonic waves travel perpendicular to the path of the light; the deflection of the light beam results from the alternating compressions and rarefactions, which alter the density and the index of refraction of the liquid. Figure 6.3 shows the experimental apparatus of De Maria and co-workers [5]. The reflectors are not positioned parallel, one of the reflectors being offset from its "true" Fabry-Perot orientation by a small angle of the order of 1 mrad. This deviation from parallelism is such that laser oscillations do not occur with the available optical pumping. When the overpopulation of the upper level reaches its desired value, the ultrasonic cell is shock-excited with a pulse of short duration. As the spontaneous radiation from the ruby rod passes through the ultrasonic field, the beam is refracted and, for a short period of time, radiation will be directed

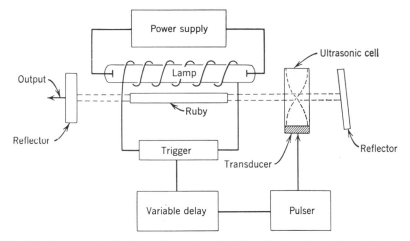

FIG. 6.3 Laser controlled by ultrasonic cell. The offset angle of the right reflector is exaggerated.

perpendicular to the offset reflector. At this time the light path is open and the giant pulse is formed.

The ultrasonic shutter can be used not only for the generation of a single giant pulse but also for the synchronization of the random output spikes of a laser to a suitable external signal frequency.

External Q-switching techniques enable the experimenter to turn on the laser at an instant chosen by him. The most precise control is available by means of the Kerr cell. This is, of course, the most expensive method for producing giant pulses. Somewhat less precise control is exercised by means of the rotating switch and the acoustic deflector. It is possible to generate giant pulses by means of a passive device, the saturable absorption cell, for a small fraction of the cost involved in the construction of externally controlled devices. The starting time of the giant-pulse laser, switched by a saturable absorber, is not subject to control as close as is available with externally controlled switches.

6.3 SATURABLE ABSORBERS

Although the starting of a giant pulse is blocked when a suitable obstacle is placed in the optical cavity of a laser, the radiation intensity within the cavity rises to very high values as the excitation of the active material progresses. It is possible to use this rising radiation as a tool to puncture the obstacle and thus to initiate the starting of the giant pulse. This idea of obstacle-piercing first led to the construction of "single shot" Q-switches, in which a thin metal layer was vaporized or a colored glass was bleached by the rising spontaneous radiation. A natural extension of this idea led to the discovery of Q-switches consisting of *reversible* bleachable glass filters and dye solutions, whose absorptivity is temporarily decreased when they are subjected to very intense radiation of the proper frequency. Such saturable absorber switches were developed simultaneously and independently in several laboratories, all reaching their goals in the early spring of 1964 [6—9]. Dye solutions have proven to be superior to glasses as saturable absorber materials. A dye switch consists of a glass cell of the order of 1-cm thickness placed between the crystal and one of the mirrors and filled with a solution of a suitable dye capable of absorption in the spectral region of the laser output. To avoid unwanted reflections, the end surfaces of the cell may be coated, or the cell may be placed at an angle to the laser axis. The dye cryptocyanine dissolved in methanol or several phthalocyanines dissolved in nitrobenzene ($10^{-6}M$ solution) may be employed as switches for rubies. Other dyes are available for the switching of neodymium lasers [10]. The concentration and the length of the absorption cell are so arranged that the cell transmits approximately 50% when the radiation density is low.

The bleaching of the dye is based on the saturation of a special transition by the intense radiation of ruby (or other active material). The absorption spectrum for a metal phthalocyanine in solution is characterized by intense band groups. One of these bands is shown in Fig. 6.4. The absorption lines correspond to electronic transitions of the phthalocyanine ring. The metal ion occupying the center of the molecule affects the exact location of the absorption peaks and the relaxation rates which govern the return of the dye to its normal absorbent condition. The relaxation time of the phthalocyanine dyes is of the order of 5×10^{-9} sec, whereas for carbocyanines and polymethines this time is of the order of 10^{-11} sec.

The giant pulse obtainable from a laser switched by means of a bleachable absorber is very similar to that obtainable by other switching methods. The duration of the pulse is usually somewhat shorter when the pulse is switched by means of a bleachable absorber.

A Kerr cell in combination with a Glan-Thomson prism and a bleachable absorption cell may be used to produce pulse-transmission-mode (PTM) operation. A schematic diagram for achieving this operation is shown in Fig. 6.5. Initially the Kerr cell is not energized; the formation of the pulse begins when the absorber is bleached as in the ordinary giant-pulse operation. When the radiation leaking out through mirror 2

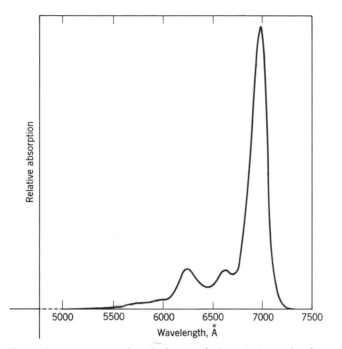

FIG. 6.4 Absorption versus wavelength for a solution of chloroaluminum phthalocyanine in chloronaphthalene.

begins to rise, the photocell activates the power supply of the Kerr cell. With the Kerr cell activated, the polarization of the radiation passing toward the Glan-Thomson prism is rotated by 90°, causing the radiation to be diverted to the output path.

Saturable absorbers have many interesting applications in laser technology. They selectively enhance strong signals; that is, they act as

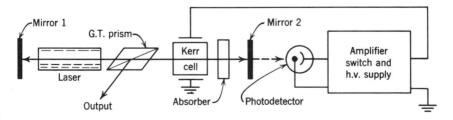

FIG. 6.5 Realization of the PTM mode with bleachable absorber and a Kerr cell controlled by a photodetector.

expander elements and they are employed in the production of ultra-short pulses–a subject discussed in Section 6.7.

6.4 LASER AMPLIFIERS

Although the basic process underlying the laser art is the amplification of light by stimulated emission of radiation, amplifiers of light have a subordinate role in laser technology. Attention is focused on light generators which are in fact regenerative, frequency-selective amplifiers.

Having reached a reasonable level of understanding of the operation of light generators, it is time to consider devices based on stimulated emission whose sole function is to amplify a signal provided by an other source. Before entering into the description of such devices, it is desirable to understand the reasons for the subordinate role of lasers as amplifiers. Most particularly, we must clarify why lasers are not used as low-level signal amplifiers in the manner their relatives, the masers, are.

Some comparisons were already made between masers and lasers in Section 2.2. It was found that masers in the wavelength range of 1 cm and longer may be constructed as single-mode, low-noise devices, but lasers operating with wavelengths of 10^{-3} cm and shorter are intrinsically multimode, high-noise devices. The fundamental difference in the noise characteristics of these devices derives from the fact that the ratio of Einstein's coefficients, which determine the rates of spontaneous and stimulated emission, varies as ν^3. Because of this ν^3 factor, the rate of spontaneous emission of radiation in the visible- and the near-infrared region exceeds the rate of stimulated emission until a very high radiation density is reached. Lasers are therefore useless as amplifiers of a low-level signal because their signal is drowned out by the spontaneous radiation. The situation is so extreme, that lasers are not only unsuitable as amplifiers of a weak signal received from a distant light source but equally unsuitable for operation when the input signal reaches the level normally obtainable from a continuously operating He—Ne laser. The outputs of pulsed solid lasers, however, are sufficiently powerful input signals for laser amplifiers. In fact, the output of an ordinary ruby laser is just marginally suitable for amplification by another ruby, as was demonstrated by the experiments of Kisliuk and Boyle [11], who built the first laser amplifier in 1961. Laser amplifiers attain practical significance when used in conjunction with a giant-pulse laser. This is the reason for the inclusion of ampliers in the chapter on giant pulses.

The use of a laser amplifier—oscillator combination offers an advantage over a stronger laser oscillator that is the same as in conventional electronics: It is possible to design and control the low-power oscillator more precisely than an oscillator of large power-generating capability. A low-level laser oscillator may be built with confinement of excitation to relatively few modes. The mode selection, in turn, is reflected in restricted beamwidth, polarization, and higher frequency selectivity. An oscillator that handles only low power is relatively unaffected by undesirable fluctuations caused by temperature changes. The amplifier that follows the oscillator need not be highly selective, but it must provide amplification in the frequency range containing the oscillator output. For stable operation of the oscillator—amplifier system it is necessary to prevent radiation emitted by the amplifier from reaching the oscillator.

The gain of a laser amplifier is limited because a very long amplifier, or one very highly excited, may turn itself into a pulse generator. Therefore, when a laser system with a very large gain is required, it must be built using several amplifiers in series, and the passage of radiation in the reverse direction must be inhibited.

The unidirectional device that prevents the passage of a light pulse in the reverse direction, that is, toward the laser oscillator, is called an *isolator*. A laser system of large gain consists of an oscillator followed by a series of amplifiers, each separated from the preceding element by an isolator, as shown in Fig. 6.6. Such a system was first designed in 1962 by Geusic and Scovil [12]. The isolators incorporated in such a traveling-wave laser amplifier are *nonreciprocal* devices which permit the passage of light in one direction with very little attenuation and which attenuate light propagating in the opposite direction. The construction of an isolator is schematically represented in Fig. 6.7. The active material of the medium is a substance that is transparent in the useful frequency region and is such that it exhibits a sizable Faraday effect. Lead oxide glass is such a material for ruby- and neodymium-laser radiation. When subjected to a magnetic field parallel to the direction of light propagation, it rotates the plane of polarization by an amount proportional to the distance traveled and to the intensity of the applied magnetic field. This medium forms the optical rotator. The magnetic field is adjusted so that the plane of polarization is rotated by 45° as the light passes once through the instrument. Depending on the direction of the magnetic field, the rotator turns either clockwise or

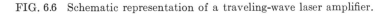

FIG. 6.6 Schematic representation of a traveling-wave laser amplifier.

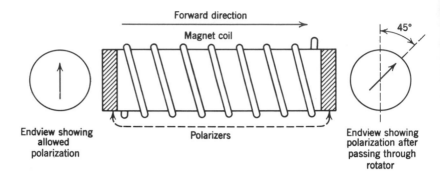

FIG. 6.7 Schematic diagram of the Faraday-rotation isolator.

counter-clockwise. On passage through the rotator in one direction and subsequent reflection and passage in the opposite direction, the plane of polarization is not restored to the original orientation, but rotated 90°. On both ends of the rotator there are polarizers, whose polarization directions are offset by 45°. Light of the proper polarization (vertical), incident from the left, passes through both polarizers unhindered. Light incident from the right is first selectively polarized, then its plane of polarization is rotated so that it arrives at the left polarizer with a horizontal polarization and is extinguished. The isolator works most efficiently when the laser light from the oscillator is already plane-polarized. Most giant-pulse laser oscillators are designed with a preferred plane of polarizations. The ruby crystals have a natural polarization preference; the neodymium glass lasers are usually terminated at Brewster's angle to minimize reflection for one polarization orientation. This determines the polarization of the output.

The amplifying lasers are constructed without mirrors, but some reflection will occur from their terminal surfaces and from the front surfaces of the isolators surrounding the laser. In order to ensure stability of the system, the maximum gain G in each section must be limited so that $Gr < 1$, where r is the largest reflection (intensity) into the amplifier section. For rubies cut into straight cylinders the reflectivity of the air—ruby interface has a reflection coefficient of about 0.07. Therefore the gain of one section must not exceed 14. Larger gains can be obtained by reducing the reflectivity at the ends by antireflection-coating or by cutting the laser rods at Brewster's angle. Laser rods cut at Brewster's angle introduce a certain amount of inconvenience in alignment and aggravate the design of a compact instrument. The difficulties are illustrated in Fig. 6.8.

The operation of laser oscillator—amplifier systems present a number of subtle problems in engineering physics. The oscillator and the amplifiers are excited by flashlamps. The flashes must be timed so that the signal arrives at the amplifiers when their material reaches maximum inversion. This inversion is greater than that reached in the oscillator because the regeneration in the amplifier is kept low. It is therefore necessary to advance the excitation of the amplifier by about 50 μsec compared to that of the oscillator.

The frequency region in which the amplifier is most effective must obviously contain the frequency of the signal to be amplified. This condition is not so easily satisfied in the case of ruby lasers which have a very narrow linewidth and whose line shifts rapidly with temperature. The temperatures of all laser elements in the system must be rigidly controlled to ensure spectral overlap. The situation is not so critical in the case of Nd-glass lasers which have a much broader linewidth 200–300 Å).

Nd-glass lasers with amplifiers are, in fact, more common than the corresponding ruby systems. Their popularity is to a large extent due to their easier adjustment and greater stability in the presence of temperature variations. Since the primary purpose of laser amplifiers is to produce energetic pulses, their operation contrasts greatly with that of the amplifiers in conventional electronics. In the case of lasers we are not dealing with small signal amplification, but with the amplification of signals so large that their passage often exhausts the amplifying capabilities of the medium. In fact, if they do not exhaust the medium, the system will not operate with the greatest possible efficiency.

The exhaustion of the medium produces a signal distortion quite different from the distortion observed in an ordinary nonlinear medium or nonlinear amplifier. The latter is characterized by the fact that the response of the system, although invariant in time, is not proportional to the signal. In a laser amplifier loaded near its capacity the amplifica-

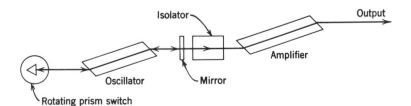

FIG. 6.8 Arrangement of components of an oscillator—amplifier system with Brewster-angle lasers.

tion decreases in time as the inversion in the medium is exhausted during the passage of the signal. The result of this gain reduction is a distortion of the pulse shape. The variation of the signal on passage through a saturable amplifier is discussed quantitatively in Section 6.6.

Somewhat related to laser amplifiers are the *optical avalanche lasers* that operate on a principle similar to the nuclear fission bomb. Two or more pieces of material are optically separated and excited to a high degree of inversion. Then they are suddenly connected optically by means of a mirror or an electrooptical device. When connected, the device is long enough for radiation starting at one end from spontaneous emission to build up to a powerful pulse that exhausts the inversion of the device. Such a device is usually constructed with a mirror at one end so that the pulse travels twice the length of the laser and exhausts all the inversion in the system [13].

6.5   GIANT-PULSE DYNAMICS

The problem of giant-pulse dynamics is to deduce the evolution of the giant pulse in time from a model representing the laser. This model consists of a material with at least two energy levels among which radiative transitions can take place and in which a population inversion can be created by means of an external agent, the pump. The material is confined between mirrors and provision is made for switching, that is, for the variation of the loss of radiation from the structure. When switching is accomplished externally, the model contains a parameter which is a prescribed function of time. When switching takes place by means of a bleachable absorber, the model includes an additional two-level system whose transitions are dependent on the radiation density.

The following quantities are of principal interest in connection with the giant pulse:

1. The total energy radiated and the efficiency of energy utilization.
2. The peak power radiated.
3. The time of formation of the giant pulse from switching to peak.
4. The rates of rise and fall.

Only under the most drastic simplifying assumptions is it possible to obtain tractable mathematical expressions for these variables. The problem is highly nonlinear, even under the most restrictive assumptions, and a mathematical attempt at linearization is self-defeating because it eliminates the essential physical features which make the evolution of the pulse possible.

We will confine ourselves here to the formulation of the problem in the simplest special case of *external fast switching*, to the calculation of some of the quantities of interest, and to a summary of the complications arising in more general situations.

External fast switching means that a shutter is provided that changes the loss rate γ of the laser from a relatively high value to a lower value in a time so short that no significant change in population inversion takes place during the switching process. The fast-switching calculations are certainly applicable when the giant pulse is produced with a Kerr cell switch or a fast-acting mechanical switch. For the sake of mathematical simplicity we assume, contrary to fact, that the laser material is homogeneous, isotropic, and uniformly excited. With these assumptions the best we can hope for is a rough agreement with results obtained on real materials.

In formulating the equations that govern the process of stimulated emission in the period following the switching, we shall neglect the effects of processes which are slow in comparison with the formation of the giant pulse. In particular, we shall neglect the effects of continued pumping and of spontaneous emission on the population inversion.

The essential physical variables of the problem are photon density and population inversion per unit volume. At the start of the process the inversion is high and the photon density is low. As the laser is switched on, the photon density rises, slowly at first, then more and more rapidly, since the rate of photon production is proportional to the photon density already present. Photons are produced at the expense of the stored inversion, which decreases ever more rapidly until the inversion remaining is no longer sufficient to maintain the rate of photon creation at the level of the photon-loss rate. At this time the photon density begins to decline and the giant pulse dies out at a rate determined by the rate of escape of photons from the laser.

The *laser material* is characterized by the following parameters:

1. N_0, the number of active ions per unit volume,
2. α_0, the absorption coefficient of the unexcited laser material. The parameter α_0 is a function of the frequency; we shall use its peak value at the center of the fluorescent line.

The *laser geometry* is characterized by the following variables:

1. V, the volume of the laser material,
2. l, the length of the laser material,
3. L, the optical distance between the reflectors calculated with due regard for the refractive indices of the materials situated between reflectors.

The *physical state of the laser* is characterized by the following variables

1. Φ, the number of photons between the reflectors divided by V. (Only photons of the laser output frequency are to be counted. Moreover, they must be in one of the axial modes of the laser.)

2. $N = N_2 g_1/g_2 - N_1$, the population inversion per unit volume of the laser material. (The density of available active atoms is denoted by N_0).

The reflectivities r_1 and r_2 of the reflectors are relevant device parameters of the laser. The loss coefficient γ is the fractional loss of photons in a single passage. When only reflection losses are significant, $\gamma = -\frac{1}{2} \log r_1 r_2$, but, in general, other losses will be present in a laser equipped for Q-switching and consequently its loss coefficient will be larger than that calculated from reflection losses alone.

Both photon density and inversion are functions of position as well as time. There are two reasons for the spatial variation. First, the initial excitation varies with the radial distance from the geometrical axis of the laser. Secondly, the relevant photon density is the sum of two densities; that of a wave increasing exponentially to the right, and that of a wave increasing similarly to the left. The total photon density is then a convex function of the displacement along the laser axis; hence its largest values are assumed at the ends with a dip located near the middle. This variation of the photon density creates a corresponding and opposite variation of the inversion along the axial direction. Nevertheless, an approximate solution of the problem may be obtained by neglecting the spatial variation of N and introducing the fictitious photon density $\Phi(t)$ in place of a true photon density which varies from point to point. The approximation will be good for short lasers and bad for long ones. We shall re-examine the validity of the solution at the conclusion of calculations.

A useful auxiliary laser parameter in this problem is the *lifetime T of a photon*. This is related to the time of a single passage $t_1 = L/c$, and to the fractional loss γ in a single passage as follows: $T = t_1/\gamma$. It is the fundamental unit of time characteristic of the laser.

The initial state for the formation of the pulse is achieved by pumping the laser with an optical source and keeping the loss coefficient at a value much higher than γ. During this period of excitation the population inversion rises from $-N_0$ to a positive value N_i; the photon density also rises to a value Φ_i. The subscript i indicates that the values are "initial" values for the giant pulse. At time $t = 0$ the loss coefficient is reduced to γ, and the formation of the pulse begins.

It was shown in Chapter 1 that α, the coefficient of amplification, is proportional to the population inversion N. According to (3.26)

$$\alpha = \frac{\alpha_0 N}{N_0},\tag{5.1}$$

where α_0 is the absorption in the unexcited material. To simplify matters, we neglect the spectral distribution of energy and calculate as if the entire phenomenon took place at the frequency that corresponds to the peak of the line. Then α and α_0 denote coefficients measured at the center of the line.

The intensity of a photon packet, traveling along the x-axis, varies as $e^{\alpha x}$; therefore, starting with ΦV photons, the increase in the number of photons from amplification in the laser of active length l is approximately $\alpha l \Phi V$, and this increase takes place in time t_1. Photons are lost at the rate of $\Phi V / T$; therefore, neglecting photons created by spontaneous emission, the variation of Φ with time is described by the equation

$$\frac{d\Phi}{dt} = \left(\frac{\alpha l}{t_1} - \frac{1}{T}\right)\Phi.\tag{5.2}$$

In writing (5.2), we have implicitly assumed that αl and $\alpha l - \gamma$ are both small compared to one; otherwise it would not be legitimate to calculate the nonlinear growth of photon density averaged over the entire laser cavity.

If the contribution of continued pumping is neglected, the density of population inversion varies at the rate

$$\frac{dN}{dt} = -\left(1 + \frac{g_1}{g_2}\right)\frac{\alpha_1}{t_1}\Phi,\tag{5.3}$$

because the emission of one photon causes N to decrease by $1 + g_1/g_2$.

We eliminate α by means of (5.1), introduce the normalized variables $n = N/N_0$, $\varphi = \Phi/N_0$, and change the timescale to make $T = t_1/\gamma$ the unit of time. Then

$$\frac{d\varphi}{dt} = \left(\frac{\alpha_0 l}{\gamma} n - 1\right)\varphi; \quad \frac{dn}{dt} = -\left(1 + \frac{g_1}{g_2}\right)\frac{\alpha_0 l}{\gamma} n\varphi.\tag{5.4}$$

We shall now introduce the constant n_p, defined by the equation $\alpha_0 l n_p = \gamma$. The final form of the differential equations is

$$\frac{d\varphi}{dt} = \left(\frac{n}{n_p} - 1\right)\varphi; \quad \frac{dn}{dt} = -\left(1 + \frac{g_1}{g_2}\right)\frac{n\varphi}{n_p}.\tag{5.5}$$

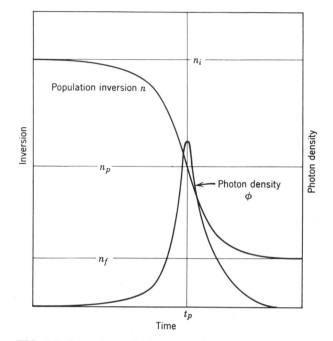

FIG. 6.9 Inversion and photon density in the giant pulse.

Once the timescale has been adapted to the laser by making $T = 1$ there remains only a single parameter characteristic of the laser, namely n_p, the inversion at the instant of the peak photon density.

At the time of switching the photon density φ is very low. It rises from φ_i, reaches a peak φ_p, generally many orders of magnitude higher than φ_i, and then declines to zero. The population inversion n is a monotone-decreasing function of time, starting at n_i and ending at n_f. Figure 6.9 illustrates the typical curves traced out by these variables. The initial rise of φ is exponential; it proceeds with a time constant $\tau = n_p T / (n_i - n_p)$ $= t_1 / (\alpha_i l - \gamma)$, in accordance with the calculations of Hellwarth [1].

It is relatively easy to obtain a first integral of the system of differential equations (5.5). The time is readily eliminated from this system and the following differential equation connects the variables φ and n

$$\frac{d\varphi}{dn} = \frac{g_2}{g_1 + g_2} \left(\frac{n_p}{n} - 1 \right).$$ (5.6)

Hence by integration

$$\varphi = \varphi_i + \frac{g_2}{g_1 + g_2} \left[n_p \log \frac{n}{n_i} - (n - n_i) \right]$$ (5.7)

results. The population inversion n_f remaining at the end of the pulse may be found by noting that the photon density φ is negligibly small both at the start and at the end of the pulse. Thus, with $\varphi_i \approx \varphi_f \approx 0$, we obtain

$$n_p \log \frac{n_f}{n_i} = n_f - n_i. \tag{5.8}$$

Equation 5.8 may be stated in the form

$$\frac{n_f}{n_i} = \exp\left\{\left(\frac{n_i}{n_p}\right)\left[\frac{n_f}{n_i} - 1\right]\right\}. \tag{5.9}$$

Figure 6.10 shows the relationship between n_f/n_i and n_i/n_p. The graph enables one to determine the fraction of inversion remaining from the initial conditions. The energy utilization factor is $(n_i - n_f)/n_i$.

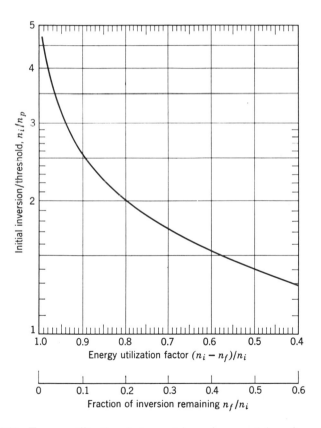

FIG. 6.10   Energy utilization factor and inversion remaining after pulse.

The total radiative energy generated in the laser is

$$E = \frac{g_2}{g_1 + g_2} (n_i - n_f) VN_0 h\nu. \tag{5.10}$$

Not all this energy represents useful output, however. There are radiation loss mechanisms in the laser cavity other than coupling to the output. Therefore we should write $\gamma = \gamma_c + \gamma_i$, where γ_c represents radiation loss due to coupling to the output and γ_i represents incidental losses due to other causes, such as diffraction, scattering, absorption at the mirrors, etc. The useful output energy of the pulse is $E_0 = E\gamma_c/\gamma$.

Other things being equal, the efficiency of the energy utilization process is proportional to the factor $n_i - n_f$. It is seen from Fig. 6.10 that a significant fraction of the inversion always remains at the end of the pulse. Efficient operation of the giant-pulse laser requires that the ratio of initial inversion to threshold inversion be large (at least 2.0) because otherwise the pulse will come to an end with much of the stored energy unutilized.

The peak power is calculated from (5.7); note that the peak is reached when $n = n_p$. Therefore, neglecting φ_i, we have

$$\varphi_p = \frac{g_2}{g_1 + g_2} \left[n_p \log \frac{n_p}{n_i} - (n_p - n_i) \right]. \tag{5.11}$$

Figure 6.11 shows φ_p/n_p as a function of the variable n_i/n_p for the case $g_1 = g_2$. It is a condensation of the numerical data published elsewhere [14].

The peak number of photons in the laser is $\varphi_p N_0 V$. The photons carry an energy $h\nu$; they decay with a lifetime $T = t_1/\gamma$. Again, only the fraction γ_c/γ of the power dissipation represents useful output. Taking all these factors into consideration, we find for the peak power radiated by the laser

$$P = \frac{\varphi_p N_0 V h\nu \gamma_c}{t_1}. \tag{5.12}$$

The calculation of n and φ as functions of time requires one more integration. This can only be carried out approximately or by numerical methods. The essential results of the calculations [14], which are too lengthy to be included here, are as follows:

The initial rise of the pulse is determined by the presence of the radiation emitted spontaneously, the evolution of the central part of the pulse is determined by the excitation parameter n_i/n_p, and the decay of the pulse only by the photon lifetime in the laser.

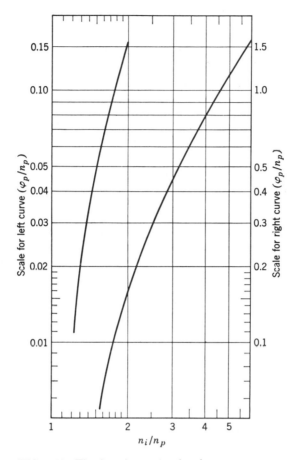

FIG. 6.11 The function φ_p/n_p for the case $g_1 = g_2$.

A numerical example may serve to illustrate the theory:

For pink ruby $N_0 h\nu = 4.65$ J/cm^3, $g_1 = g_2$, and $\alpha_0 = 0.28$ cm^{-1} at room temperature. In a typical situation the incidental internal loss rate γ_i is around 0.05. The loss rate due to coupling to the output with the shutter open varies between 0.05 and 0.20. The optical length of a short laser is between 20 and 30 cm and the initial inversion between 0.1 and 0.3.

We now choose the following specific design values to illustrate the application of the theory developed: $l = 5$ cm, $L = 30$ cm, $\gamma_i = 0.04$, $\gamma_c = 0.1$. From these data it follows that $\gamma = 0.14$, $n_p = \gamma/\alpha_0 l = 0.1$, $t = 10^{-9}$ sec, $T = 5 \times 10^{-9}$ sec. The initial excitation n_i must exceed n_p.

We choose for illustration $n_i = 0.2 = 2n_p$. The conditions required for the validity of the theory are fulfilled since, with the above choice of the parameters, we have $\alpha l = n_i$, $\alpha_0 l = 0.28$ when the pulse starts, and less, as it evolves. Moreover, $\alpha l - \gamma < 0.14$ for the entire pulse. From the graph of Fig. 6.10 we find that the energy utilization factor is 0.8, the inversion remaining at the end of the pulse is $n_f = 0.04$. The energy generated during the pulse is calculated from (5.10):

$$E = \tfrac{1}{2} \times (0.20 - 0.04) \times 4.65 \text{ J/cm}^3 = 0.372 \text{ J/cm}^3.$$

The energy of the output pulse is therefore $E_0 = 0.7E = 0.26$ J/cm^3. The peak power output is calculated by means of (5.11) which gives $\varphi_p = 0.0178$. Then with the given values of $\gamma_c = 0.1$, $t_1 = 10^{-9}$ sec and $N_0 h\nu = 4.65$ J/cm^3 one obtains $P = 7.14 \times 10^6$ W/cm^3.*

The calculations assume homogeneous and isotropic crystals uniformly excited. Since such conditions do not prevail in practice, the peak power output of an actual laser will remain well below the theoretical limit calculated in these assumed ideal conditions.

It should be stressed in conclusion that the many simplifying assumptions introduced limit the applicability of the theory introduced here. It has already been stated that spatial variations of photon density were neglected, and, therefore, the results apply only to short lasers. Moreover, the calculations assumed instantaneous switching. This limitation may be removed at the expense of moderate mathematical complications. The subject was further developed by Wang [15], who generalized the present method for the case of gradual switching. Szabo and Stein [16] obtained solutions for a laser switched with a saturable absorber.

6.6 AMPLIFIER DYNAMICS

The equations of giant-pulse dynamics were obtained in Section 6.5 as a result of assumptions which greatly restrict their validity. Perhaps the most drastic was the assumption that one may neglect the effects of the spatial variation of photon density and inversion within the laser. Such an assumption is certainly not proper for laser amplifiers so designed that the signal passes through the amplifier only in one direction. At the input end, the signal is generally small, at the output end, it is as large as is consistent with the design of the amplifier. These amplifiers are generally long, in the sense that the quantity αl is larger than

---

* Since the diameter of the crystal was not specified, the results are given per unit volume of the ruby.

one. The opposite is usually the case for laser oscillators. In these the large gain is achieved over many passages through the material while in the amplifier only one passage is contemplated. We now set aside special regenerative amplifiers, which are provided with partial reflectors, and concentrate our attention on nonregenerative types, which are so designed that essentially no reflection takes place.

First, we formulate the differential equations that relate the inversion and photon densities in an amplifying medium. The shape of the amplifier is assumed to be cylindrical, with the x-axis chosen along the cylinder axis. The amplifier is terminated by the planes $x = 0$ and $x = l$. Within the medium, the photons travel with a velocity $v = c/\eta$, where η is the refractive index.

We restrict our attention to photons traveling in the $+x$ and in the $-x$ directions and denote the densities of these photons by Φ_+ and Φ_-, respectively. The density of inversion is denoted by N. These quantities are functions of the position x and the time t. For simplicity, we carry out the calculation for the case $g_1 = g_2$ and assume that the rate at which inversions are created by pumping and the rate at which they are annihilated by spontaneous emission are negligible compared to the rate at which stimulated processes take place.

The differential equations governing the variation of photon densities are*

$$\frac{\partial \Phi_+}{\partial t} + v \frac{\partial \Phi_+}{\partial x} = \alpha v \Phi_+, \tag{6.1}$$

$$\frac{\partial \Phi_-}{\partial t} - v \frac{\partial \Phi_-}{\partial x} - \alpha v \Phi_-. \tag{6.2}$$

The amplification α is related to inversion according to (3.26) of Chapter 1:

$$\alpha = \alpha_0 N / N_0. \tag{6.3}$$

The variation of inversion density is governed by the equation

$$\frac{\partial N}{\partial t} = -2\alpha v(\Phi_+ + \Phi_-). \tag{6.4}$$

---

* The first of these equations is found by noting that the net outflow of photons traveling along the x-axis through a cylinder of cross-sectional area A and height h in time dt is $Ah(\partial \Phi_+/\partial x)v\, dt$. The number of additional photons created during the period dt is $Ah\alpha\Phi_+ v\, dt$ because in time dt the photons advance by $dx = v\, dt$, and α is the amplification factor per unit distance. The rate of change of photons (of the proper type) contained in the cylinder is $Ah(\partial \Phi_+/\partial t) = -Ahv(\partial \Phi_+/\partial x) + Ah\alpha v \Phi_+$. Reversal of the x-direction gives the second equation.

Equations 6.1 to 6.4 govern the variation of the essential densities in an active (amplifying) material, in which wave propagation is possible in the $+x$- and $-x$-directions only. These equations are applicable to the giant-pulse laser as well as to regenerative and nonregenerative amplifiers. These devices are distinguished by the boundary conditions. When reflections are present, the values of Φ_+ and Φ_- are tied together at the ends of the device, that is, $\Phi_+(0,t) = r_1\Phi_-(0,t)$ and $\Phi_-(l,t) = r_2\Phi_+(l,t)$. The simplified treatment of the giant pulse, given in Section 6.5, pertains to an average of the function $\Phi_+ + \Phi_-$ over the entire device.

The nonregenerative amplifier is characterized by the absence of a backward wave, that is, $\Phi_- = 0$. The remaining equations in this case are

$$\frac{\partial \Phi_+}{\partial t} + v\frac{\partial \Phi_+}{\partial x} = \frac{\alpha_0 v N \Phi_+}{N_0}, \tag{6.5}$$

$$\frac{\partial N}{\partial t} = -\frac{2\alpha_0 v N \Phi_+}{N_0}. \tag{6.6}$$

On introducing the normalized densities $\varphi = \Phi_+/N_0$ and $n = N/N_0$, we obtain

$$\frac{\partial \varphi}{\partial t} + v\frac{\partial \varphi}{\partial x} = \alpha_0 v n \varphi, \tag{6.7}$$

$$\frac{\partial n}{\partial t} = -2\alpha_0 v n \varphi. \tag{6.8}$$

The boundary and initial conditions applicable in case of an amplifier are as follows: the photon density is 0 for $t < 0$. Then a signal $\varphi_i(t)$ is impressed at $x = 0$; that is, $\varphi(0,t) = \varphi_i(t)$, where φ_i is a prescribed (arbitrary) function of t for $t \geq 0$, and 0 for $t < 0$. The initial inversion is a prescribed function of the position; that is, $n(x,0) = n_i(x)$. In the most common situation the initial inversion is uniform, $n_i(x) = n_0$.

The object of the theory is to predict the output signal $\varphi_o(t) = \varphi(l,t)$. To attain this objective we must solve (6.7) and (6.8), using the prescribed functions $\varphi_i(t)$ and $n_i(x)$.

A quick examination of simple special cases gives hints at what to expect and what mathematical procedures to follow. In the absence of amplification, $\alpha_0 = 0$, the equations are decoupled and the solution of (6.7) is an arbitrary function of $t - x/v$. Hence in this case the solution is

$$\varphi_1(x,t) = \varphi_1\left(t - \frac{x}{v}\right). \tag{6.9}*$$

* Note that $\varphi_i = 0$ for negative arguments!

In this case, the signal propagates through without any change in shape. It is simply delayed in proportion to the distance traveled. Slightly more complicated is the case of constant amplification. This occurs when α_0 is not zero, but when the signal is so small that n suffers no noticeable decrease during its passage through the amplifier. Then we have $n = n_0$. The solution of (6.7) in this case is

$$\varphi_2(x,t) = e^{\alpha_0 n_0 x} \varphi_i \left(t - \frac{x}{v} \right).$$ (6.10)

The solution of the amplifier equations in the general case is obtained most directly if new variables are introduced and the photon density is renormalized to bring out the intrinsic symmetry of the problem. The special solutions shown in (6.9) and (6.10) suggest that the "natural" independent variables of the problem are $\xi = x/v$ and $\tau = t - x/v$. In fact, when the variables $x = v\xi$ and $t = \tau + \xi$ are eliminated and the photon density is expressed as a function φ' of ξ and τ, we find that

$$\frac{\partial \varphi'}{\partial \xi} = \frac{\partial \varphi}{\partial t} + v \frac{\partial \varphi}{\partial x},$$ (6.11)

which is the left side of (6.7). To eliminate the factor 2 from (6.8) we introduce not φ' but the function

$$p(\xi,\tau) = 2\varphi(x,t),$$ (6.12)

and

$$q(\xi,\tau) = n(x,t).$$ (6.13)

The differential equations 6.7 and 6.8 are then replaced by

$$\frac{\partial p}{\partial \xi} = \alpha_0 v p q,$$ (6.14)

$$\frac{\partial q}{\partial \tau} = -\alpha_0 v p q.$$ (6.15)

The boundary conditions have to be rephrased since the initial region $0 \le x \le l, 0 \le t < \infty$ is now replaced by $0 \le \xi \le l/v$ and $-\xi \le \tau < \infty$. When $\xi = 0$ then $t = \tau$ so that the condition $\varphi(0,t) = \varphi_i(t)$ becomes $p(0,\tau) = 2\varphi_i(\tau)$. The given initial inversion $n_i(x)$ will remain unchanged until $t = x/v$ because no signal will reach the point x prior to that time. We may therefore write $q(\xi,0) = n_i(v\xi)$ as the initial condition of the transformed system.

Since (6.14) and (6.15) imply that

$$\frac{\partial p}{\partial \xi} = -\frac{\partial q}{\partial \tau},$$ (6.16)

there is a function $U(\xi,\tau)$ such that

$$p = \frac{\partial U}{\partial \tau} \text{ and } q = -\frac{\partial U}{\partial \xi}, \tag{6.17}$$

moreover, this function satisfies the differential equation

$$\frac{\partial^2 U}{\partial \xi\, \partial \tau} + \alpha_0 v \frac{\partial U}{\partial \xi} \frac{\partial U}{\partial \tau} = 0. \tag{6.18}$$

We shift our attention from finding p and q to finding U. Let $V = e^{kU}$, then

$$\frac{\partial^2 V}{\partial \xi\, \partial \tau} = k e^{kU} \left(\frac{\partial^2 U}{\partial \xi\, \partial \tau} + k \frac{\partial U}{\partial \xi} \frac{\partial U}{\partial \tau} \right). \tag{6.19}$$

Therefore in the present case the function V satisfies the differential equation

$$\frac{\partial^2 V}{\partial \xi\, \partial \tau} = 0, \tag{6.20}$$

and $k = \alpha_0 v$. The general solution of (6.20) is $V = f(\xi) + g(\tau)$, where f and g are arbitrary differentiable functions. Thus

$$\alpha_0 v U = \log \left[f(\xi) + g(\tau) \right]. \tag{6.21}$$

Now we proceed to determine $f(\xi)$ and $g(\tau)$ from the boundary conditions, noting that an additive constant in U may be assigned arbitrarily. We will choose this constant so that $U(0,0) = 0$ and therefore

$$f(0) + g(0) = 1. \tag{6.22}$$

From (6.17) and the boundary conditions it follows that

$$\left(\frac{\partial U}{\partial \tau} \right)_{\xi=0} = p(0,\tau) = 2\varphi_i(\tau), \tag{6.23}$$

$$\left(\frac{\partial U}{\partial \xi} \right)_{\tau=0} = -q(\xi,0) = -n_i(v\xi). \tag{6.24}$$

After integration according to τ and ξ, noting that $U(0,0) = 0$,

$$U(0,\tau) = 2 \int_0^\tau \varphi_i(\tau')\, d\tau', \tag{6.25}$$

$$U(\xi,0) = - \int_0^\xi n_i(v\xi')\, d\xi'. \tag{6.26}$$

Introducing the value of U from (6.21), these become

$$\log \left[f(0) + g(\tau) \right] = 2\alpha_0 v \int_0^\tau \varphi_i(\tau')\, d\tau', \tag{6.27}$$

$$\log \left[f(\xi) + g(0) \right] = -\alpha_0 \int_0^\xi n_i(v\xi')v\, d\xi'. \tag{6.28}$$

Hence

$$g(\tau) = -f(0) + \exp 2\alpha_0 v \int_0^\tau \varphi_i(\tau') \, d\tau', \tag{6.29}$$

$$f(\xi) = - g(0) + \exp - \alpha_0 \int_0^{v\xi} n_i(x') \, dx'. \tag{6.30}$$

Now we introduce the saturation–and gain functions which are defined as the reciprocals of the exponential expressions in (6.29) and (6.30); that is,

$$S(\tau) = \exp - 2\alpha_0 v \int_0^\tau \varphi_i(\tau') \, d\tau', \tag{6.31}$$

$$G(x) = \exp \alpha_0 \int_0^x n_i(x') \, dx' \qquad (x = v\xi). \tag{6.32}$$

Consequently, after adding (6.31) and (6.32), and noting (6.22), we get

$$U = \frac{1}{\alpha_0 v} \log (S^{-1} + G^{-1} \quad 1). \tag{6.33}$$

The solutions are now obtained by differentiation, noting that S^{-1} is a function of τ and G^{-1} is a function of ξ; moreover,

$$\frac{dS^{-1}}{d\tau} = 2\alpha_0 v\varphi_i(\tau)S^{-1}; \frac{dG^{-1}}{d\xi} = - \alpha_0 v n_i(v\xi)G^{-1}. \tag{6.34}$$

Hence

$$p = \frac{2S^{-1}(\tau)\varphi_i(\tau)}{S^{-1} + G^{-1} - 1}, \qquad q = \frac{G^{-1}(v\xi)n_i(v\xi)}{S^{-1} + G^{-1} - 1}, \tag{6.35}$$

or in terms of the original variables

$$\varphi(x,t) = \frac{S^{-1}(t - x/v)\varphi_i(t - x/v)}{S^{-1} + G^{-1} - 1}, \tag{6.36}$$

$$n(x,t) = \frac{G^{-1}(x)n_i(x)}{S^{-1} + G^{-1} - 1}. \tag{6.37}$$

These results correctly reflect the fact that there is no signal when $t < x/v$, nor is there depletion of inversion.

It is to be noted that the gain function does not depend on time. It governs the function of the amplifier for very small signals; that is, in the case in which $S(\tau)$ is nearly equal to one. In this case we have *approximately*

$$\varphi(x,t) = G(x)\varphi_i \left(t - \frac{x}{v} \right).$$

For large signals S becomes small after a considerable portion of the signal has passed (for large τ), and then S^{-1} is the determining factor. Because of this dependence on τ, the shape of the signal will be progressively distorted as it travels toward the output.

The effective gain of the amplifier is the function

$$G_E(x,t) = \frac{S^{-1}}{S^{-1} + G^{-1} - 1}. \tag{6.38}$$

It is a function of the signal amplitude because of its dependence on S. In deriving the above expressions for amplification of a signal, we assumed that it is legitimate to deal with the electromagnetic field as a transport phenomenon. We considered the transport of photons across a material and we regarded the interaction of the electromagnetic field with the atomic system in the simplest approximation, characterized by a gain or absorption coefficient. The limitations of such an approach should be noted because a correct description of the electromagnetic field ought to be based on Maxwell's equations rather than the transport equations, and a correct description of the interaction of the field with an ensemble of atomic systems requires the use of Schrödinger's equation. It is likely that the present development is valid as long as the pulse length is greater than the reciprocal linewidth and as long as the pulse passing over the atomic system is not so large as to invalidate a linear approximation in the perturbation calculations.

Many calculations have been published pertaining to the total gain obtainable from amplifiers and the shape distortion suffered by signals of various types. The works of Schulz-De Bois [17], Siegman [18], and Steele [19] are excellent sources for such material.

Because amplification is greater in the early phases of the pulse, most pulses tend to become shorter when they are sent through a saturable amplifier. To produce shortening of such a signal, it may be necessary to remove its slowly rising portion. Pulse sharpening by means of saturable amplification was investigated in great detail by a group of investigators at the Lebedev Institute in Moscow. Their publications contain a wealth of material on pulse propagation in gain-saturated amplifiers [20, 21]. They have demonstrated the rather remarkable phenomenon that the exhaustion of inversion in the material, and the peak of the pulse, may propagate at a speed greater than that of light.

6.7 ULTRASHORT PULSES

Ordinary giant-pulse techniques produce laser pulses of 10-to-200-nsec duration. It is possible to produce much shorter pulses than these by breaking up the giant pulse into a sequence of equally spaced, very sharp pulses. The technique that accomplishes this makes use of the fact that the laser, when excited well above threshold, oscillates not in a single mode but in a series of modes. The production of a series of sharp pulses is achieved by exciting a large number of modes in proper phase and amplitude relationship.

Let us confine our attention to the axial modes of a laser assuming that the off-axial ones have been eliminated, or reduced in excitation, by insuring proper symmetry of excitation and by proper choice of all geometrical factors. The distribution of the electromagnetic field in a transverse plane will be the same for all the axial modes, but these modes are distinguished from one another by the location of their nodal planes and by the oscillation frequencies associated with them. It was shown in Chapter 3 that the axial modes are equidistant in frequency, with a separation of $\Delta\nu = c/2L'$ between adjacent modes. Here L' is the optical distance between the reflectors. The time of passage from one reflector to the other is L'/c, therefore

$$\Delta\nu = \frac{1}{T}, \tag{7.1}$$

where T is the time required for a full, round-trip passage of light through the laser.*

The number of axial modes excited simultaneously varies with the linewidth and with the degree of excitation over threshold. It may be quite large in giant-pulse operation when the inversion is driven way above the minimum necessary for threshold. In large neodymium glass lasers, where L' is of the order of 1 m and where the linewidth is large, several thousand of these modes may be excited.

Let us consider the type of output that would result if a large number of modes, equidistant in frequency, could be excited with approximately equal amplitude and if these modes could somehow be locked together in a constant phase relationship. Each oscillatory mode gives rise to a wave with an amplitude described by the (real part of the) function

$$e^{i\omega(t-x/c)},$$

where $\omega = \omega_0 + k\Delta\omega$ and $\Delta\omega = 2\pi\Delta\nu$. Then the addition of $2n + 1$ of such waves with frequencies centered around ω_0, all having equal ampli-

---

* The letter T is used here to denote a quantity different from the lifetime of photon which was designated by the same symbol in Section 6.5.

tudes, leads to a wave described by the function

$$\sum_{k=-n}^{n} e^{i(\omega_0 + k\Delta\omega)t}.$$

This function is the product of the function $e^{i\omega_0 t}$ and an amplitude function

$$F(t) = \frac{\sin (n + \frac{1}{2}) \Delta\omega t}{\sin \frac{1}{2} \Delta\omega t}. \tag{7.2}$$

The result of this synthesis is an amplitude-modulated wave of frequency ω_0 whose intensity varies at the rate $F(t)^2$. Figure 6.12 illustrates a typical such function. The function $F(t)$ is periodic with the period $\tau = 2\pi/\Delta\omega = T$. In the neighborhood of the primary maxima the intensity variation is similar to that of the function

$$\left[\frac{\sin (n + \frac{1}{2}) \Delta\omega t}{\frac{1}{2}\Delta\omega t} \right]^2.$$

Thus the peak intensity reaches $(2n + 1)^2$, and the first zero of $F(t)$ is separated from the peak by a time interval $\tau' = T/(2n + 1)$.

For a laser of optical length $L' = 60$ cm $T = 4 \times 10^{-9}$ sec. If 100 adjacent axial mode oscillations of such a laser are locked together,

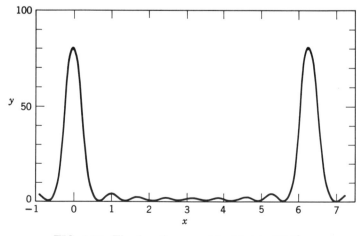

FIG. 6.12   The function $y = (\sin 4.5x/\sin 0.5x)^2$.

one could obtain a peak intensity 10^4 times that of the individual modes, that is, 100 times the sum of all intensities. The peaks would repeat at the rate of $T^{-1} = 250$ MHz and the peak pulses would have a half-width of 4×10^{-11} sec.

Experimentally, phase-locking of the axial modes was first realized by means of an acoustic modulator incorporated in a He—Ne laser [22]. The expected changes in the output waveform were observed with pulses of the order of 2.5×10^{-9}-sec width produced at the rate of 56 MHz.

The number of modes available in a gas laser is relatively limited because of the narrow linewidth. Therefore the method of phase-locking attains its real significance in the case of solid lasers with broad lines. Mode-locking in ruby was demonstrated in 1965 by Deutsch [23], who obtained spikes of 2 to 4-nsec duration by acoustic modulation at the rate determined by the transit time. The sharp spikes occur during a single spike of the ordinary laser operation ($\frac{1}{2}$ to 1μsec) or during the emission of the giant pulse. Greater enhancement of the peaks could be obtained with a neodymium glass laser. Working with an acoustic modulator incorporated in a flash-excited neodymium laser, in 1966 De Maria and associates [24] obtained regular pulsations with peaks only 1 nsec wide. They also established that the modulation frequency must be adjusted quite accurately to the mode-difference frequency. Even finer peaks were obtained by a group of scientists at Bell Telephone laboratories [25], who obtained pulses about 0.1 nsec wide from an YAG:Nd^{3+} laser excited to continuous operation at a low power level. A careful analysis of their data showed that the 300 MHz modulation locked approximately 40 modes.

Further experimentation with mode-locking revealed that it can be accomplished without the use of an externally driven modulator. Incorporation of a suitable bleachable absorber cell produces self-locking of the longitudinal modes [26, 27]. The self-locking of modes and the production of highly-peaked pulses, spaced at intervals equal to the travel time across the cavity, may be comprehended in terms of an electrical circuit analogy illustrated in Fig. 6.13. Part (a) is an illustration of the laser with the saturable dye cell, part (b) of the figure is the representation of an analog electrical pulse-generating network containing an expandor, a filter, an amplifier, and a feedback through a delay line. The expandor is a circuit element that provides less attenuation for high-level signals than for low-level ones. It prevents the degradation of the signal by noise as the pulse is recirculated around the feedback loop. The expandor reduces signals of low amplitude and acts so as to shorten the recirculating pulse until the pulsewidth is limited by the frequency response of the filter. The regenerative pulse oscillator produces pulses spaced at

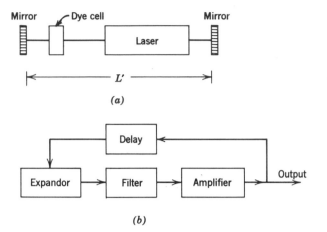

FIG. 6.13 Schematic representation of a repetitive optical pulse generator (a) and its analogy in conventional electronics—the Cutler pulse generator (b).

the circulation time of the generator. The pulsewidth generated is the reciprocal system bandwidth.

In the laser system illustrated, the dye cell is the expandor, the laser crystal is the amplifier—filter combination, and the time required for the optical signal to travel through the system provides the delay. A signal originates from noise; it is amplified, filtered, clipped of its low amplitude components, and returned to the system [26].

Although the above analogy helps to obtain an intuitive grasp of the self-locking mode operation, a detailed mathematical analysis is necessary to explain its quantitative features, in particular the fact that self-locking is dependent on the position of the dye cell within the laser cavity. The modes are all locked in phase when the dye cell is relatively short and is located near one of the mirrors. The mode locking is accomplished by the nonlinear interaction of radiation of differing frequencies within the dye cell. The nonlinearity of the interaction is the consequence of the fact that the dye cell is operated in a region of intensity where the transitions are nearly saturated [28].

The highest peak powers achieved by lasers are obtained from solid lasers in the giant-pulse mode, in combination with repetitive ultrashort pulse techniques. The generation and the application of ultrashort pulses and pulse trains are among the rapidly developing fields at present. A comprehensive review article by De Maria and associates [29] provides the reader an insight into the state of this art as of the end of 1968.

REFERENCES

1. R. W. Hellwarth, Control of fluorescent pulsations, *Advances in Quantum Electronics,* J. R. Singer, Ed., Columbia University Press, New York, 1961, pp. 334–341.
2. A. A. Vuylsteke, Theory of laser regeneration switching, *J. Appl. Phys.,* **34,** 1615–1622 (1963).
3. F. J. McClung and R. W. Hellwarth, Giant optical pulsations from ruby, *J. Appl. Phys.,* **33,** 828–829 (1962).
4. R. J. Collins and P. P. Kisliuk, Control of population inversion in pulsed optical masers by feedback modulation, *J. Appl. Phys.,* **33,** 2009–2011 (1962).
5. A. J. De Maria, R. Gagosz and G. Barnard, Ultrasonic-refraction shutter for optical maser oscillators, *J. Appl. Phys.,* **34,** 453–456 (1963).
6. G. Bret and F. Gires, Giant-pulse laser and light amplifier using variable transmission coefficient glasses as light switches, *Appl. Phys. Letters,* **4,** 175–176 (1964).
7. P. P. Sorokin, J. J. Luzzi, J. R. Lankard, and G. D. Pettit, Ruby laser Q-switching elements using phthalocyanine molecules in solution, *IBM J. Res. Dev.,* **8,** 182–184 (1964)
8. P. Kafalas, J. I. Masters, and E. M. E. Murray, Photosensitive liquid used as a nondestructive passive Q-switch in a ruby laser, *J. Appl. Phys.,* **35,** 2349–2350 (1964).
9. B. H. Soffer, Giant pulse laser operation by a passive reversibly bleachable absorber, *J. Appl. Phys.,* **35,** 2551 (1964).
10. B. H. Soffer and R. H. Hoskins, Generation of giant pulses from a Nd-laser by a reversible bleachable absorber, *Nature,* **204,** 276 (1964).
11. P. P. Kisliuk and W. S. Boyle, The pulsed ruby laser as a light amplifier, *Proc. IRE,* **49,** 1635–1639 (1961).
12. J. E. Geusic and H. E. D. Scovil, A unidirectional traveling-wave optical maser, *Bell System Tech. J.,* **41,** 1371–1397 (1962).
13. C. G. Young, J. W. Kantorski, and E. O. Dixon, Optical avalanche laser, *J. Appl. Phys.,* **37,** 4319–4324 (1966).
14. W. G. Wagner and B. A. Lengyel, Evolution of the giant pulse in a laser, *J. Appl. Phys.,* **34,** 2040–2046 (1963).
15. C. C. Wang, Optical giant pulses from a Q-switched laser, *Proc. IEEE,* **51,** 1767 (1963).
16. A. Szabo and R. A. Stein, Theory of laser giant pulsing by a saturable absorber, *J. Appl. Phys.,* **36,** 1562–1566 (1965).
17. E. O. Schulz-Du Bois, Pulse sharpening and gain saturation in traveling-wave masers, *Bell System Tech. J.,* **43,** 625–658 (1964).
18. A. E. Siegman, Design considerations for laser pulse amplifiers, *J. Appl. Phys.,* **35,** 460 (1964).
19. E. L. Steele, *Optical Lasers in Electronics,* Wiley, New York, 1968.
20. N. G. Basov and R. V. Ambartsumian et al., Nonlinear amplification of light pulses, *Soviet Phys. JETP,* **23,** 16–22 (1966), [**50,** 23–34 (1966)].
21. N. G. Basov and V. S. Letokhov, Change of light pulse shape by nonlinear amplification, *Soviet Phys. Dokl.* **11,** 222–224, (1966), [**167,** 73–76, (1966)].
22. L. E. Hargrove, R. L. Fork, and M. A. Pollack, Locking of He-Ne laser modes induced by synchronous intra-cavity modulation, *Appl. Phys. Letters,* **5,** 4–5 (1964).

23. T. Deutsch, Mode-locking effects in an internally modulated ruby laser, *Appl. Phys. Letters,* **7,** 80–82 (1965).
24. A. J. De Maria, C. M. Ferrar, and G. E. Danielson Jr., Mode-locking of a Nd^{3+}-doped glass laser, *Appl. Phys. Letters,* **8,** 22–24 (1966).
25. M. Di Domenico, J. E. Geusic, H. M. Marcos, and R. G. Smith, Generation of ultrashort optical pulses by mode-locking the YAG laser, *Appl. Phys. Letters,* **8,** 180–183 (1966).
26. A. J. De Maria, D. A. Stetser, and H. Heynau, Self mode-locking of lasers with saturable absorbers, *Appl. Phys. Letters,* **8,** 174–176 (1966).
27. H. W. Mocker and R. J. Collins, Mode competition and self-locking effects in a Q-switched ruby laser, *Appl. Phys. Letters,* **7,** 270–273 (1965).
28. C. A. Sacchi, G. Soncini, and O. Svelto, Self-locking of modes in a passive Q-switched laser, *Nuovo Cimento,* **48,** 58–72 (1967).
29. A. J. De Maria, W. H. Glenn, M. J. Brienza, and M. E. Mack, Picosecond laser pulses, *Proc. IEEE,* **57,** 2–25 (1969).

Chapter 7   Semiconductor lasers

7.1 INTRODUCTION

The coherent radiation emitted from a laser usually derives its energy from an electric power supply. Ordinarily, the energy undergoes a long chain of transformations before it assumes the form of coherent radiation. In the optically excited lasers the electrical energy is first converted into kinetic energy of charged particles in an electric discharge, then atoms in the discharge are excited and emit radiation. The incoherent radiation is transported to an active material where it is absorbed, and where other processes take place before the active material is placed into an amplifying condition. In the electrically excited gas lasers the processes are also complex, and the energy stored in the inverted population distribution of the gas appears as the end product of a long chain of events, each event being relatively inefficient in the sense that it also permits energy transformations in other directions than the one that results in laser action. In semiconductors, however, it is possible to convert electrical energy directly into coherent light. Such conversion takes place in the diode injection lasers in which excitation is the *immediate* result of work done by an imposed electric field on the charge carriers in the material. This carrier-injection process is the most efficient process for the conversion of electrical energy to coherent radiation. It is not the only process available for the excitation of semiconductors; in fact, semiconductor lasers may be produced by optical excitation, electron bombardment, and avalanche breakdown as well.

Semiconductor lasers differ from other solid lasers in most of their physical and geometric characteristics. Most conspicuous, is the difference in size. The semiconductor laser is two orders of magnitude smaller than a typical crystal or glass laser. The largest dimension of a common semiconductor laser is at most 1 mm. The relevant physical properties of semiconductors and their variations with external parameters, such as pressure and temperature, are vastly different from those of an ionic crystal or glass. As a result of these differences, the analysis of oscillation conditions, mode structure, and related performance data requires calculations different from those encountered in connection with other solid lasers.

In order to prepare for the discussion of semiconductor lasers, a brief summary is first given of the relevant properties of the semiconductors with special emphasis on those properties that relate directly to the absorption and emission of radiation. The sections that follow contain descriptions of the various semiconductor laser types, the injection laser being given the most attention because of its practical significance.

The manifold and complex energy-exchange processes in a semiconductor attracted the attention of many investigators in search of laser materials. A considerable amount of theoretical work was done in a relatively short period of time in various widely separated institutions, whose common purpose was to sift out the processes and the semiconductor materials which might be used for laser devices. Some of this speculation and calculation started before 1960 and it continued until after the first semiconductor lasers made their appearance in late 1962. The relation of the theoretical speculative work to the invention of lasers is briefly discussed in the concluding section.

7.2 BASIC PROPERTIES OF SEMICONDUCTORS

Semiconductors are crystalline solids whose electrical conductivity is 7 to 14 orders of magnitude less than that of metals. They are typified by elements such as Si and Ge, and compounds such as GaAs, InP, PbTe, and PbS.

Semiconductors differ radically from ionic crystals in their energy-level structure and the restraints imposed on the electrons in the material. In the ionic crystals electrons "belong" to individual ions. Although the energy levels of the ions are influenced by the forces exerted by neighboring ions, one may in good approximation talk about energy levels of individual ions. When a semiconductor crystal is formed, the atoms retain individual possession of only those electrons which are in their inner shells. The outer, less tightly bound electrons become the collective property of the crystal as a whole. The physical properties of the semiconductors are related to the distribution of these collectively owned electrons among the energy levels of the crystal.

The main characteristic of the energy-level scheme of an ideal (pure) semiconductor crystal is illustrated in Fig. 7.1. The energy spectrum consists of broad bands of permitted levels: the *valence band* V and the *conduction band* C separated by a forbidden band (gap) of width E_g.

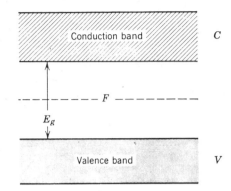

FIG. 7.1 Energy levels of an ideal semiconductor.

The density of states in the valence and conduction bands is a function of the energy and each state may be occupied by at most one electron. The probability that a given state will be occupied is given by the *Fermi-Dirac distribution function*

$$f = [1 + e^{(E-F)/kT}]^{-1}, \tag{2.1}$$

where F is the Fermi level shown in Fig. 7.1, E is the energy of the electron in the given state, and T is the absolute temperature. At 0°K the valence band of an ideal semiconductor is completely filled and the conduction band is empty. In this situation the semiconductor cannot conduct current; it becomes an insulator. Above 0°K some electrons are always present in the conduction band; consequently electric current may flow both as a result of the motion of electrons in the conduction band and as a result of the motion of vacancies (holes) in the valence band created by the promotion of electrons to the conduction band. These holes in the valence band are completely equivalent to particles with a positive charge provided with a mass which is in general different from the mass of a free electron. In an ideal semiconductor the number of electrons in the conduction band is exactly equal to the number of holes in the valence band.

The energy of an electron is actually a function of its momentum $\mathbf{p}$ or the related quantum-mechanical wave vector $\mathbf{k} = \mathbf{p}/h$. In one dimension the energy E may be represented as a function of a momentum component, say p_x, and it can be shown that E is a function of p_x^2 so that E has an extremum for $p_x = 0$. The energy in the conduction band and in the valence band may be represented by curves such as shown in Figs. 7.2a and b. The highest energy in the valence band is represented

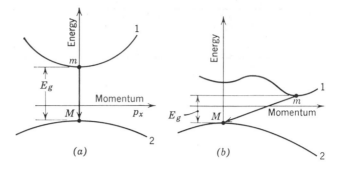

FIG. 7.2 Energy structure of direct (a) and indirect (b) semiconductors; line 1, conduction band, line 2, valence band.

by the points M and the lowest energy in the conduction band by the points m. The energy gap, or the width of the forbidden zone E_g, is represented by the vertical component of the vector directed from M to m. In Fig. 7.2a the energy surfaces are such that lowest minimum in the conduction band is attained for the same value of the momentum p for which the energy in the valence band has its maximum. When a material is such that this is the case it is called a *direct semiconductor*. The opposite situation prevails in Fig. 7.2b where the *lowest* minimum in the conduction band is attained at the point m which does not correspond to the same value of the momentum as the point M, the highest peak in the valence band. This type of a semiconductor is called *indirect*. GaAs is a direct semiconductor while GaP, Si, and Ge are indirect.

The significance of a distinction between these two types of semiconductors arises from the fact that in a direct semiconductor an electron may return from the bottom of the conduction band, represented by the point m, to the top of the valence band, represented by the point M, without significant change in momentum, whereas in an indirect semiconductor a large change of momentum is necessary. The following simple calculation shows that the photon created in the transition cannot provide this momentum change:

According to quantum theory, the range of variability of momentum component in a periodic structure of period a is restricted from $- h/2a$ to $+h/2a$. Thus the change of momentum, corresponding to a displacement from the center of Fig. 7.2 to its edge, is $p_{max} = h/2a$. On the other hand, the momentum carried by a photon of frequency ν is $P = h\nu/c = h/\lambda$. The ratio of these momenta is $P/p_{max} = 2a/\lambda$. Since a is only a few Ångströms and the wavelength of the emitted light is

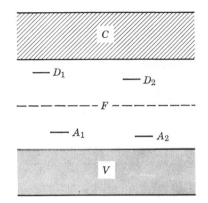

FIG. 7.3 Energy levels of a semiconductor with impurities.

of the order of 10^4 Å, the photon can provide only a very small momentum change compared to the range of the momentum component shown in the figures. Hence, according to the principle of the conservation of momentum, a transition which is accompanied *only* by the emission of a photon must be such that it can be represented in Fig. 7.2 by a nearly vertical line. If a transition such as mM in Fig. 7.2b occurs, it must be accompanied by a process that provides the difference in momentum. This process is usually the emission of a phonon, which takes up the momentum and carries away a part of the energy as well.

The simple energy-level schemes described are applicable to pure and ideally perfect semiconductors. In a real crystal additional energy levels may arise, such as those designated by D and A in Fig. 7.3. They are created by the presence of crystal imperfections, that is, impurities, vacancies, and dislocations. When the imperfections are few, these levels are localized in the neighborhoods of the corresponding imperfections, in contradistinction to the permitted zones which belong to the crystal as a whole.

In a real crystal the number of particles that serve as current carriers is determined primarily by the presence of impurities which may be classified in the following three categories: donor impurities, whose energy levels generally lie near the conduction band, acceptor impurities, whose levels normally lie near the valence band, and finally impurities, whose levels lie deep in the forbidden zone. The first two kinds are most important; they are deliberately introduced in many instances to produce an n-type or a p-type semiconductor. An *n-type semiconductor* is obtained when one introduces into the crystal lattice an element that has more valence electrons than the lattice site ordinarily calls for. Such an atom or ion becomes an *electron donor;* it provides an electron that can move about relatively freely. When an element with a lower than appropriate valence is incorporated into the lattice, it becomes an *electron acceptor;* it robs the valence band of one of the electrons that belong there in the ideal crystal. In this manner a hole is created in the material, which is then called a *p-type semiconductor.* For many applications semiconductors are doped with impurities of both kinds, the donor atoms prevailing in one part of the semiconductor, the acceptors in the other. The transition region separating these parts is called the *p-n junction.*

As the number of donor impurities in an *n*-type semiconductor is increased, their effect does not remain localized. Acting collectively at concentrations of 10^{16} cm^{-3} and above, the donor atoms distort the entire energy-level structure of the semiconductor. By providing electrons that can easily pass into the conduction band, they cause a shift of the Fermi level upward. As their concentration is further increased, their energy levels broaden and merge into the conduction band so that the Fermi level is shifted entirely into that band, as illustrated on the left side of Fig. 7.4. In a *p*-type semiconductor the presence of acceptor atoms depresses the Fermi level, which may then lie below the top edge of the valence band. This is shown on the right side of the figure. When the Fermi level penetrates the valence band or the conduction band, the semiconductor is called *degenerate.*

It is shown by a simple thermodynamic argument that when solids are in contact in thermodynamic equilibrium their Fermi levels are equal. The Fermi energy plays the role of the electrochemical potential. When two dissimilar solids are placed in contact, a transfer of charge takes place between them until the Fermi levels are equalized. This transfer of charge also accounts for the contact potential differences observed between metals.

As a consequence of the equality of the Fermi levels, the distribution of the electrons in a *p-n* junction in thermal equilibrium will take the form shown in Fig. 7.4. The Fermi level is shown as separating the filled from the unfilled states; actually, the probability of a state being filled is given by (2.1), and there are always some electrons above the Fermi level. The absolute temperature determines how sharp the separation of the filled and unfilled regions is. When a forward voltage is applied to a *p-n* junction by connecting the *p* and *n* regions to the positive and negative terminals of a battery, electrons will be driven from the *n* region to the *p* region, and holes from the *p* region to the

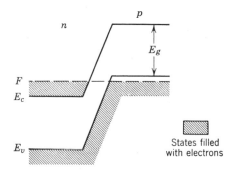

FIG. 7.4   Energy-level diagram of a p-n junction.

n region. Electrons and holes are then present in the same region; their recombination results in emission of radiation. The energy-level structure of a p-n junction with a forward bias is shown in Fig. 7.5. The energy levels are now displaced with respect to each other, so that $eV = F_u - F_l$, where V is the impressed electromotive force less the potential drop caused by ohmic losses in the semiconductor material. The situation represented in Fig. 7.5 corresponds to a degenerate semiconductor. The separate Fermi levels drawn in the n and p regions, which are not in thermal equilibrium with each other, have the following meaning: Within each band a near-thermal equilibrium is rapidly established, and the distribution of the electrons within the band may be described by a Fermi-Dirac function with a parameter F characteristic of that band.

It should be kept in mind that a p-n semiconductor with a forward potential applied is not in thermodynamic equilibrium. A large current flows through the junction and the junction region itself is far from equilibrium. The concept of a Fermi level is not applicable in the transition region.

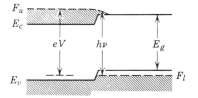

FIG. 7.5   Energy-level diagram of a p-n junction with a forward bias.

7.3 EMISSION AND ABSORPTION OF RADIATION IN SEMICONDUCTORS

When an electron recombines with a hole, energy becomes free which is partly or wholly converted into radiation. In a direct semiconductor the entire energy may be converted into a photon whose frequency is then determined by the equation

$$h\nu = E_2 - E_1, \tag{3.1}$$

where E_2 and E_1 are the energies of the electron in the initial and the final levels. In an indirect semiconductor, (3.1) is replaced by

$$h\nu + \Delta = E_2 - E_1, \tag{3.2}$$

where Δ is the energy of the phonon emitted. The inverse processes can also take place, incident radiation may be absorbed by a semiconductor and an electron may thereby be raised from the valence band into an empty level in the conduction band. Again, such a phenomenon may take place with or without the simultaneous absorption or creation of a phonon. Transitions may also occur in which the entire energy difference is converted into phonons and no photon is emitted.

Electrons in the valence band and holes in the conduction band may move around from one level to another. Electrons drift rapidly to the lowest available level in the conduction band, holes drift to the highest available level in the valence band. These changes within each band take place much faster than the recombination processes and, therefore, the frequency of light observed on emission is nearly such as would correspond to the band-gap energy. On the other hand, the absorption spectrum of the semiconductor is broad, with the lower limit of the frequency being determined by $h\nu = E_g$ because more energetic photons can be absorbed. They may create a hole below the top of the valence band and they may raise an electron to a level above the bottom of the conduction band. The energy evolved in the process of carrier drift within the band will appear as heat.

The mechanism described here is similar to the funneling mechanism noted in ruby and other crystal lasers. This similarity suggests that it should be possible to excite semiconductors by optical pumping. It is possible indeed to do so, at least in the case of some semiconductors, but there is a more convenient and more efficient method of creating the necessary electron–hole pairs, and we shall give most of our attention to that method. It is called the *carrier injection method*. Lasers produced by this method are called *injection lasers*.

Electrons are present in the conduction band of a heavily doped n-type semiconductor; holes are present in the valence band of a similar p-type material. When a semiconductor diode is formed consisting of a single crystal containing a p-type and an n-type region, there is a transition layer between these regions. It is called the *p-n junction.* Carriers of opposite types are present on opposite sides of the junction. Recombination can take place when electrons and holes are brought into the same region in space. This is accomplished by applying a forward voltage (positive on the p-side) to the diode. A sizeable current will result, consisting of the flow of holes from the p-side and a flow of electrons from the n-side into the junction. They recombine in (or around) the junction region giving rise to *recombination radiation,* which is emitted normally from a very thin transition region. For moderate current densities this radiation has the characteristics of spontaneously emitted radiation. As the current density is increased, the stimulated emission of radiation may become the dominant process in certain semiconductors, of which GaAs is the best known example.

The most important property of GaAs is that it is a direct semiconductor, in contrast to Ge and Si which are *indirect* semiconductors and which also emit recombination radiation. We shall return to the examination of the significance of this difference from the viewpoint of laser action.

Gallium arsenide occupies the same role among semiconductors that ruby occupies among ionic crystals. It is the first and the most used semiconductor laser material. Deferring the detailed description of the GaAs injection laser until Section 7.4, we note the following properties for illustration: The band-gap energy of GaAs varies with temperature, impurity content, and pressure. At 77°K the band gap, E_g, of pure GaAs, is 1.51 eV; at 300°K it is only around 1.41 eV. The presence of impurities tends to decrease the width of the forbidden region. It must be emphasized that the sharp, well-defined band structure is predicated on a strictly periodic lattice structure. The presence of a large number of foreign ions scattered in a random manner among the regularly placed Ga and As ions represents a departure from the model which results in the band structure. This causes irregularities at the band edges and as a consequence transitions may take place with less energy change than E_g in the pure material. In fact, recombination radiation is obtained from heavily doped GaAs diodes at 77°K, with a spectral distribution that has a peak between 8400 and 8500 Å; this corresponds to a photon energy between 1.46 and 1.48 eV. These figures are necessarily vague because the location of the peak depends on doping and the current density. As the current density increases, the peak of the recombination radiation shifts toward shorter wavelengths and the characteristics of stimulated radiation begin to appear. These can become dominant if the emitter of radiation is provided with proper reflecting surfaces which give it a laser structure.

Since the general laws of emission and absorption of radiation are as valid for recombinations as they are for ordinary intra-atomic transitions, it is expected that stimulated emission and absorption will occur whenever spontaneous emission occurs. The question is: Under what circumstances does the rate of stimulated emission exceed the rate of absorption in the p-n junction of a direct semiconductor?

Emission and absorption in a narrow spectral interval is associated with transitions between a group of levels at the bottom of the conduction band and a group at the top of the valence band. The rates at which these processes occur are proportional to the existing radiation density (per unit frequency interval), u_ν, and to Einstein's transition coefficient B, which is the same for the up–and the down transitions. In addition, stimulated emission is proportional to the number of transitions possible between filled upper and empty lower states, while absorption is proportional to the number of transitions possible between filled lower–and empty upper states.

Let $f_c(E)$ and $f_v(E)$ be the probabilities that a state with energy E in the conduction band and the valence band, respectively, is occupied. Then the probability that a state with energy E_2 in the conduction band is occupied, while the state with energy E_1 in the valence band is empty, is $f_c(E_2)[1 - f_v(E_1)]$, or briefly $f_c(1 - f_v)$. The rate of downward transition with a change of energy from E_2 to E_1 is therefore proportional to $Bu_\nu f_c(1 - f_v)$. The rate of the reverse process is proportional to $Bu_\nu f_v(1 - f_c)$. Hence stimulated emission will exceed absorption if, and only if,

$$f_c(1 - f_v) > f_v(1 - f_c). \tag{3.3}$$

This inequality simplifies to

$$f_c > f_v. \tag{3.4}$$

An interesting conclusion can be drawn from (3.4) in the case when the electrons in both bands have reached thermal equilibrium with the crystal lattice. Such a condition of equilibrium may be reached in a time short compared with the carrier lifetime. Then, when electrons are injected into the conduction band or holes in the valence band, the probability that a level in either band is occupied can be written in the form (2.1), namely,

$$f_c^{-1} = 1 + e^{(E_2 - F_c)/kT}; \quad f_v^{-1} = 1 + e^{(E_1 - F_v)/kT},$$

where F_c and F_v are the parameters of the distribution in the two bands. They are called the *quasi-Fermi levels* of the conduction and valence bands, respectively; E_2 and E_1 are the energies of the initial and final states. From (3.4) it follows that

$$e^{(E_1-F_v)/kT} > e^{(E_2-F_c)/kT},$$

and hence

$$F_c - F_v > E_2 - E_1 = h\nu, \tag{3.5}$$

where ν is the frequency of the emitted photon. Since $E_2 - E_1$ is not less than E_g, it also follows that

$$F_c - F_v > E_g. \tag{3.6}$$

This inequality is equivalent to the condition for population inversion in ordinary lasers. It was derived independently and simultaneously by Basov, Krokhin, and Popov [1] at the Lebedev Institute (Moscow) and by Bernard and Duraffourg [2] at the French National Telecommunications Establishment. This result, obtained shortly before the first semiconductor laser was successfully constructed, implies the following two requirements:

1. The laser material must be provided with enough impurities to make at least one side of the junction degenerate, otherwise the Fermi levels would both lie in the energy gap of width E_g.

2. The forward bias V must exceed E_g/e.

It should be stressed that the above calculations apply only to direct semiconductors. In the case of indirect semiconductors the calculation is more complicated because one has to take into account the statistics of the phonons required for the transitions. At low temperature the processes which require the absorption of a phonon from the lattice may be neglected, and then, in place of (3.6), one obtains the inequality

$$F_c - F_v > E_g - \Delta, \tag{3.7}$$

where Δ is the energy of the phonon created by the transition.

When comparing (3.6) and (3.7), the case of the indirect semiconductor looks superficially favorable. It must be remembered, however, that population inversion is not sufficient to produce a laser. The threshold condition must be satisfied, which requires that in one mode at least the rate of photon production should exceed the rate of photon loss *due to all causes*. In the case of indirect semiconductors the transition rates (up and down) are so slow in comparison to other processes responsible for losses that a net gain can not be achieved in spite of population inversion. The essential difference between the direct and the indirect semiconductor is the fact that the radiative transition rates in the former are high while they are low in the latter.

At first sight it might be puzzling that the low transition rates could prevent laser action altogether because, when a similar situation occurs in the case of ion crystal lasers, the consequence is merely that the laser has to be made longer or the terminations more reflective. It is possible to operate an ion laser with a small gain. Here the situation is different because the significant radiation loss mechanisms in semiconductors are different from those encountered in three- and four-level lasers. In the latter types the principal losses occur at the terminating surfaces; therefore the threshold condition can be found by equating the end losses with the amplification, calculated from the value of the (negative) absorption coefficient appropriate to the population inversion. In semiconductor materials the volume losses dominate; they are caused by the absorption of coherent radiation in the semiconductor without a corresponding increase in the capability of the semiconductor to re-emit radiation of the same frequency. Such events do not take place in ruby and similar crystals because the relevant emission and absorption phenomena take place in connection with transitions between two sharply defined levels in an essentially homogeneous material. As we have already stated, the semiconductor contains energy bands and—as long as there are vacancies below—an electron may drift to a lower level within a band without the emission of radiation. It is therefore possible for an electron in either band to absorb radiation by transferring from a low-energy state to an empty high-energy state within that band. This phenomenon is called *free carrier absorption*. The electron may then return to a lower state by the release of the excess energy in the form of heat.

Free carrier absorption is a very serious limiting factor in semiconductor laser operation because it not only causes a diminution of the effective amplification, but it causes the crystal to heat up with consequent further deterioration of its properties as an amplifier. Other phenomena, such as impurity absorption and scattering, also reduce the effective amplification. The combined effect of these phenomena is so large that in the first approximation we can neglect the effect of the terminations entirely and calculate the threshold current from the condition that the rate of carrier injection must be large enough to make the effective amplification at least zero.

The difficulty is illustrated by the fact that the measured volume-loss coefficient in various GaAs lasers is of the order of 10 cm^{-1}. This contrasts with the maximum amplification rate theoretically obtainable from a fully excited ruby, which is 0.4 cm^{-1}. Operation of a semiconductor laser requires a material with very large intrinsic gain before free carrier absorption and other unavoidable losses.

Exclusion of indirect semiconductors because of their low gain unfortunately eliminates the popular semiconductors Ge and Si as laser materials. It also eliminates pure GaP, which would be more desirable than GaAs, because GaP has a large bandgap which could produce visible radiation.

Injection of carriers into the junction region is a highly efficient method of creating conditions favorable for amplification (population inversion), but it is not the only method suitable for the excitation of semiconductors. It was already noted that optical excitation may create the necessary hole-electron pairs. Efficient as carrier injection is, it produces amplification only in a very thin layer.* It is therefore of interest to develop methods which result in excitation of the semiconductor material in bulk. In addition to optical pumping, such excitation has been successfully accomplished by means of irradiation by fast electrons and by the so-called avalanche breakdown method which consists of the imposition of a sudden pulse of very high electric field. This field accelerates carriers to an extent that, on collision, they produce new carriers by ionization, creating a vast plasma-like sea of carriers in the semiconductor. Lasers based on all these unorthodox methods will be briefly described, but most attention will be given to the injection lasers.

7.4 THE GaAs INJECTION LASER

Description of the Device. Gallium arsenide is the most commonly used material in injection lasers. Since the injection lasers are quite similar to one another, it is expedient to describe the GaAs laser in some detail and then in the following section to note special considerations applicable to other injection lasers.

As already noted, an injection laser consists of a single crystal having a p-type region and an n-type region. A device of this type is often used in electronics and is called a diode.

Gallium arsenide laser diodes are fabricated from single crystal wafers of n-type GaAs containing 10^{17} to 10^{18} donors (Te or Se) per cm^3. An acceptor element, such as zinc, is then diffused into the top layer of the wafer to a depth of 10 to 100 μm until the concentration of acceptors in excess of the donors reaches a value between 10^{18} and 5×10^{19} per cm^3. Then the top of the wafer is p-type, the main body n-type, with a transition region, or junction, between these two types.

---

* For an exception to this rule see p. 228.

The single crystal is usually so oriented that the surface of the wafer is perpendicular to the (001) crystallographic direction. Then the (110) and (1$\bar{1}$0) cleavage planes are perpendicular to each other and to the plane of the junction. The wafer is cut or rather cleaved along these planes into small rectangular segments. These are terminated by good, partially reflecting surfaces which are naturally aligned. One pair of the side surfaces is roughened to reduce specular reflection, the other pair provides the partial reflectors which form the Fabry-Perot cavity. The refractive index of GaAs is around 3.6, therefore the terminal faces have a reflectivity of about 35% at normal incidence. It is usually not necessary to provide reflective coatings on these surfaces.

The diode segments are mounted on a solid metal base to provide good electrical and thermal contact. An electrode is soldered to a metal contact deposited on top of the p-layer. The size of such a diode laser is surprisingly small. Typically, a GaAs diode measures 0.5 mm in length and 0.2 mm in its short dimensions.

To produce recombination radiation, an electric field is impressed across the diode in the form of a pulse lasting about 1 μsec. The electromotive force required is approximately the size of the bandgap, that is, 1.5 V. The current through the junction is a highly nonlinear function of the impressed voltage; it increases rapidly as the voltage approaches the band gap. The emission of light (infrared) is noticeable even at low current levels. The radiation emitted under such circumstances is essentially omnidirectional in the plane of the junction. It has a broad spectral content with a maximum between 8300 and 8500 Å. As the current density is increased, the character of the emitted radiation changes drastically. This change is best described in the words of the original observers of laser action in GaAs. In reporting what can be seen with a detector located along the beam axis, that is, in a direction perpendicular to the polished cleaved surfaces, the team of discoverers wrote [3]:

Below 5000 A/cm^2 the light intensity varied linearly with current density. Near 8500 A/cm^2 the intensity increased very rapidly with current, reaching a value about ten times the extrapolated low-current intensity at 20,000 A/cm^2. Such a current threshold is characteristic of the onset of stimulated light emission and it is significant that the azimuthal maxima of the radiation patterns make their appearance at this threshold.

We must stress that the rapid rise of the detector current in a preferred direction is in large part due to the narrowing of the emitted beam. The actual increase in detector response depends on the solid angle intercepted by the detector.

The discoverers, Hall and associates at the General Electric Laboratories, also observed the spectral distribution of the emitted radiation with a relatively low resolution spectrometer:

Below threshold the spectral width at half maximum is 125 Å. . . . As the current is increased through threshold, the spectral width decreases suddenly to 15 Å in a manner, which is again characteristic of the onset of stimulated emission.

As they immediately suspected, the 15 Å width was not an intrinsic property of GaAs. Subsequent investigations revealed that the output of the GaAs diode has a mode structure, as have other solid lasers, and that extremely narrow band outputs may be obtained in single-mode operation.

The wavelength of the GaAs laser is approximately 8400 Å. Its value depends on several variables, mainly the concentration of impurities, the temperature, and the current through the diode.

Stimulated radiation is produced only in a very thin layer about 2 μm in thickness. This layer is located on the p-side of the junction. The spread of the radiation in the far field is appropriate to an aperture about 2 μm high. Thus the light emanating from a GaAs diode is not as well collimated as that obtainable from rubies and many other ionic crystals.

Several hundred watts of peak power may be obtained from GaAs diodes in intermittent (pulsed) operation at 77°K; around 15 W has been reported at room temperatures. Specially designed GaAs diodes have operated continuously at liquid hydrogen temperature.

The operational characteristics of GaAs lasers depend on several variables. They will now be described separately.

Spectrum and Mode Structure. The gross spectral distribution of light emitted from a GaAs diode is shown in Fig. 7.6. The curves illustrate the strong dependence of the spectral distribution on the current passing through the junction. They were obtained by Burns and Nathan [4] on a particular diode whose temperature was maintained at 77°K. Measurements on different diodes or at different temperatures will result in similar, but not identical, curves. The peak wavelength depends on the impurity content of the material, on the temperature, and even on the pressure.

Measurements with higher spectral resolution reveal the mode structure of the laser output. This is illustrated in Fig. 7.7. The center of the emission spectrum is approximately at the peak of the fluorescent line, and the fine structure of the emitted radiation is determined by the resonator modes available in the spectral region where the intensity

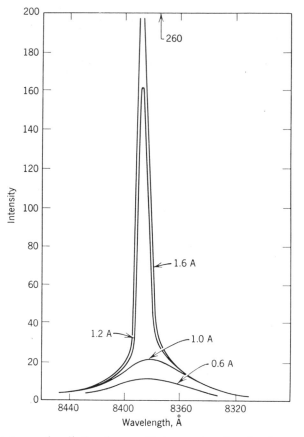

FIG. 7.6 Spectrum of radiation from a GaAs p-n junction for four values of the current. (After Burns and Nathan [4].)

of the fluorescent line is significant. The spectrum of the fluorescent line is a property of the material while the resonator modes are dependent on the construction of the device.

In a semiconductor laser the separation of the simplest cavity modes is uneven because the laser is operated in a wavelength region in which the index of refraction varies rapidly, whereas in other solid lasers the variation of the refractive index with frequency may be neglected. In the plane wave approximation the longitudinal modes of a Fabry-Perot interferometer satisfy (4.2) of Chapter 3, which may be put in the form

$$\eta\nu = nc/2L. \tag{4.1}$$

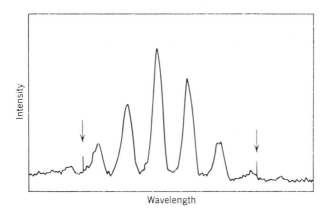

FIG. 7.7 Mode structure observed in the spectrum of a GaAs laser. At lower current more modes of approximately equal intensity are observed, and as the current
is increased the central mode becomes dominant. The markers (arrows) are separated by 11.26 Å.

Consequently, in first approximation the frequency difference $\Delta\nu$ of
consecutive modes must satisfy the equation

$$\left(\eta + \nu \frac{\partial\eta}{\partial\nu}\right)\Delta\nu = c/2L. \tag{4.2}$$

Let us introduce

$$\eta_0 = \eta + \nu \frac{\partial\eta}{\partial\nu}. \tag{4.3}$$

Then

$$\eta_0 \Delta\nu = \frac{c}{2L} \tag{4.4}$$

takes the place of $\eta\Delta\nu = c/2L$ valid for a constant η. The second term
on the right side of (4.3) is not negligible. In fact, it may reach one-half
of the value of the first term [5,6].

The frequency of recombination radiation is determined by the energy
that becomes free upon the recombination of a hole with an electron.
This energy is approximately equal to the band-gap energy E_g, which
is uniquely determined for a pure semiconductor at a given temperature
and pressure. Sturge [7] determined E_g of pure GaAs at 1 atm pressure
in the temperature range of 10 to 294°K. In this region, a rough approximation of the experimental values is given by

$$E_g(T) = E_g(0) - \beta T^2, \tag{4.5}$$

where $E_g(0) = 1.521$ eV, and $\beta = 1.21 \times 10^{-6}$ eV/$°K^2$. Experimental studies of recombination radiation in doped GaAs crystals show that the peak of the emission spectrum depends on the concentration of the carriers (doping) and on the current passing through the crystal. Currents well below threshold produce a broad emission band which shifts to higher energy and becomes narrower as the current is increased.

These and other related observations indicate that the recombination in heavily doped GaAs does not take place by a direct passage from the bottom of the conduction band to the top of the valence band, but that the donor– and acceptor levels are involved in a rather complex manner. They extend the edges of these bands into the forbidden region, and they thus make it possible to obtain radiation with photon energy less than the band-gap energy.

In discussing the temperature dependence of coherent radiation it is practical to specify the situation by requiring that the spectrum is to be measured just above threshold. When this is done, a curve of the type shown in Fig. 7.8 is obtained for the peak of the coherent radiation versus temperature. The points in the figure are observations made on a single GaAs diode operated at different temperatures. Correction was made for the rise of temperature during the pulse. The parabola represents the variation of the band-gap energy displaced downward by 41.6 meV.

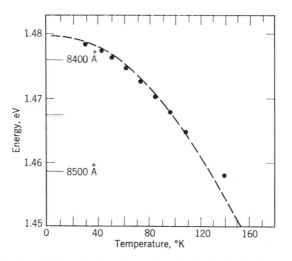

FIG. 7.8 Spectral variation of the coherent output of a GaAs laser with temperature. The dots represent measured photon energies with the laser excited just above threshold; the curve represents the variation of the band gap of pure GaAs less 0.0416 eV. (After Engeler and Garfinkel [8].)

It fits the observed points reasonably well. Observations on other GaAs diodes lead to similar figures differing only in the displacement of the parabola. In this matter there is a variation of about 5 meV from one diode to another. The important matter is that frequency of the output varies with the temperature in the same manner as the bandgap energy [8]. The spectrum shifts to lower frequencies as the temperature increases. It was observed that the individual peaks in the resolved mode structure of the stimulated radiation shift only about one-third as fast as the entire group; the modes at higher energies fall behind as the temperature increases. They drop out and may be replaced by new modes at the long wavelength end.

The temperature dependence of the frequency of a given mode can be deduced from (4.1) by taking into account that the refractive index depends on the temperature as well as on the frequency. By differentiation of (4.1) we obtain

$$ -\frac{d(\log \nu)}{dT} = \frac{d(\log L)}{dT} + \frac{d(\log \eta)}{dT}. \tag{4.6} $$

The first term on the right is the thermal expansion coefficient, whose value does not exceed 6×10^{-6} per °K, whereas the observed temperature variation of cavity modes is around 5×10^{-5} per °K. The effects of thermal expansion are therefore neglected; the major contribution to the frequency shift arises from the variation of η with the temperature. Then (4.6) is replaced by

$$ -\frac{1}{\nu}\frac{d\nu}{dT} = \frac{1}{\eta}\left(\frac{\partial \eta}{\partial T} + \frac{\partial \eta}{\partial \nu}\frac{d\nu}{dT}\right), \tag{4.7} $$

which combined with (4.3) results in

$$ \frac{d\nu}{dT} = -\frac{\nu}{\eta_0}\frac{\partial \eta}{\partial T}. \tag{4.8} $$

The left side of (4.8) is the rate of change of the frequency of a given mode with temperature. The partial derivative on the right is the rate of change of the index of refraction with the frequency held fixed. This quantity can be determined by measurement.

There is a necessary vagueness in the description of the spectral properties of injection lasers, which contrasts with the precision common in the description of ion crystal lasers. The spectrum of the output of an injection laser varies from one crystal to another because of variations of impurity content. In addition to this variation from crystal to crystal, the properties of the semiconductor lasers are rather sensitive functions of environmental variables such as temperature, pressure, and magnetic field. Notwithstanding all this variability, it is possible to produce a highly precise signal from an injection laser operating under properly controlled circumstances. A GaAs laser operated in the continuous regime in a single mode at 77°K had a frequency spread of only 150 kHz—a spread of less than one part in 10^9 [9].

Threshold Current and Efficiency. The temperature dependence of the threshold current density of a GaAs diode laser is shown in Fig. 7.9. In the vicinity of 77°K the threshold current varies approximately as T^3. The efficiency of the diode laser decreases with increasing temperature, and the heat dissipated in an inefficient laser contributes to further increase of the temperature. For these reasons most semiconductor lasers are operated at liquid nitrogen temperature (77°K) or lower.

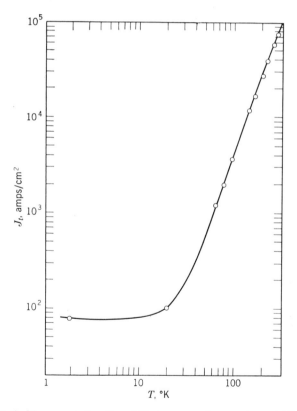

FIG. 7.9 Threshold current density of GaAs versus temperature. (After Burns and Nathan [5].)

Uniaxial pressure perpendicular to the junction plane lowers the threshold current of GaAs and shifts the spectrum of the output toward shorter wavelengths. Hydrostatic (omnidirectional) pressure does not affect GaAs diodes, but it does affect some other diode lasers. The imposition of a magnetic field H shifts the energy of photons upward by an amount proportional to H^2.* These effects permit the tuning of a GaAs diode laser over a significant range of frequencies.

High conversion efficiency (from electrical energy to light) is one of the most important properties of injection lasers. The figure of merit used to describe such a laser is usually the *external efficiency*, defined as the number of photons emitted by the laser divided by the number of carriers injected into the junction. The *internal efficiency* is the number of photons generated per carrier injected. This is not directly measurable because many of the photons get absorbed. External efficiencies around 0.5 have been measured in GaAs lasers operating at 77°K. At room temperature a more usual figure for the external efficiency is 0.1. The efficiencies quoted apply to laser operation, that is, above threshold.

Other Properties of the Laser Output. Light emitted from a GaAs diode is usually plane polarized, but the polarization varies from one diode to another. The factors determining polarization are not known.

The spatial distribution of the radiated energy has the general appearance of the diffraction pattern from a rectangular aperture, complicated by distortions attributable to the simultaneous presence of several modes of oscillation. The vertical spread of the pattern is indicative of an active region of thickness of around 2 μm. It corresponds to a beamwidth of about 10°—much broader than the beams radiated from ion crystal lasers.

Several hundred watts of peak power may be obtained from GaAs diodes in intermittent (pulsed) operation at 77°K; around 15 W is a good figure for room temperature operation. Specially designed GaAs diodes have operated continuously at liquid hydrogen temperature and produced an output (steady) of 3.2 W with an external efficiency of nearly 50% [10]. Such high-output diode lasers are designed with very massive heat sinks in order to maintain the diode at the necessary low temperature.

It has been observed that high-power GaAs diode lasers emit not only in the near-infrared region around 8400 Å, but also in the blue region at twice the frequency of the infrared radiation. The appearance of this blue light is the result of harmonic generation within the GaAs

---

* This quadratic dependence on the magnetic field is a specific property of GaAs.

crystal. In keeping with the nonlinear nature of the frequency-doubling phenomenon, the intensity of the blue light is proportional to the square of the intensity of the fundamental radiation.

The detailed spectral characteristics of GaAs luminescence depend on the concentration of impurities, on the temperature, and on the current. The mechanism of recombination is actually more complicated than here described. A quantitative theory of the spectrum takes into account the existence of banded donor– and acceptor states adjacent to, and partially overlapping, the conduction– and valence bands. Calculations have shown that most of the observed features of the spectrum can be accounted for by assuming that recombination transitions take place mainly between the impurity bands. This explanation is consistent with the observed quadratic dependence of the photon energy on the external magnetic field. The interested reader is referred to the work of Lucovsky [11].

The properties of GaAs diode lasers, as well as those other semiconductor lasers, are discussed in greater detail in several excellent review articles which all contain numerous references [5, 12, 13].

7.5 INJECTION LASERS OTHER THAN GaAs

Semiconductor junction lasers cover a large wavelength range because a number of III–V compounds similar to GaAs can be used as basic materials. The energy gaps of such materials with laser capabilities are known to vary from about 0.16 to about 1.5 eV with a corresponding variation in the frequency of the emitted radiation. Such materials are shown in Table 7.1.

Gallium arsenide has the highest band-gap energy of the simple III–V junction laser compounds. Ordinarily the light emitted from a GaAs diode has a wavelength of around 0.84 μm. Gallium phosphide has a higher energy gap than GaAs. The photon whose energy corresponds to this gap is in the green part of the spectrum, whereas the radiation from the other simple III–V compounds is in the infrared. Gallium phosphide would make an excellent injection laser material were it not for the fact that it is an indirect semiconductor. Fortunately GaAs and GaP can be combined in any proportion, and the transition from the direct semiconductor GaAs to the indirect GaP takes place when the proportion of GaP reaches 0.4. The energy gap E_g of the mixed semiconductor $GaAs_xP_{1-x}$ lies between the band gaps of the two pure materials. As a consequence, junctions of the mixed semiconductor $GaAs_xP_{1-x}$ operate as lasers when x exceeds 0.6. By varying the composition, lasers can be constructed from this material covering the wavelength range from 0.84 μm (pure GaAs) to about 0.64 μm. (40% GaP).

The behavior in InP is very similar to that of GaAs. The band gap in InP is slightly lower than in GaAs; consequently, the spectrum of the emitted radiation is shifted to longer wavelengths (0.90 to 0.91 μm). Still farther in the infrared region lie the outputs of GaSb, InAs, and InSb lasers. Their wavelengths are 1.56, 3.11, and 5.18 μm, respectively. Such lasers must be operated at very low temperatures. By using mixtures, or alloys, of these semiconductor materials, lasers can be constructed whose emission lies in the wavelength range between the pure

TABLE 7.1
INJECTION LASER MATERIALS AND
THEIR WAVELENGTHS

Material	Wavelength (μm)
GaAs	0.84
Ga(AsP)	0.64 − 0.84
(Ga,In)As	0.84 − 3.11
(Ga,Al)As	0.64 − 0.84
InAs	3.11
In(As,P)	0.90 − 3.11
InP	0.90
In(As,Sb)	3.11 − 5.18
InSb	5.18
GaSb	1.56
PbS	4.32
PbTe	6.5 (at 12°K)
PbSe	8.5 (at 12°K) 7.3 (at 77°K)
(Pb,Sn)Te	6 − 28
(Pb,Sn)Se	8 − 31

materials. Further "tuning" of the laser frequency is possible by varying the concentration of the carriers as well as by varying the temperature and pressure, or by impressing on the semiconductor an external magnetic field. A moderate magnetic field, applied in the direction perpendicular to the current, tends to decrease the thresholds of InAs and InSb diodes. Unlike the GaAs diodes, the diodes of these materials change their output frequencies linearly with an applied magnetic field. This variation indicates that the recombination mechanism in InAs and in InSb is different from that encountered in GaAs [13].

The mode structure of the emitted radiation, the variation of the spectrum with temperature and current, and the variation of the threshold current with temperature are very similar in nature in all these injection lasers. The general principles explained apply to all these types.

Injection lasers may be made not only of the III–V compounds discussed so far, but also of some IV–VI compounds, appearing at the end of Table 7.1. The preparation of p-n junctions from such materials is different in this case because an excess of the IV element is used to provide the acceptors, and an excess of the VI element provides the donors. First an n-type PbTe is prepared by annealing the crystal under Te vapor pressure, then Pb vapor is infused by exposing the wafer to lead vapor at a suitable temperature [14, 15].

While the energy gap of GaAs is around 1.5 eV, this gap is almost an order of magnitude smaller for PbTe and PbSe. Consequently the radiation emitted by the lead telluride–and selenide lasers have a wavelength longer by about a factor of ten than that of the gallium compounds. The lead compound lasers should be operated at, or below, liquid nitrogen temperature. Their outputs may be tuned by means of an impressed magnetic field [16] and also by hydrostatic pressure. The wavelength of a single PbSe laser can be tuned from 7.5 μm to 11.0 μm by varying the pressure from 0 to 7000 bars. The variation of the photon energy (frequency) is linear with the pressure, as shown in Fig. 7.10 [17].

Mixed lead–tin tellurides and selenides offer very interesting possibilities for laser design. The unusual properties of the band structures of these materials were discovered in connection with laser studies. The materials in question have compositions of the form $Pb_{1-x}Sn_xTe$ and $Pb_{1-x}Sn_xSe$. They were explored in the region $0 \leq x \leq 0.28$. Using such materials, lasers can be constructed capable of providing coherent radiation in a frequency band located anywhere within a very broad range by choosing the composition and the operating temperature. This range extends from about 6 μm upward to around 28 μm without a sharp limit on the long wavelength end. The reason for this great flexibility is that in these mixed lead compounds the band gap E_g becomes zero for some value of x, which depends on the temperature. The vanishing of the band gap is believed to come about as follows: the bands which are conduction and valence bands of pure PbTe are valence and conduction bands for SnTe, but the roles of the bands are interchanged. The interchange of the bands is inferred from the following experimental facts: The variation of the band-gap energy E_g with temperature is such that $\partial E_g/\partial T$ is positive for PbTe and negative for SnTe. For the mixed compounds $Pb_{1-x}Sn_xTe$ the slope of the $E_g(T)$ curve is intermediate between those of the pure compounds. Moreover, the band gap energy

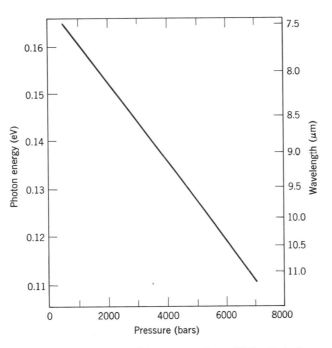

FIG. 7.10 Variation of wavelength with pressure for a PbSe diode laser at 77°K. Current density 4800 A/cm^{-2}. (After Besson et al. [17].)

of the mixed compounds is less than those of the pure compounds. Starting with PbTe, as the tin content is increased, E_g decreases nearly linearly toward zero, and then it increases to the value appropriate for SnTe.

According to the interpretation of Dimmock, Melngailis, and Strauss [18], the energy gap between two *fixed* bands, characterized by their symmetry properties and quantum mechanical labels, decreases smoothly and nearly linearly with increasing tin content, in the manner shown in Fig. 7.11. For an appropriate molar proportion the bands pass each other. The point of passing depends on the temperature. The band-gap measurements are naturally not made at this passing point, but the location of the passing point is deduced from the line fitted to observational points on either side. A similar situation prevails for the mixed selenides of lead and tin [19]. These materials may then be used to design lasers emitting in a specific region of a broad infrared range, provided the composition and the operating temperature are properly chosen. When the temperature of such a laser is carefully controlled, the frequency of the semiconductor laser may be maintained to a

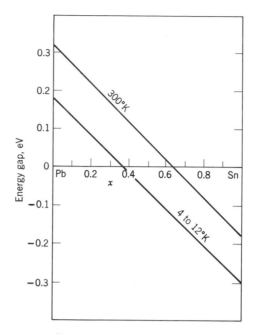

FIG. 7.11 Energy gap of $Pb_{1-x}Sn_xTe$ as a function of x. *Upper Curve:* data for $300°K$; *lower curve:* data for 4 to $12°K$. Lines represent straight-line fit to the data collected by Dimmock et al. [18].

remarkable degree of stability, as demonstrated by the observation of beat frequencies between a CO_2 gas laser and a $Pb_{0.88}Sn_{0.12}Te$ diode laser [20].

In the injection lasers described so far, recombination takes place only in a very thin sheet. It is possible, however, to produce an injection laser in which recombination takes place over a larger volume. Such a volume laser may be constructed of three layers of InSb forming a so-called n^+pp^+ structure. The $+$ signs indicate heavily doped regions while the center layer (p) is lightly doped [21]. An interesting modification of such a volume injection laser is shown in Fig. 7.12. The unusual feature of this laser is that the emission of coherent light takes place in a direction parallel to the current. The application of a magnetic field of the order of a kilogauss facilitates the operation of this device [22]. Three-layer structures of GaAs have also been investigated and rather complex phenomena were observed including stimulated emission from a center region 2 to 20 μm thick [23]. These investigations may yet produce injection lasers capable of large power-handling capacity.

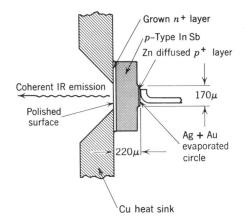

FIG. 7.12 InSb n^+pp^+ laser structure designed for coherent emission parallel to the current. (After Melngailis [22].)

7.6 OPTICALLY PUMPED SEMICONDUCTOR LASERS

Generation of stimulated radiation in injection lasers is normally restricted to a very thin layer. This fact imposes limitations on the power handling capacity of the laser and on the collimation of the beam emerging from the laser. Optical excitation makes it possible to use a somewhat larger volume for the generation of radiation, but it does not provide as great advantages as were hoped for. One of the main disadvantages of optical pumping is that, in the case of a semiconductor, the pump must be another laser.

Well before optically pumped semiconductor lasers were successfully constructed, Basov [24] reached the conclusion that such lasers must be excited by relatively monochromatic light sources, and that the excitation must be of short duration otherwise the absorption of light energy, not properly used in the semiconductor, causes an intolerable increase in temperature. He suggested excitation with giant-pulse lasers.

The first successful demonstration of stimulated emission from an optically excited semiconductor took place at the end of 1964 with a p-type GaAs crystal excited by means of a focused beam from an ordinary ruby laser [25]. The external efficiency of the system was extremely low; about 0.5%.

Much higher conversion efficiency is attained by means of a giant-pulse ruby, especially when the frequency of the ruby radiation is shifted

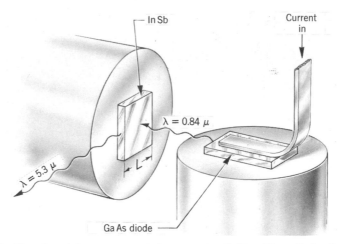

FIG. 7.13 Experimental arrangement for pumping InSb with a GaAs diode laser. (After Phelan and Rediker [27].)

down by the use of the Raman effect so that the wavelength of the exciting radiation is 8281 Å, not much below the wavelength that corresponds to the energy gap of GaAs. With this radiation incident on the GaAs crystal wafer, stimulated radiation can be observed both at 77°K and at room temperature. Basov and co-workers obtained in this manner a laser output in the wavelength region between 8330 and 8360 Å with a quantum efficiency (in the GaAs) of around 4% [26]. The energy of the photons so obtained from GaAs is higher than the energy of the photons obtained from diode injection lasers whose emission at 77°K peaks around 8460 Å. Such a difference between the photons obtained by optical excitation and by carrier injection into the same basic material is not unique. It is observed on materials other than GaAs as well. The homogeneous semiconductors used in the optical excitation experiments are quite different from the heavily doped diode materials. The mechanisms of recombination in these materials are also different. It is easier to correlate the results obtained on optically excited semiconductors with the properties of the pure material than to interpret the results of the diode laser experiments.

Optically excited semiconductor lasers have been useful mainly as tools of research for the exploration of semiconductor materials. They permit the study of recombination and stimulated emission under much simpler and cleaner circumstances than those that prevail in diodes.

The optical excitation of a semiconductor laser is often accomplished with the output of another semiconductor (diode) laser. An arrangement for pumping InSb with the output of a GaAs diode is shown in Fig. 7.13. It represents the arrangement used in the experiments of Phelan and Rediker [27], who were the first to obtain success with semiconductor laser combinations. The laser action in the optically excited InSb is demonstrated by the sudden increase of radiative output at threshold and by the narrowing of the luminescence of InSb, conspicuously seen in Fig. 7.14. Both curves were obtained with the same configuration at 4°K. The broad, spontaneous emission curve results when the GaAs laser is pulsed with 10 A, 5-μsec pulses and the narrow laser emission curve with 15 A, 50-nsec pulses. The peak intensity of the laser pulse is much higher than that of the spontaneous radiation, but the curves in the figure are normalized for equal peak ordinates.

Some of the optically excited semiconductor lasers are listed in Table 7.2 on p. 232.

An interesting property of some of these optically excited lasers is that they may be pumped by a source whose photons have less energy than the band gap. Excitation then occurs with the simultaneous absorption of two photons [28, 29].

Although the layers of semiconductors that can be excited optically are thicker than the active regions of ordinary junction diodes, these layers are still quite thin. They are of the order of 20 μm, in contrast to the 2-μm thickness typical for diodes. So far it has not been possible to excite optically significant volumes of semiconductor material.

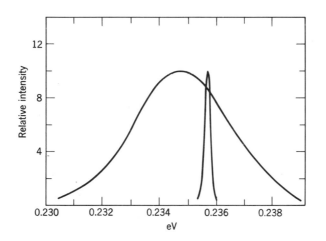

FIG. 7.14 Spontaneous and laser emission from InSb. The peak intensities of the broad spontaneous emission and the narrow laser line are adjusted to compare linewidths. (After Phelan and Rediker [27].)

TABLE 7.2

OPTICALLY EXCITED SEMICONDUCTOR LASERS

Laser material	Wavelength (μm)	Exciter	Note
GaAs	0.836	Ruby	
GaAs	0.836	Nd glass	Two photons
InAs	3.1	GaAs	
InSb	5.3	GaAs	
PbTe	6.5	GaAs	
CdS	0.495	Ruby	Two photons
CdSe	0.697	Nd glass	Two photons
CdSe	0.69	Ga(AsP)	

7.7 SEMICONDUCTORS EXCITED BY ELECTRON BEAMS

The phenomenon of cathodoluminescence has been known for some time. It consists of a chain of events which begins with the impact of fast electrons on a semiconductor and ends with the emission of light. The intermediate processes include the creation of carrier pairs by the fast electrons and the slowing down of the energetic carriers by collisions, until finally recombination takes place with the emission of light.

It has been anticipated for several years that this process may be utilized for the excitation of some semiconductors to a degree necessary for laser action. In order that this be accomplished, it is necessary to bombard materials with pulses of sufficient intensity to attain population inversion. The duration of the pulses must be short so that the heat evolved in intermediate processes does not increase the temperature of the semiconductor significantly. We have noted that the attainment of threshold in injection lasers is progressively more difficult at elevated temperatures. The reasons for the increase of the threshold inversion with temperature are independent of the process of excitation. They apply equally in the present case. Calculations have shown that in the process of creating a carrier pair approximately $3E_g$ energy is removed from the electron beam. Of this energy only about E_g is recovered as radiation, the rest is converted into heat. The system will operate as a laser only when the heat generated is promptly removed, and when the rate at which the electrons lose energy and settle at the bottom of the conduction band is fast compared to the rate of recombination.

The technique of electron beam excitation is illustrated in Fig. 7.15, which shows the apparatus of Rediker [30]. A high-energy beam is focused on a small semiconductor sample whose dimensions are fractions of a millimeter. As in the case of diode lasers, the single crystals are

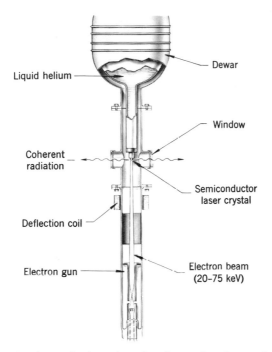

Liquid helium

Dewar

Window

Coherent radiation

Semiconductor laser crystal

Deflection coil

Electron gun

Electron beam (20–75 keV)

FIG. 7.15 Apparatus for excitation of semiconductors by electron beams. (Courtesy Lincoln Laboratories.)

cleaved or polished to form a pair of parallel surfaces, thus forming a Fabry-Perot resonator, whose axis is perpendicular to the electron beam. The other surfaces are roughly finished. As the electron beam penetrates the surface layer to about 10-μm depth, it creates electron— hole pairs. These drift ultimately (and hopefully fast) to the bottom of the conduction band and to the top of the valence band, respectively. Their presence in sufficient numbers provides the capability for amplification. Laser oscillations build up when the threshold condition is satisfied. When very energetic (over 100 kV) electron beams are used, the actual apparatus is more complicated than shown in Fig. 7.15.

A different type of arrangement proposed by Basov and Bogdankevich [31] is shown in Fig. 7.16. In this arrangement a thin layer of an extended semiconductor film is irradiated. This film, combined with an external mirror, forms the Fabry-Perot cavity. A vacuum envelope, not shown in the figure, is also necessary for operation.

Table 7.3 shows some of the materials excitable to laser action by

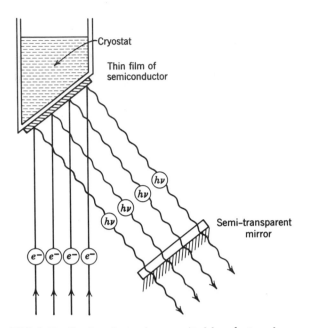

FIG. 7.16 Semiconductor laser excited by electron beam.

TABLE 7.3
SEMICONDUCTOR LASER MATERIALS
EXCITABLE BY FAST ELECTRONS

Material	Wavelength (μm)
GaAs	0.84
GaSb	1.53
InAs	3.00
InSb	4.95
PbS	4.27
PbSe	8.5
PbTe	6.5
Te	3.72
CdS	0.495
CdSe	0.69
CdTe	0.78
ZnO	0.38
ZnS	0.33

electron bombardment. The common diode laser materials can all be excited by fast electrons. In addition to these, the table contains materials, such as Te, as well as the cadmium and zinc compounds, which are not suitable for injection lasers because p-n junctions cannot be fabricated from these materials. The conditions of excitation of these materials are quite variable. Although InAs and InSb are excited by 15-keV electrons, excitation of the cadmium compounds requires 200 keV. The electron beams are pulsed with a duration of 1 to 2 μsec at a repetition rate of a few thousand per second. The current densities vary between 0.1 and 1.0 A/cm^2. For details the reader is advised to consult the collection of articles entitled *Radiative Recombination in Semiconductors* in the proceedings of the Seventh International Conference on the Physics of Semiconductors [31], and other pertinent works listed in the review articles [12, 32].

Besides being the only elemental semiconductor laser, Te is interesting because its output at 3.7 μm is in an infrared atmospheric window.

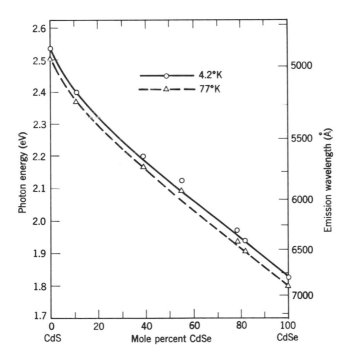

FIG. 7.17 Photon energy and corresponding wavelength of CdS$_x$Se$_{1-x}$ lasers at 4.2 and 77°K, as a function of crystal composition. (After Hurwitz [36].)

This element is easily excited to laser oscillations, not only at 4°K, but also at 77°K. According to the experience of Benoit à la Guillaume and Debevre, the discoverers of this laser, 15-to-20-keV electrons and 1-μsec pulses with 50 mA/cm^2 provide adequate excitation at 4°K, and ten times this current is sufficient at 77°K [33].

Cadmium salts are interesting partly because CdS is the first material in which several indications of stimulated emission were observed upon electron bombardment [34]. Subsequently, excellent lasers were made of CdS, CdSe, and mixed compounds of composition CdS$_x$Se$_{1-x}$. By varying the proportion of S to Se it is possible to cover a large region of the visible spectrum from 4900 to 6900 Å. The lasers constructed by Hurwitz [35, 36] were capable of producing up to 20 W of peak power with up to 11% power efficiency. The variation of the wavelength of these lasers with composition and temperature is shown in Fig. 7.17. Table 7.4 lists other relevant parameters of such lasers.

TABLE 7.4[a]

THRESHOLD CURRENT DENSITY, PEAK OUTPUT POWER, AND MAXIMUM OBSERVED EFFICIENCY OF CdS$_x$Se$_{1-x}$ LASERS AT 4.2°K (UPPER FIGURES) AND 77°K (LOWER FIGURES)[b]

CdSe (mol-%)	Threshold Current Density (mA/cm^2)	Output Power (W)	Efficiency (%)
0	1000	10	0.7
	1200	5	0.4
11	80	10	7
	400	6	4
39	30	20	11
	80	8	9
55	40	11	9
	400	7	5
79	120	3	3
	600	1	2
82	100	5	4
	450	3	3
100	50	16	8
	200	14	7

[a] After Hurwitz [36].

[b] Sample dimensions: 0.25 mm × 1.5 mm × 4 to 20 μm; beam voltage: 50 kV.

Zinc oxide and sulfide lasers extend the range of semiconductor lasers into the ultraviolet region. The material preparation for these lasers is rather demanding. It requires thin (3 to 15 μm) platelets of vapor-grown perfect crystals. Laser action is achieved both at liquid helium and at liquid nitrogen temperature, but with lesser efficiency at the latter [37, 38].

7.8 LASERS EXCITED BY AVALANCHE BREAKDOWN

One of the earliest schemes proposed for the excitation of semiconductors is the creation of pairs of opposite carriers by the application of an electric field large enough to produce impact ionization within the semiconductor. This method was proposed by Basov, Vul, and Popov as early as 1959 [39]. Although the scheme was much discussed theoretically, it was difficult to put into practice. When the semiconductor breaks down under the influence of an electric field of several thousand volts per centimeter, a hot plasma is formed in which population inversion is established only after the field is removed and some cooling takes place. Whether a population inversion is established or not depends on the rates of the processes with which the carriers return to an equilibrium situation.

The clearest evidence for excitation of a bulk semiconductor by avalanche breakdown was obtained by Southgate [40], who obtained stimulated emission from n-type GaAs. The experiments are performed with Te-doped crystals at 77°K, subjected to pulsed fields in excess of 2200 V/cm and of 10-nsec duration. Outputs of 3 W have been measured with 3×10^4 W input power indicating a pitifully low power efficiency, at least in the early stages of the development of this type of laser.

7.9 HISTORICAL REMARKS ON SEMICONDUCTOR LASERS

There is considerable controversy over priorities of discovery and invention in all branches of quantum electronics. The situation is perhaps most involved in the field of semiconductor lasers. Given almost any such device, several investigators and inventors can claim credit for its discovery or invention with some justification. Conflicting claims arose partly because the proponents of a new method or principle for the excitation of semiconductors were seldom able to carry out the construction of a device based on their principle and partly because groups in several laboratories worked feverishly and independently to complete the development of very similar devices.

It is an interesting historical sidelight that in a set of notes sent to Edward Teller in September 1953, John von Neumann calculated that, if electrons and holes are injected into a p-n junction, it is possible to upset the equilibrium distribution of the carriers to such an extent that light amplification may be obtained as a result of the stimulated recombination of the carriers. This correspondence did not come to light until sometime in 1963, and it did not have an effect on the invention of the semiconductor laser. Other similar speculations in the early sixties received considerable publicity. Many schemes and calculations were discussed in various seminars, published in journals and in reports of restricted circulation, or debated at conferences. These discussions and publications often guided or aided subsequent experimental developments even when the lasers that were constructed did not operate according to the concepts first proposed.

Speculations concerning semiconductor lasers were favorite topics at the Second International Conference on Quantum Electronics held in Berkeley in March 1961, and at the third such conference, named congress, held in Paris in February 1963. The proceedings of these meetings contain ample material of historic interest. Many semiconductor schemes that eventually worked, and some that have since been found impractical, were introduced and debated. The scientists at the Lebedev Institute in Moscow, under the leadership of Basov, introduced and analyzed many of the semiconductor laser schemes. From 1959 on they published calculations pertaining to the use of avalanche breakdown as an excitation method for indirect semiconductors, and in 1961 they proposed the p-n junction laser [41]. They derived the necessary condition for amplification at the same time that the French investigators Bernard and Duraffourg also obtained this condition [2]. Basov and his co-workers contributed significantly to the optically excited semiconductor laser [24]. The major credit for the development of the semiconductor laser excited by electron beam excitation also belongs to this Russian group, although the first operating laser of this type was constructed in France.

A research team at Lincoln Laboratories (Massachusetts Institute of Technology), led by B. Lax, has made a number of notable contributions in diverse areas. Although their much discussed cyclotron resonance scheme did not lead to a practical laser, their investigations greatly clarified the structures of the potential laser materials. They pioneered work on the tuning of diode lasers and made many significant improvements in several laser types, in particular, in the diode lasers operating in the middle-infrared region [13–20] and the electron-beam-excited lasers [35, 36, 38].

The development of the individual laser types progressed roughly as follows: the development of injection lasers got seriously under way in 1961 when Basov, Krokhin, and Popov [41] proposed the injection of carriers in a semiconductor junction as a means for the production of a light-amplifying material and published detailed calculations concerning the feasibility of such a device. Shortly thereafter Dumke at IBM made an essential contribution by demonstrating on theoretical grounds that direct semiconductors have a much greater potential than the indirect ones favored by the Russian group. The actual construction of the first injection lasers took place simultaneously and independently in three American laboratories: General Electric [3], International Business Machines [42], and Lincoln Laboratories [43] announced the successful operation of their GaAs junction lasers in the early fall of 1962. The extension of this work to other injection lasers was mostly carried out at Lincoln Laboratories where the techniques of magnetic- and pressure tuning were also developed.

Basov was first to propose and analyze theoretically the optically excited semiconductor laser [24]. The first such operative device was made by Schlickman, Fitzgerald, and Kingston at M.I.T., in 1964 [25]. Shortly afterward the group at the Lebedev Institute has scored many experimental successes with these devices [26, 28, 29].

The first suggestions pertaining to excitation by electron beams were also made by Basov and his co-workers, who almost achieved full experimental success. Their experiments on CdS showed signs of stimulated emission [34], but the first such operating laser was built by Benoit à la Guillaume and Debevre [33]. As was the case with the injection lasers, much subsequent progress was made at Lincoln Laboratories [35, 36, 38].

The history of the excitation by avalanche breakdown started in 1959 when Basov, Vul, and Popov [39] proposed this method for the excitation of pure germanium and silicon. When Southgate [40] finally succeeded in 1968, he obtained laser action in a very different material, namely Te-doped GaAs.

Because of the active interchange of ideas and because of the influence that the experiments with unsuccessful schemes provided for the later successful ones, it is not fair to assign credit only to those who produced the first working laser. Neither is it fair to regard the first proponent of an excitation scheme the inventor of a laser, because usually additional ideas and experimentation were necessary for the construction of an operating device. In several instances proponents of a scheme failed to bring it to successful conclusion, in spite of a determined and sometimes frantic effort. Numerous controversies and lawsuits in progress indicate that it is difficult to assess objectively the role of the contributions made by the theorist and the experimentors. They also reflect the fact that in the field under discussion several of the most important phenomena were discovered independently and nearly simultaneously.

REFERENCES

1. N. G. Basov, O. N. Krokhin, and Yu. M. Popov, Generation, amplification and detection of infrared and optical radiation by quantum mechanical systems, *Soviet Phys. Uspekhi.*, **3**, 702–728 (1961): **72**, 161–209 (1960).
2. M. Bernard and G. Duraffourg, Possibilitiés de lasers à semiconducteurs, *J. Phys. Radium,* **22**, 836–837 (1961).
3. R. N. Hall, G. E. Fenner, J. D. Kingsley, T. J. Soltys, and R. O. Carlson, Coherent light emission from GaAs junctions, *Phys. Rev. Letters,* **9**, 366–368 (1962).
4. G. Burns and M. I. Nathan, Line shape in GaAs injection lasers, *Proc. IEEE,* **51**, 471–472 (1963).
5. G. Burns and M. I. Nathan, P-n junction lasers, *Proc. IEEE,* **52**, 760–794 (1964).
6. P. P. Sorokin, J. D. Axe, and J. R. Lankard, Spectral characteristics of GaAs lasers operating in Fabry-Perot modes, *J. Appl. Phys.,* **34**, 2553–2556 (1963).
7. M. D. Sturge, Optical absorption of GaAs between 0.6 and 2.75 eV, *Phys. Rev.,* **127**, 768–773 (1962).
8. W. E. Engeler and M. Garfinkel Temperature effects in coherent GaAs diodes, *J. Appl. Phys.,* **34**, 2746–2750 (1963).
9. W. E. Ahearn and J. W. Crowe, Linewidth measurements of cw GaAs lasers at 77°K, *IEEE J. Quant. Electr.,* **QE-2**, 597–602 (1966).
10. W. Engeler and M. Garfinkel, Characteristics of a continuous high-power GaAs junction laser, *J. Appl. Phys.,* **35**, 1734–1741 (1964).
11. G. Lucovsky, Mechanism for radiative recombination in GaAs p-n junctions, *Physics of Quantum Electronics,* P. L. Kelley, B. Lax, and P. E. Tannenwald, Eds., McGraw-Hill, New York, 1966, (pp. 467–475.)
12. M. I. Nathan, Semiconductor lasers, *Appl. Opt.,* **5**, 1514–1528 (1966).
13. I. Melngailis and R. H. Rediker, Properties of InAs lasers, *J. Appl. Phys.,* **37**, 899–911 (1966).
14. J. F. Butler, A. R. Calawa, R. J. Phelan, T. C. Harman, A. J. Strauss, and R. H. Rediker, PbTe diode laser, *Appl. Phys. Letters,* **5**, 75–76 (1964).
15. J. F. Butler, A. R. Calawa, and R. H. Rediker, Properties of the PbSe diode laser, *IEEE J. Quant. Electr.,* **QE-1**, 4–7 (1965).
16. J. F. Butler and A. R. Calawa, Magnetoemission studies of PbS, PbTe, and PbSe diode lasers. *Physics of Quantum Electronics,* P. L. Kelley, B. Lax, and P. E. Tannenwald, Eds., McGraw-Hill, New York, 1966.
17. J. M. Besson, J. F. Butler, A. R. Calawa, W. Paul, and R. H. Rediker, Pressure-tuned PbSe diode laser *Appl. Phys. Letters,* **7**, 206–208 (1965).
18. J. O. Dimmock, I. Melngailis, and A. J. Strauss, Band structure and laser action in Pb_xSn_{1-x} Te, *Phys. Rev. Letters,* **16**, 1193–1196 (1966).
19. T. C. Harman, A. R. Calawa, I. Melngailis, and J. O. Dimmock, Temperature and compositional dependence of laser emission in $Pb_{1-x}Sn_xSe$, *Appl. Phys. Letters,* **14**, 333–334 (1969).

20. E. D. Hinkley, T. C. Harman, and C. Freed, Optical heterodyne detection at 10.6μm of the beat frequency between a tunable $Pb_{0.88}$ $Sn_{0.12}Te$ diode laser and
21. I. Melngailis, R. J. Phelan, and R. H. Redliker, Luminescence and coherent a CO_2 gas laser, *Appl. Phys. Letters,* **13,** 49–51 (1968).
    emission in a large-volume injection plasma in InSb, *Appl. Phys. Letters,* **5,** 99–100 (1964).
22. I. Melngailis, Longitudinal injection-plasma laser of InSb, *Appl. Phys. Letters,* **6,** 59–60 (1965).
23. M. Pilkuhn and H. Rupprecht, Spontaneous and stimulated emission from GaAs diodes with three-layer structure, *J. Appl. Phys.,* **37,** 3621–3628 (1966).
24. N. G. Basov, Inverted populations in semiconductors, *Quantum Electronics III,* P. Grivet and N. Bloembergen, Eds., Columbia University Press, New York, 1964, pp. 1769–1785.
25. J. J. Schlickman, M. E. Fitzgerald, and R. H. Kingston, Evidence of stimulated emission in ruby-laser-pumped GaAs, *Proc. IEEE,* **52,** 1739–1740 (1964).
26. N. G. Basov, A. Z. Grasyuk, V. A. Katulin, Stimulated emission in GaAs with optical excitation, *Soviet Phys.-Dokl.,* **10,** 343–344 (1965); **161,** 1306–1307 (1965).
27. R. J. Phelan and R. H. Rediker, Optically pumped semiconductor laser, *Appl. Phys. Letters,* **6,** 70–61 (1965).
28. N. G. Basov, A. Z. Grasyuk, I. G. Zabarev, and V. A. Katulin, Generation in GaAs under two-photon optical excitation of Nd-glass laser emission, *JETP Letters* **1,** 118–120 (1965); **1,** 29–33 (1965).
29. L. A. Kulevsky and A. M. Prokhorov, The nature of the laser transition in CdS crystal at 90°K with two-photon excitation, *IEEE J. Quant. Electr.,* **QE-2,** 584–586 (1966).
30. R. H. Rediker, Semiconductor lasers, *Physics Today,* **18,** 42–54 (1965).
31. N. G. Basov and O. V. Bogdankevich, Excitation of semiconductor lasers by a beam of fast electrons, *Radiative Recombination in Semiconductors,* Academic Press, New York and Dunod Editeur, Paris, 1965.
32. H. F. Ivey, Electroluminescence and semiconductor lasers, *IEEE J. Quantum Electr.* **QE-2,** 713–726 (1966).
33. C. Benoit à la Guillaume and J. M. Debevre, Electron-beam excitation of semiconductor lasers, *Physics of Quantum Electronics,* P. L. Kelley, B. Lax, and P. E. Tannenwald, Eds., McGraw-Hill, New York, 1966.
34. N. G. Basov, O. V. Bogdankevich, and A. G. Devyatkov, Cadmium sulfide laser with fast electron excitation, *Soviet Phys.-JETP,* **20,** 1067–1068 (1965); **47,** 1588–1590 (1964).
35. C. E. Hurwitz, Electron-beam pumped lasers of CdSe and CdS, *Appl. Phys. Letters,* **8,** 121–124 (1966).
36. C. E. Hurwitz, Efficient visible lasers of CdS_xSe_{1-x} by electron-beam excitation, *Appl. Phys. Letters,* **8,** 243–245 (1966).
37. F. H. Nicoll, Ultraviolet ZnO laser pumped by an electron beam, *Appl. Phys. Letters,* **9,** 13–15 (1966).
38. C. E. Hurwitz, Efficient ultraviolet laser emission in electron-beam-excited ZnS, *Appl. Phys. Letters,* **9,** 116–118 (1966).
39. N. G. Basov, B. M. Vul, and Yu. M. Popov, Quantummechanical semiconductor generators and amplifiers of electromagnetic oscillations, *Soviet Phys.-JETP,* **10,** 416 (1960); **37,** 587–588 (1959).

40. P. D. Southgate, Laser action in field-ionized bulk GaAs, *Appl. Phys. Letters*, **12,** 61–63 (1968).
41. N. G. Basov, O. N. Krokhin, and Yu. M. Popov, Production of negative temperature states in *p-n* junctions of degenerate semiconductors, *Soviet Phys.-JETP*, **13,** 1320–1321 1961, [**40,** 1879–1880 (1961)].
42. M. I. Nathan, W. P. Dumke, G. Burns, F. H. Dill, and G. J. Lasher, Stimulated emission of radiation from GaAs *p-n* junctions, *Appl. Phys. Letters*, **1,** 61–63 (1962).
43. T. M. Quist, R. H. Rediker, R. J. Keyes, W. E. Krag, B. Lax, A. L. McWhorter, and H. J. Zeiger, Semiconductor maser of GaAs, *Appl. Phys. Letters*, **1,** 91–92 (1962).

Chapter 8   Liquid lasers

8.1 EXPLORATION OF LIQUIDS AS LASER MATERIALS

Although the initial successes in the laser field were achieved with solids and gases, liquids appear to have interesting possibilities as laser materials. It has not been stressed sufficiently that when laser action is based on ions imbedded in a crystal lattice, only single crystals unusually free of imperfections can serve as solid laser materials. Crystal imperfections, strains, and inhomogeneities impair coherent amplification. The crystals must be cut and polished accurately, and often attention must be paid to the orientation of the crystal axes. Preparation of solid laser materials is time consuming and expensive. Once the laser has been built, the concentration of active ions cannot be changed. When the laser is overloaded, cracks or other defects develop which greatly degrade laser performance. Glass as a host material replacing crystals is useful in many respects, but its poor thermal conductivity aggravates the problem of heat removal.

Liquids and gases allow complete freedom from the problems of single-crystal-growing and shaping. They allow change of shape and concentration and they will not develop the defects that make experimentation with solid lasers so costly. The use of liquids and gases makes the cooling of the active medium quite simple compared to the cooling of solid lasers, because fluids may be circulated.

Although gases are ideal laser materials in many ways, they suffer from a serious disadvantage: Their density is low, consequently their performance is limited because of the low concentration of active atoms. Search for suitable liquid laser materials is motivated by the fact that liquids may possess the advantages of gaseous laser materials and at the same time permit a reasonable concentration of active material in a given volume. The exploration of liquid materials for their suitability is simplified by the fact that, within limits, the concentration of the ingredients can be changed quickly and easily.

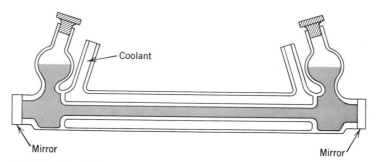

FIG. 8.1 Liquid laser cell with expansion volume and cooling attachment.

There are a number of obstacles to the construction of a laser based on the use of liquids and solutions. These were overcome, in some cases, by careful selection of the materials, following an extensive analysis of the relevant spectroscopic properties.

On comparing fluorescence in liquids to fluorescence in solids, it becomes apparent that the following circumstances prevail:

1. Spectral lines are broader in liquids (solutions) because of perturbing effects of the constantly changing environment.

2. Interaction with the solvent may lead to nonradiative deactivation of excited atoms or molecules, causing the quantum efficiency to decrease.

A minor disadvantage of liquids as laser materials is their large coefficient of thermal expansion. The linear expansion coefficient of the liquids suitable for lasers is of the order of one part in one thousand per degree centigrade. This value is much greater than the expansion coefficients of solid materials from which a vessel might be constructed. Since the temperature of the active material changes during laser operation, it is essential to make provision for a change in volume. This may be accomplished by attaching expansion volumes, as shown in Fig. 8.1, or, in the case of lasers with small diameters, by confining the liquid between moveable pistons as shown in Fig. 8.2. Thermal changes

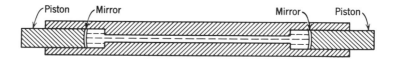

FIG. 8.2 Small-diameter liquid laser with moveable pistons.

and the motion of the liquid introduce inhomogeneities in the refractive index and degrade the laser performance.

The fluorescent and phosphorescent organic materials are natural starting materials for laser development because they possess the attributes necessary for laser action, namely, they can be excited by irradiation and they are capable of transferring the energy obtained by optical excitation to an energy level which might become the starting level for stimulated emission.

The number of organic compounds which manifest fluorescence in the liquid form or in solutions is large. Many of these have been examined as potential liquid laser materials. Since most of these compounds must be in a solution to serve as working liquids, they must be studied taking into account their interaction with the solvents. The number of possible combinations is staggering, therefore experimentation must be preceded by an analysis of the physical chemistry and spectroscopy of the possible materials. Through such analysis it is possible to select groups of materials and material combinations which offer some promise of success. The reader interested in this type of analysis is referred to the excellent review article of Lempicki and Samelson [1]. The same authors have also written a well-illustrated popular article [2].

Three types of liquid laser will be discussed in detail:

1. Organic chelate lasers, whose active material is a rare-earth ion coupled with a group of organic radicals.
2. The nonorganic neodymium–selenium oxychloride laser.
3. The purely organic "dye" laser

All these laser types are excited optically, the first two types by means of flashlamps, the last type by solid lasers or by flashlamps. The first two types of liquid lasers are only of academic interest; the third type has found application in scientific experiments.

8.2   RARE-EARTH CHELATE LASERS

The chelates are metallo-organic compounds. In the chelates to be considered here, a rare-earth ion is surrounded by oxygen atoms from the ligand or chelate group. The fluorescence of a central metallic ion may be excited by the absorption of radiation in the complex organic groups which surround it and shield it from the environment. A typical chelate is formed when benzoylacetonate ions enter into a combination with a trivalent metal ion. Benzoylacetonate (B) arises by the removal of a proton from the benzoylacetone molecule; the excess negative charge is distributed between the two oxygen atoms which are not essentially

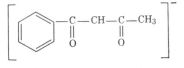

FIG. 8.3 Benzoylacetonate ion.

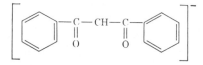

FIG. 8.4 Dibenzoylmethide ion.

different. The structure of benzoylacetonate is shown in Fig. 8.3 and that of the similar dibenzoylmethide (D) in Fig. 8.4. When three benzoylacetonate ions combine with a tripositive metal ion, the six oxygen atoms form a regular octahedron around the metal, as illustrated in Fig. 8.5. Similar configurations may be constructed with three dibenzoylmethide (D) ions or with three benzoyltrifiuoroacetone (BTF) ions. The resulting compounds are called "tris" chelates.

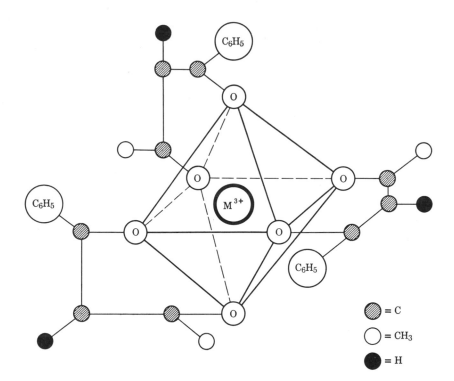

FIG. 8.5 Perspective drawing of a rare-earth trisbenzoylacetonate chelate (MB_3). The lines outlining the octahedron do not represent bonds. The sizes of the circles do not indicate atomic dimensions.

Compounds involving four ligand groups; that is, eight oxygen atoms around a metal ion are called "tetrakis" chelates. In trivalent europium we have compounds denoted by $EuKe_3$ and ions like $(EuKe_4)^-$. (Here Ke stands for the chelate ion.) This last ion is used in the form of the compound $EuKe_4P$, where P is a monovalent positive ion, most frequently piperidinium.

The chelate ions belong to a class of organic molecules noted for their interesting luminescent behavior. The ground state of the radical is a singlet state, $S = 0$. Absorption of ultraviolet radiation causes excitation from the ground state S_0 to one of the lower singlet excited states S_1, S_2, etc. By dissipating energy through vibrations the molecule may change without the emission of radiation from the states S_2, S_3, etc., to the lowest singlet excited state S_1, or to one of the low triplet states T_1 or T_2. The transition $S_1 \rightarrow S_0$ is permitted by the selection rules; it takes place rapidly—the lifetimes vary between 10^{-6} and 10^{-9} sec—and it manifests itself as fluorescence. Radiative transitions from the triplet states to the ground state are forbidden by the selection rules; therefore molecules may be trapped in a triplet state with a lifetime of the order of 10^{-3} sec.* Rare-earth chelates differ from the radicals in that the chelate possesses, in addition to the above energy levels, a fairly isolated set of low-lying energy states derived from the $4f$ electronic configuration of the metal ion. The excited electronic states of a chelate are shown schematically in Fig. 8.6. Under favorable circum-

* In a liquid rich in a triplet-quenching agent, such as O_2, this time may be considerably reduced.

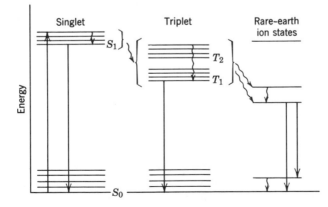

FIG. 8.6 Schematic energy-level diagram for a rare-earth chelate possessing low-lying $4f$ electronic states: $\rightarrow$ radiative transitions; $\rightsquigarrow$ radiationless transitions.

stances, the luminescence cycle may be completed by a transfer of excitation from level T_1 (or T_2) to one of the levels of the metal ion and a subsequent return to the ground with accompanying emission of lines characteristic of the ion. In this manner, the broad absorption characteristics of the organic radical combine with the narrow emission lines of the rare-earth ion to form a situation favorable for the construction of an optically pumped laser.

The potential value of the chelates as laser materials was noted shortly after stimulated emission was first accomplished in ruby, and the systematic investigation of the luminescence of chelates was intensified. Recognition of the essential features of the fluorescence mechanism pointed to europium as the rare-earth ion most likely to be a suitable laser material and caused an intensive investigation of europium in benzoylacetonate [3] and thenoyltrifluoroacetonate (TTA) [4], combined with estimation of the pumping energies and concentrations required to produce laser action in solutions of such molecules [4, 5]. The results of these investigations were published early in 1963, about the same time as successful operation of europium chelate lasers was reported simultaneously from two laboratories. Wolff and Pressley [6] at RCA obtained stimulated emission from a long coil of methylmethacrylate fiber containing $Eu(TTA)_3$, while Lempicki and Samelson [7, 8] at General Telephone and Electronics obtained the same in solutions of europium benzoylacetonate and europium dibenzoylmethide in a mixture of ethyl and methyl alcohols. Both groups of experiments were carried out at temperatures between 110° and 140°K in order to work with as narrow lines as possible and thus to reduce the energy required for excitation. At these temperatures the solutions resemble solids more than ordinary liquids; they flow like tar or thick honey. Eventually the range of the operating temperature was extended to room temperature. This success was attained by replacing the mixture of alcohols by acetonitrile, a solvent which does not cause a significant dissociation of the tetrakis chelate [9, 10].

Before proceeding to the description of the composition, structure, and operational characteristics of the chelate lasers we should like to concern ourselves with the spectroscopy and the physical chemistry of the materials involved. The europium chelate laser operates at a wavelength around 6130 Å; the radiation corresponds to a transition in the Eu^{3+} ion from the $^5D_0$ level to one of the $^7F_2$ sublevels located about 950 cm^{-1} above the ground level of the ion. The splitting of the $^7F_2$ levels is determined by the fields surrounding the Eu ion. The situation varies, depending on the number and nature of the ligands and possibly even on other ions coordinated with the Eu ion.

The fluorescence properties of europium benzoylacetonate and europium dibenzoylmethide were studied in great detail. These laser materials were dissolved at first in a standard solvent consisting of a 3:1 mixture of ethanol and methanol. Later a 3:1:1 mixture of ethanol, methanol, and dimethylformamide (DMF) was used as a solvent, and finally special other solvents were tried. When the lasers were first discovered, it was thought that the active materials were of the form EuB_3 and EuD_3; that is, they were assumed to be tris chelates (six-fold coordinated). By careful comparison of the spectral lines Brecher, Lempicki, and Samelson proved that the active materials were really of the form EuB_4 and EuD_4, the charge on each of these ions being compensated for by a positive piperidinium ion [11, 12]. This work was complicated by the fact that the solutions contained both the tris and the tetrakis chelate because the solvent caused a dissociation of the chelate, which was originally in the tetrakis form. The identification of the lines was accomplished by taking into account the different symmetries prevailing at the sites of a six-fold and an eight-fold coordinated ion. The degree of dissociation varies with the chelate, with the solvent and, of course, with the temperature. Quantitative data on this dissociation are available in the publications of the General Telephone and Electronics group [1, 12 (III)]. It is interesting to note that at 93°K temperature the dissociation of the tetrakis B and D chelates in alcohol is partial, but that of the BTF chelate is complete.

The chelates containing four ligands have sharper fluorescent lines, and this is probably the reason why they are easier excited to laser action. In any case, it is clear that unintended dissociation of the four-ligand chelate has two deleterious effects: It reduces the concentration of the species available for laser action. Moreover, the products of dissociation, that is, the free enolate ion and the tris chelate, absorb the exciting light in the same region as the active material does. The presence of dimethyl-formamide in the solvent inhibits the dissociation of the $Eu(BTF)_4$ and it is, in fact, possible to obtain laser action with solvent containing DMF, while the same chelate does not lase in the standard alcohol solution. The dissociation of this chelate is even less in the solvents acetonitrile and propionitrile. With these solvents $Eu(BTF)$, is a suitable laser material even at room temperature [9, 10].

Excitation of the europium chelate laser is accomplished with a powerful ultraviolet source with an output in the absorption band of the radical. This is generally in the 3000-to-4000-Å range; for europium benzoylacetonate the peak is at 3900 Å. Europium ions are present in concentrations varying between 10^{18} and 2×10^{19} per cm^3. The lower limit is set by the threshold condition, which requires a minimum gain to offset the losses at the ends of the laser; the upper limit is set by the solubility. It is interesting to note that the absorption constant of a EuB solution containing 1.2×10^{19} ions cm^{-3} is about 250 times greater than that of pink ruby of similar ion concentration. As a result of the intense absorption, the excitation of EuB will be extremely nonuniform, and the threshold depends strongly on cell geometry. In fact, the high absorbance of the chelates limits the thickness of the active material that may be used effectively.

Chelate lasers have an active material diameter of the order of 1 mm. They are excited by a flash in a manner similar to that used for ruby. An experimental arrangement for the flash excitation of a chelate laser is shown in Fig. 8.7. The liquid is located in a quartz cylinder, which is held within a dewar flask, and is cooled by a stream of cold nitrogen gas. Cooling of this laser is more critical than the cooling of ruby. The threshold pump energy varies greatly with the solvent used for the chelate and with the temperature. The dependence of pump energy on temperature is illustrated in Fig. 8.8, which exhibits the data obtained by Schimitschek [10] on a laser designed for room-temperature operation. The active material of this laser is a 0.005 M solution in acetonitrile of the chelate $Eu(BTF)_4H$ Pyrr, where Pyrr stands for pyrrolidine. The solution is confined in a capillary tube located in the focal line of an elliptic cylinder. It is clear from the graph displayed in Fig. 8.8 that even a moderate increase of temperature causes a substantial increase in threshold energy.

The typical range of the overall efficiency of a chelate laser is around 10^{-6}. This low efficiency is in part due to the complex excitation process that leads to traps and paths of return which do not contribute to laser action, and in part to the high absorbance of the material that prevents uniform excitation of the active medium.

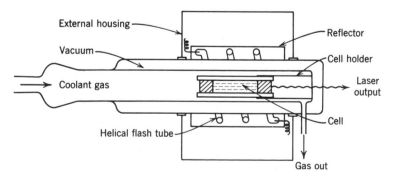

FIG. 8.7 A chelate laser.

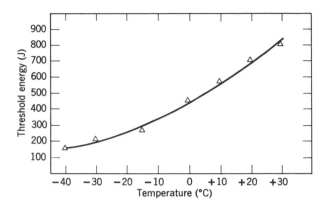

FIG. 8.8   The threshold pump energy of a europium chelate laser versus temperature. (After Schimitschek [10].)

8.3   THE NEODYMIUM-SELENIUM OXYCHLORIDE LASER

The performance of the metal–chelate lasers is limited by the high absorbance of the organic ligand and by the dissipation of excitation energy through processes that do not contribute to laser action. To circumvent the limitations of metal–chelate lasers, one is tempted to experiment with aqueous solutions of a rare-earth ion and to excite them to fluorescence by optical pumping, as these ions are excited in a four-level solid laser. One could hope that trivalent neodymium—an excellent four-level laser ion when incorporated in solids—might work even in aqueous solutions. Experience has shown, however, that excited Nd^{3+} ions relax rapidly in aqueous solution without the emission of radiation. The nonradiative relaxation process results in the transformation of the excitation energy of the ions into vibrational energy of the molecules of the solvent. The probability of such a process depends on the number of vibrational quanta that must be created in order to carry away the excitation energy of the ion. A process requiring the simultaneous creation of several quanta is highly unlikely; therefore, in those cases in which the available vibrational energy quantum is small compared to excitation energy of the ion a nonradiative energy transfer will seldom take place. On the other hand, relaxation will take place more frequently when an available vibrational quantum is nearly equal to the excitation energy of the ion. Experimental studies of fluorescence have shown that when heavy water was substituted for ordinary water, the rate of the nonradiative decay of Nd fluorescence was lowered. This effect is attributed to the increased mass, hence the decreased frequency, of molecular vibrations. The highest energy of vibrations in a molecule is determined by the lightest constituent.

Using this clue about the role of the atomic masses, Heller and co-workers [13, 14] solved the problem of creating an efficient liquid laser by dissolving Nd in a liquid system which contains no atoms lighter than oxygen. This system has no vibrations of sufficient energy to carry away the energy corresponding to the gap between the excited and the ground multiplet levels of Nd^{3+}.

Neodymium oxyde and neodymium chloride dissolve in selenium oxychloride to a limited extent. Solubility improves if the solution is acidified, but for reasons already explained above hydrogen-containing acids should not be used. Hydrogen-free acidification may be obtained by adding tin tetrachloride to the solvent because of the formation of the powerful acid $(SeOCl^+)_2SnCl_6^{2-}$. This acid then produces the required Nd^{3+} from either Nd_2O_3 or $NdCl_3$.

The absorption spectrum of Nd^{3+} in Heller's solvent is similar to that of Nd^{3+} in an ionic crystal. The excitation of the liquid neodymium laser can be accomplished by flashtubes placed side by side with the laser cell and surrounded by a cylindrical reflector. Stimulated emission with the wavelength 1.056 μm is readily obtained, even without the use of aligned mirrors because strong internal reflection takes place at the surface of the liquid. (The solvent has a refractive index of 1.65.) A liquid neodymium laser with no mirrors has a limited usefulness because the stimulated radiation spreads out in all directions and does not form a collimated beam. Such beam can be produced using cells with properly aligned mirrors.

Liquid neodymium lasers have good efficiency and comparatively low threshold. In the initial experiments Heller reached threshold in a small laser with an electrical input of 5 J and obtained a 1-J output from a 15-cm tube of 0.6-mm inner diameter with an electrical input of 1000 J. These figures compare very favorably with the corresponding data for chelate lasers. An important added advantage of the liquid neodymium laser is that it need not be refrigerated, whereas most of the chelate lasers must be operated well below room temperature. A serious disadvantage of this laser is that its solvent is a very dangerous material. It is highly corrosive and very toxic; on contact with the skin it will cause burns and poisoning.

8.4 ORGANIC DYE LASERS

The active medium of an organic dye laser is an organic fluorescent material dissolved in a common solvent. Typical materials are rhodamines dissolved in alcohol and fluorescein dissolved in water. These substances derive their colors from strong absorption bands in the visible region. Their excitation is accomplished by optical pumping, using solid lasers in giant-pulse operation or flashlamps that deliver a short pulse with a fast risetime. The output of dye lasers is a short pulse of broad spectral content. The spectral distribution depends on the solvent used, on the concentration of the dye, and on several other parameters. Because of this dependence, the spectral output is variable and only rough wavelength data can be given. The gain achievable in these dye solutions is high; the active material of an operating laser seldom must exceed 2 cm in length. Dye lasers operate at room temperature.

The excitation–emission cycle of organic dyes varies somewhat from one material to another, but all these materials have certain structural properties in common that determine the main features of their relevant optical processes. The polymethine and cyanine dyes have a pair of electrons, which possess a certain freedom of motion within the molecule. The motion of this electron pair determines the electronic configuration of the molecule. The lowest electronic configuration S_0 is a singlet state (spins opposite). There are excited singlet (S_1, S_2, . . .) and triplet (T_1, T_2, . . .) states. Vibrations of the atoms contribute energy which is small compared to the energy difference between the lowest S electronic states. The energy of the molecule is a function of the interatomic distances. On plotting the energy of the molecule in the S_0 and S_1 electronic states as a function of a typical (or generalized) distance coordinate, one obtains curves as shown in Fig. 8.9. The horizontal bands within each electronic energy curve represent molecular vibration bands. Absorption and emission of energy takes place when the molecule passes from one band to another. In the case of the dyes, the passage from the lowest lying bands of the S_0 level to one of the lower bands of the S_1 level requires the absorption of one quantum of visible light.

The interatomic distances that correspond to the minima m_0 and m_1 of the S_0 and S_1 curves are generally somewhat different, and this difference affects the absorption and emission spectra of these materials. Of particular interest to laser technology are the fluorescent dyes which emit light after a short delay and whose emission spectrum is shifted with respect to their absorption spectrum.

The cause and nature of this shift is comprehended by means of the Franck-Condon principle, which states that the electronic transitions take place so rapidly that the atoms of the molecule cannot alter significantly their relative positions and their relative velocities during the time of such an electronic transition [15]. Consequently those transitions will take place at the highest rate which require the minimal rearrangement of interatomic distances and velocities. When the situation is such as shown in Fig. 8.9, the $v' = 0 \rightarrow v'' = 1$ transition may be more likely than the $v' = 0 \rightarrow v'' = 0$ transition. The opposite transition may be very weak in absorption because the $v'' = 1$ level is not populated. The starting level of the transition $v' = 0 \rightarrow v'' = 1$ may become populated when the molecule is optically pumped into the $v' = 1$ level because the energy is rapidly redistributed among adjacent vibrational levels and the population tends to favor the lowest of the levels.

Dye lasers are generally pumped from the lowest S_0 vibrational band ($v'' = 0$) into some of the S_1 bands. Stimulated radiation is emitted on return from $v' = 0$ into one of the bands *above* $v'' = 0$. Thus the laser operates on the four-level scheme.

The presence of molecules in one of the metastable triplet levels may be detrimental for laser action in several ways. First, the molecules stored in a triplet level are eliminated from participating in the laser cycle. Second, when some of the lowest triplet levels are populated, the

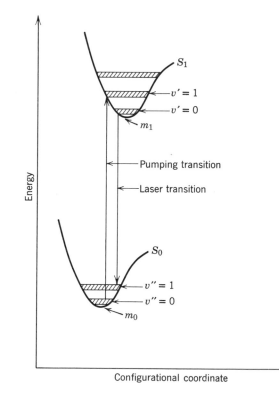

FIG. 8.9 Partial energy-level diagram of a dye laser material.

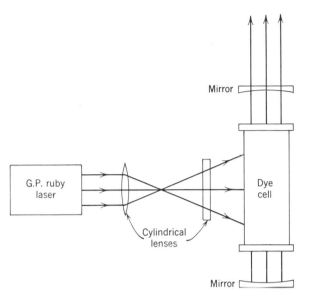

FIG. 8.10 Dye laser configuration with transverse pumping. (After Sorokin et al. [17].)

material may become absorbent for the laser radiation because of triplet—triplet transitions that overlap the laser line. It should be noted that in comparison to atomic levels the molecular levels are broad and numerous. Therefore spectral overlaps occur relatively often among molecular lines.

The first dye lasers were discovered by Sorokin and associates [16, 17] at the IBM Laboratories in the course of the extension of an investigation pertaining to dyes used in passive Q-switching of ruby lasers. Laser action was obtained in 1965 from a solution of chloro-aluminum phthalocyanine (CAP), excited by the radiation obtained from a giant-pulse ruby laser. The configuration of this dye laser is shown in Fig. 8.10. The dye solution is contained in a cell placed between external mirrors. While the original arrangement provided illumination from the side, end-on illumination is also possible by the use of a selective trans-mitting mirror. The dye concentration is about 10^{-3} mole/liter; the optimum concentration depends on the size of the cell and the geometry of the illumination. The coherent radiation extends over a spectral region about 40 Å wide and centered around 7560 Å. The center wavelength and the width of the spectrum vary with the concentration, the solvent and the level of excitation. The dependence of the relevant optical parameters of CAP on the solvent used are shown in Table 8.1.

TABLE 8.1

Central Wavelengths of Stimulated Emission, λ_s, and Wavelengths of Maximum Absorption, λ_A for 10^{-3} Mole Solutions of CAP

Solvent	λ_s (Å)	λ_A (Å)
1-propanol	7550 ± 5	6705 ± 10
Ethanol	7555 ± 5	6700 ± 10
Dimethyl sulfoxide	7615 ± 5	6770 ± 10
Ethylene glycol	7630 ± 5	6815 ± 10

After Sorokin [17].

This variation is typical. Similar variation has been observed for other dyes, but one must not conclude that the laser wavelength will always increase if one replaces 1-propanol by ethanol, or dimethyl sulfoxide by ethylene glycol.

The dye 3–3′ diethylthiatricarbocyanine (DTTC) iodide is one of the best-explored dye-laser materials. It is typical of the variety of cyanine dyes which have shown laser capabilities. The DTTC iodide is solvable in a large variety of solvents. Its peak fluorescent emission varies between 7960 Å for methyl alcohol solvent to 8160 Å for dimethyl sulfoxide solvent. This dye is excitable by the ruby radiation, although its absorption peak is reached at around 7600 Å.

About twenty similar dyes were shown to be suitable laser materials in a variety of solvents. By selecting the proper dye and adjusting its concentration between the limits 3×10^{-6} to 5×10^{-4} moles/liter it is possible to cover the entire spectral range from 0.71 to 1.06 μm without a gap [18].

Many dyes emit fluorescent radiation of shorter wavelength than that of the ruby laser. These dyes cannot be excited to stimulated emission by irradiation from a ruby laser. It was discovered simultaneously by Sorokin [17] and McFarland [19] that many such dyes can be made to emit coherent radiation if they are excited by the second harmonic of the output of a ruby. Thus the radiation is first converted to the wavelength 3471 Å and then made incident on the dye cell. The generation of a laser output in this manner requires very high power output in the first stage (100 MW) because the conversion losses are quite high. Table 8.2 contains some of the common dye lasers that operate in the orange–to blue region of the spectrum.

Early attempts to pump dye lasers by means of flashlamps failed. The analysis of Sorokin [17] has shown that the coherence quality of the pumping beam was not essential, but that the important features required of the pumping source are high peak intensity and short rise-time. These qualities are required because in a slow excitation process some of the lowest triplet levels become populated and inhibit the attainment of gain in the intended laser material. The effect of the triplet states is analysed with the aid of Fig. 8.11, which shows the details of the energy level structure of a dye molecule. Each of the vibrational levels has a definite spread because it has a rotational fine structure not shown on the diagram. Transitions between singlet states are "allowed" by the spectroscopic selection rules. They give rise to intense absorption and emission spectra. Transitions between different triplet states are also allowed, but singlet-triplet transitions, known as *inter-system crossings*, are forbidden. That the transition is forbidden means that it does not take place as fast as the allowed transitions, or that it does not take place fast unless accompanied by a process other than the emission of a photon.

Molecules in the S_1 group of states return to the S_0 states with the emission of radiation. The typical decay time of this fluorescence in organic dyes is 5×10^{-9} sec. They may change to the lower-lying triplet state T_1 without the emission of radiation by giving up the energy difference to the lattice. The rate for this intersystem crossing, k, in typical dyes is such that $k^{-1} \approx 5 \times 10^{-8}$ sec. Thus the radiationless intersystem

TABLE 8.2

DYE LASERS EXCITED BY THE SECOND HARMONIC
OF RUBY [17, 19]. ETHANOL SOLVENT, 10^{-3}
MOLES/LITER CONCENTRATION

Dye	Laser Wavelength (μm)
Acridone	0.437
Acriflavin hydrochloride	0.510
Fluorescein (Na)	0.527[a]
Eosin	0.540
Rhodamine 6 G	0.555
Rhodamine B	0.577
Acridine Red	0.580
Rhodamine G	0.585

[a] Water may also be used as a solvent.

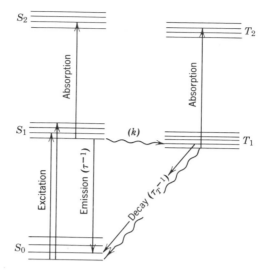

FIG. 8.11 Energy level scheme of a dye molecule. Horizontal lines represent vibrational states. Wavy lines indicate nonradiative transitions; solid lines, radiative transitions. Letters in parentheses are reciprocals of process rates.

crossing is a process that proceeds about one tenth as fast as fluorescence and is therefore only mildly competitive in the deactivation of S_1. The lifetime τ_T of the decay from the triplet state T_1 to S_0 is much longer, 10^{-3} to 10^{-4} sec being typical for a pure dye solution. Because of its relatively long lifetime, the triplet state T_1 acts as a trap for the dye molecules and depletes the supply available for the laser process. In addition to depleting the number of active molecules, the accumulation in the T_1 state has another detrimental effect. Triplet-to-triplet transitions are allowed, and the optical absorption associated with these is strong. Unfortunately, the absorption band associated with the T_1 to T_2 transitions generally overlaps the S_1 to S_0 spectrum. Hence the accumulation of molecules in the T_1 state produces a large optical loss in the singlet fluorescent spectrum. It can prevent or stop laser emission. In order to minimize the detrimental effect of triplet-to-triplet absorption it is necessary either to reduce the triplet lifetime somehow, or to excite the laser so fast that laser threshold is reached before a significant number of molecules have accumulated in the T_1 state. Fast excitation requires an optical pump of high intensity and short risetime. Hence the advantage of the giant pulse for dye-laser excitation.

Dye lasers may also be excited by means of specially constructed flashlamps with high output and short risetime [17, 20]. Rhodamine

6G is among the most frequently used dyes in flash-excited lasers. This dye is dissolved in water or in methyl– or ethyl alcohol, or alternatively it may be incorporated in a polymethyl methacrylate plastic. Concentrations between 10^{-3} M and 10^{-4} M are suitable in any solvent. The wavelength of the laser output varies between 0.57 and 0.61 μm, depending on the solvent and on the concentration [21, 22]. In a pure, outgassed solvent the laser output is quenched in about 0.3 μsec because of the accumulation of dye molecules in triplet states. When the solution is saturated with oxygen, the triplet lifetime decreases from 10^{-3} to 10^{-7} sec. With this drastic change in lifetime, laser oscillations of relatively long duration may be obtained. In fact, laser pulses lasting from 50 to 100 μsec were observed during the flat period of the exciting pulse. The laser oscillations are finally quenched, not by the accumulation of molecules in the triplet levels, but by the optical inhomogeneities induced by the heating and the nonuniform excitation of the solution [23, 24].

Dye lasers promise to provide sources of coherent light easily tunable over considerable bands of the visible spectrum. One of the tuning methods makes use of the variation of wavelength with concentration. To make use of concentration tuning it is not actually necessary to change the concentration of the dye solution, the same effect can be obtained by altering the ratio of the active length to the total length of the dye cuvette [25]. Another elegant tuning method was invented by Soffer and McFarland [26], who replaced one mirror of a laser-pumped rhodamine 6G laser with a diffraction grating. This grating was mounted in the Littrow arrangement adjusted so that the first-order reflection of the desired wavelength was reflected back upon itself along the axis of the laser. This wavelength depends on the orientation of the grating. Tuning is thus accomplished by rotating the grating. The tuning range of the rhodamine 6G laser was found to cover the region 5550 to 5950 Å. Replacement of the broadband dielectric mirror by the selective grating results in spectacular narrowing of the spectrum of the laser output. Soffer and McFarland observed a change of spectral width from 60 to 0.6 Å. This sharpening of the laser output in the frequency domain more than compensates for the 40% loss of total output energy that takes place as a consequence of the introduction of the grating into the system.

There are a number of other ingenious methods available for the tuning of the dye lasers. These and other interesting techniques pertaining to the dye lasers are described in some detail in the excellent review article of Snavely [23], which contains many references to the literature of the dye-laser field published in 1967 and 1968.

REFERENCES

1. A. Lempicki and H. Samelson, Organic laser systems, *Lasers*, A. K. Levine, Ed., Dekker, New York, 1966, pp. 181–252.
2. A. Lempicki and H. Samelson, Liquid lasers, *Scientific American*, **216**(6), 81–90 (1967).
3. V. A. Voloshin, A. G. Goryushko, and V. A. Kulchitskii, Spectral study of polymethylmethacrylate activated by europium benzoylacetonate, *Opt. Spectry.* (U.S.S.R.), **15**, 154–155 (1963); **15**, 286–287 (1963).
4. H. Winston, O. J. Marsh, C. K. Suzuki, and C. L. Telk, Fluorescence of europium thenoyltrifluoroacetonate. Evaluation of laser threshold parameters, *J. Chem. Phys.*, **39**, 267–271 (1963).
5. H. Lyons and M. L. Bhaumik, Rare-earth chelates and the molecular approach to lasers, *Quantum Electronics III*, Columbia University Press, New York, 1964, pp. 699–708.
6. N. E. Wolff and R. J. Pressley, Optical maser action in an Eu^{3+}-containing organic matrix, *Appl. Phys. Letters*, **2**, 152–154 (1963).
7. A. Lempicki and H. Samelson, Optical maser action in europium benzoylacetonate, *Phys. Letters*, **4**, 133–135 (1963).
8. A. Lempicki and H. Samelson, Stimulated processes in organic compounds, *Appl. Phys. Letters*, **2**, 159–161 (1963).
9. H. Samelson, A. Lempicki, C. Brecher, and V. Brophy, Room-temperature operation of a europium chelate liquid laser, *Appl. Phys. Letters*, **5**, 173–174 (1964).
10. E. J. Schimitschek, J. A. Trias, and R. B. Nehrick, Stimulated emission in a europium chelate solution at room temperature, *J. Appl. Phys.*, **36**, 867–868 (1965).
11. C. Brecher, A. Lempicki, and H. Samelson, Evidence for eightfold co-ordination in europium chelates, *J. Chem. Phys.*, **41**, 279–280 (1964).
12. H. Samelson et al., Laser phenomena in europium chelates, I, *J. Chem. Phys.*, **40**, 2547–2553; *II*, **40**, 2553–2558; *III*, **42**, 1081–1096 (1965).
13. A. Heller, A high-gain, room-temperature liquid laser: trivalent neodymium in selenium oxychloride, *Appl. Phys. Letters*, **9**, 106–108 (1966).
14. A. Lempicki and A. Heller, Characteristics of the Nd^{3+}:SeOCl$_2$ laser, *Appl. Phys. Letters*, **9**, 108–110 (1966).
15. G. Herzberg, *Molecular Spectra and Molecular Structure*, Vol. 1, Van Nostrand, New York, 1950.
16. P. P. Sorokin and J. R. Lankard, Stimulated emission from an organic dye, chloro-aluminum phthalocyanine, *IBM J. Res. Dev.*, **10**, 162–163 (1966).
17. P. P. Sorokin, J. R. Lankard, E. C. Hammond, and V. L. Moruzzi, Laser-pumped stimulated emission from organic dyes. Experimental studies and analytical comparisons, *IBM J. Res. Dev.*, **11**, 130–148 (1967).
18. Y. Miyazoe and M. Maeda, Stimulated emission from 19 polymethine dyes—Laser action over the continuous range 710–1060 mμ, *Appl. Phys. Letters*, **12**, 206–208 (1968).
19. B. B. McFarland, Laser second-harmonic-induced stimulated emission of organic dyes, *Appl. Phys. Letters*, **10**, 208–209 (1967).
20. B. B. Snavely, O. G. Peterson, and R. F. Reithel, Blue laser emission from a flashlamp-excited organic dye solution, *Appl. Phys. Letters*, **11**, 275–276 (1967).

21. P. P. Sorokin, J. R. Lankard, V. L. Moruzzi, and E. C. Hammond, Flashlamp-pumped organic dye lasers, *J. Chem. Phys.*, **48,** 4726–4741 (1968).
22. O. G. Peterson and B. B. Snavely, Stimulated emission from flashlamp-excited organic dyes in polymethyl methacrylate, *Appl. Phys. Letters,* **12,** 238–240 (1968).
23. B. B. Snavely, Flashlamp-excited organic dye lasers, *Proc. IEEE,* **57,** 1374–1390 (1969).
24. B. B. Snavely and F. P. Schäfer, Feasibility of cw operation of dye lasers, *Physics Letters,* **28A,** 728–729 (1969).
25. G. I. Farmer, B. G. Huth, L. M. Taylor, and M. R. Kagan, Concentration and dye length dependence of organic dye laser spectra, *Appl. Opt.,* **8,** 363–366 (1969).
26. B. H. Soffer and B. B. McFarland, Continuously tunable, narrow-band organic dye lasers, *Appl. Phys. Letters,* **10,** 266–267 (1967).

Chapter 9 Gas lasers

In gases, as in other materials, amplification of radiation takes place only in the condition that is variously referred to as population inversion, negative temperature, or negative absorption. It is a nonequilibrium condition, characterized by a distribution of atomic systems such, that for some pairs of stationary energy levels more atoms are found in the higher than in the lower energy state. The general requirements necessary for the attainment of this condition, called population inversion, were already discussed in Chapter 3. This discussion must now be expanded to include certain special aspects of spectroscopy which are relevant to gas lasers only and which have therefore not been included in the earlier general discussion. It will be necessary to deal with spectroscopic properties of molecules and with an excitation process peculiar to gases: excitation by means of collisions.

In addition to spectroscopic material, some factors relating to the shape and size of the laser must be re-examined. These factors were introduced in Chapter 3 under such titles as threshold condition and mode structure, but the discussion was then kept general or was oriented toward the requirements of solid lasers. The contents of that earlier chapter will now be supplemented with material especially applicable to gas lasers. Only after the completion of these general topics will the description of specific gas lasers be undertaken.

9.1 EXCITATION PROCESSES IN GASES

Molecular Energy Levels. In some respects the spectroscopic problems of gas lasers are simpler than those of solid and liquid lasers. When the gas consists of free atoms, these interact with each other and with the walls of the vessel for only very short intervals of time. Therefore, the relevant energy levels will be those of a free atom in the absence of external fields. Under such circumstances, the multiple levels are not resolved, spectral lines are generally sharp, and selection rules for electric dipole radiation are valid. At pressures used in gas lasers, the width of a spectral line is determined mostly by the Doppler effect. Nonradiative energy exchanges, so frequent in solids and liquids, are confined in gases to energy conversions in collisions.

Absorption bands, so useful in other types of laser materials, are absent in atomic gases. Their absence eliminates the possibility of exciting these gases by irradiation from an ordinary light source whose energy is distributed over a broad region of the spectrum.

The inclusion of molecular gases among laser materials requires the consideration of some of the laws of molecular spectroscopy. As in atoms, we encounter a series of discrete energy levels in molecules, but in the latter the energy-level structure is more complicated because it consists of the superposition of electronic, vibrational, and rotational levels.

The electronic levels are similar to those encountered in free atoms, they are associated with electron orbits and spin orientations. The order of magnitude of these levels in molecules and in free atoms is the same. The energy differences between the lower levels of this type are ordinarily around 1 to 5 eV; the associated wavenumbers are 8000 to 40,000 cm^{-1}. Consequently, transitions between lower-lying electronic levels of the molecule result in the emission (or absorption) of radiation in– or near the visible range.

Next in the hierarchy are the vibrational levels. The vibrations are periodic variations of the distances between the atomic nuclei. In the simplest molecules that consist of only two atoms, there is only one degree of freedom involved in the vibrational motion: the distance between the two nuclei. The forces of electrostatic repulsion between the nuclei and the binding forces due to shared electrons determine a certain equilibrium distance at which the nuclei tend to remain. If the potential energy of the molecule is plotted as a function of the internuclear distance, a curve of the type shown in Fig. 9.1 is obtained. A harmonic oscillation is predicted when the potential energy curve in the neighborhood of the equilibrium distance r_0 is approximated by a parabola. The energy levels of the harmonic oscillator calculated according to quantum mechanics satisfy the formula

$$E_v = h\nu_0(v + \tfrac{1}{2}), \quad v = 0, 1, 2, \ldots . \tag{1.1}$$

Here ν_0 is the classical oscillator frequency, and the integer v is the *vibrational quantum number*. The selection rule appropriate to radiative transitions among levels of a harmonic oscillator states that transitions occur only between adjacent levels ($\Delta v = \pm 1$). Since adjacent levels of the harmonic oscillator differ in energy always by $h\nu$, the quantum me-

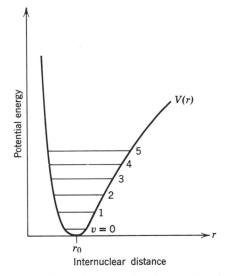

Internuclear distance

FIG. 9.1 Potential energy of a diatomic molecule as a function of nuclear separation. Horizontal lines indicate vibrational energy levels.

chanical prediction is that such an oscillator emits and absorbs only at a single frequency ν_0, which is the classical frequency of the oscillator.

When the expansion of the potential energy around r_0 is continued to include higher-order terms, the calculated oscillation will no longer be a simple harmonic motion, and the energy level scheme of the now anharmonic oscillator will deviate from the simple equidistant spacing represented by (1.1). In fact, the slowly decreasing spacing of the energy levels shown in Fig. 9.1 is intended as an illustration of the consequence of the anharmonic nature of the molecular oscillations. The emission and absorption spectrum will now contain not one vibrational line but many, with a systematic shift in frequency on passing from low to high values of v. The HCl molecule maybe used to illustrate the magnitude of the quantities involved. In this molecule the energy difference between the lowest two vibrational levels ($v = 0$ and $v = 1$) is about 2900 cm^{-1}. Passing toward higher values of v, the difference decreases by about 100 cm^{-1} for each step. For most light, diatomic molecules the separation of the lowest two vibrational levels lies between 1000 and 3000 cm^{-1} [1].

A polyatomic molecule has several vibrational degrees of freedom, each associated with a single normal coordinate. Then for *each such coordinate* the preceding discussion is applicable; one obtains in rough approximation a single frequency and, on closer examination, an entire series of vibrational frequencies.

Molecules also posses energy by virtue of their rotation. For the purpose of the most elementary calculation of this energy, a diatomic molecule may be regarded as a rigid dumbbell consisting of two mass-points at a fixed distance. The essential characteristic of such a model is the moment of inertia I about an axis through the center of mass and oriented in a direction perpendicular to the line connecting the two nuclei. In classical mechanics the rotational energy of such a structure is $E = \frac{1}{2}P^2/I$, where $P = I\omega$ is the angular momentum. In quantum mechanics it is shown that only such energy values are stationary for which

$$E = \frac{h^2}{8\pi^2 I} J(J + 1), \tag{1.2}$$

where the *rotational quantum number* J has integral values only; that is, $J = 0, 1, 2, \ldots$. Thus the rotator has a series of discrete energy levels whose energy increases quadratically with J. The selection rule requires that transitions take place only between adjacent rotational levels; that is, $\Delta J = \pm 1$. Then, when the rotator emits one quantum of radiation in changing from the level $J = J' + 1$ to the level $J = J'$, the frequency of the emitted radiation is

$$\nu = \frac{h}{8\pi^2 I} [(J' + 1)(J' + 2) - J'(J' + 1)] = \frac{h}{4\pi^2 I} (J' + 1). \tag{1.3}$$

It follows then that the entire spectrum of the rigid rotator consists of equidistant lines.*

The frequencies of these rotational lines are usually at least one order of magnitude less than the first vibrational frequency. The combination of vibrational and rotational energies leads to a partial molecular energy-level diagram, shown in Fig. 9.2. All the levels shown belong to the same electronic configuration of a hypothetical molecule.

It must be stressed that the rigid rotator model is a rough approximation. Vibrations and rotations interact, the presence of vibrations changes the moment of inertia, and the presence of a rotation introduces centrifugal forces with the resultant displacement in the average internuclear distance. More precise calculations take account of these interactions [1].

---

* This only applies when the vibrational energy of the molecule does not change. Otherwise, the transition $\Delta J = 0$ is also possible. More about such vibration—rotation transitions in Section 9.13.

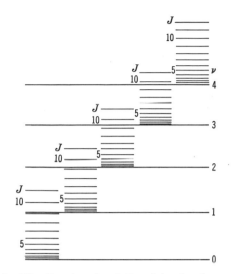

FIG. 9.2 Vibrational and rotational levels of a molecule.

A general transition in the molecule involves the change of electronic, vibrational, and rotational quantum numbers. The largest contribution to energy change comes from the change of the electronic configuration. Change in the vibrational energy contributes less, and the least is the contribution of the rotational part. For the time being we leave the contribution of the rotational energy changes out of consideration and examine in greater detail the interrelation of changes in electronic configuration and vibrational energy. The potential energy curve in Fig. 9.1 was drawn for a fixed electronic configuration. A different curve is obtained when this configuration is changed. Not only the shapes of the potential curves may change, but the internuclear distance that corresponds to the minimum of this energy is in general different for each curve. Figure 9.3a illustrates the situation when the equilibrium internuclear distance is the same for electron configurations I and II, while Fig. 9.3b illustrates the case when these equilibrium distances are different.

When a transition takes place from one electronic configuration to another, the selection rules for purely vibrational transitions do not apply. The question arises: How are the electronic and vibrational transitions interrelated? It is known experimentally from spectral intensity measurements that certain vibrational transitions will be preferred. The observed intensity distributions are explained on the basis of the *Franck-Condon principle*. The essence of this principle is the following

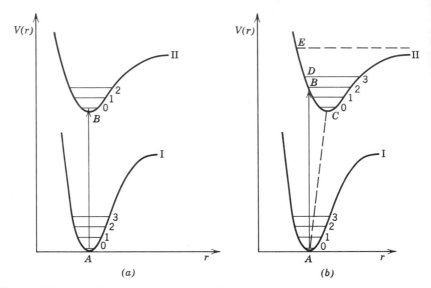

FIG. 9.3 Diagrams illustrating the Franck-Condon principle: (*a*) potential minima attained at the same internuclear distance; (*b*) potential minima attained for different internuclear distances.

observation: *The electron jump in a molecule takes place so rapidly in comparison to the vibrational motion that, immediately after the jump, the nuclei have nearly the same relative position and velocity they had before the jump.*

Let us apply this principle to discover what happens when a molecule originally in the $v = 0$ vibrational state, in the lower electron configuration absorbs radiation capable of lifting the molecule into one of the low vibrational levels of the higher electron configuration. When the situation is such as shown in Fig. 9.3*a*, the molecule is very close to the minimum of the potential curve *I*. It cannot reach the vibrational levels $v' = 2$ or 3 without changing significantly either its internuclear distance or the velocity with which the nuclei are moving toward each other. Only for the $v = 0 \rightarrow v' = 0$ transition is the requirement of small position- and momentum change satisfied. This transition will be preferred most. For the level $v' = 1$, the necessary alteration of position and momentum will be greater, but less than required for the $v' = 2$ transition. Therefore the intensity of the 0–0 transition will be largest, followed by that of the 0–1 transition, and so on.

In Fig. 9.3*b* the minimum *C* of the upper potential curve lies to the right of the minimum *A* of the lower curve. Therefore the transition from $A \to C$ is no longer the most probable since the internuclear distance must alter on that transition. The most probable transition is that from A to B; that is, the 0–2 transition. For this transition, there is no change in either the internuclear distance or velocity. A transition of the type 0–3 to point D involves a change in internuclear velocity. Therefore, under the circumstances described, the intensity of the 0–2 transition will be the largest, followed by the 0–3 and 0–1 transitions.

The Franck-Condon principle is applicable even when the excitation of the molecule occurs by an electron collision instead of the absorption of a light quantum. The reason for its validity is that the electron generally carries too little momentum to alter appreciably the momenta of the nuclei.

When the electronic configuration of a molecule is changed by the absorption of energy for an external agent, the molecule may dissociate in several ways. First, it may occur that, even though the higher electronic configuration has the general shape of curve II in Fig. 9.3b, the molecule arrives at a point so far above B that its total energy does not correspond to a closed orbit. The horizontal dotted line ending at E illustrates such an "orbit" with only one turning point. A second alternative is equally important. Only relatively few electronic configurations of the molecule are bonding configurations which exhibit a minimum potential energy for some internuclear distance. In a nonbonding configuration, the potential energy steadily decreases with increasing nuclear separation. If the interaction with an external agent brings the molecule into such a configuration the parts of the molecule fly apart since they are acted upon by repulsive forces.

When a molecule dissociates, its fragments may end up in their ground states, but this is frequently not the case. Dissociation may proceed through molecular configurations which lead ultimately to excited fragments. Thus under certain circumstances dissociation is a source of excitation. In a similar manner, chemical combination may lead to the formation of a molecule in one of its excited states.

Electron Impacts in Gaseous Discharges. From the point of view of spectroscopy, the directly observable energy-exchange processes are those in which the energy of an atomic system changes with the emission or absorption of electromagnetic radiation. The selection rules introduced earlier apply to such radiative changes. Other energy-exchange processes also take place in a gaseous discharge. They occur when particles collide with each other and with the walls of the discharge.

Collisions between fast moving electrons, atoms, and molecules are usually the prime sources of excitation for an atomic system in a discharge. Electric power is fed into the discharge mostly in the form of work done on the electrons by the accelerating electric field between the electrodes. Many elastic collisions take place between the fast-moving electrons and the atoms and molecules of the discharge. The major effect of these elastic collisions is to randomize the direction of motion of the electrons. Because of the large disparity in the masses of electrons and atoms, relatively little kinetic energy is transferred between them in elastic collisions.

The electrons may lose significant amounts of kinetic energy in *inelastic* collisions with atoms. In such collisions a conversion of kinetic energy to stored atomic energy takes place, but such conversion can take place only with the conversion of such amounts of kinetic energy as the atom (or molecule) is capable of accepting by changing from its initial state to another stationary state. Thus for an inelastic collision of this type to be possible it is necessary that the impinging electron have enough kinetic energy to raise the internal energy of the atom to the next available level. Such an inelastic collision, in which the atom gains internal energy, is called a *collision of the first kind*. Inelastic collisions between atoms and electrons may also take place, with the result that the initially excited atom gives up internal energy which is then converted into kinetic energy, with the electron gaining most of it. Naturally, the amount of internal energy that the atom is capable of giving up is determined by the available stationary states of the atom because the atom must end up in one of these states after the collision. An inelastic collision of the type in which the atom gives up energy is called a *collision of the second kind*.

That collisions of the first kind actually occur is amply demonstrated by experience. That collisions of the second kind must also occur may be deduced from the argument that it must be possible to establish thermodynamic equilibrium in a system that consists of a completely isolated plasma containing electrons and atoms. In fact, with the use of a thermodynamic argument it is possible to derive a relation between the inelastic transition rate θ_{nm}, for the passage of an atomic system from state n to m, and the transition rate θ_{mn} applicable to the reverse process, assuming that both rates are calculated for collisions with an electron gas in thermal equilibrium at a given temperature.

The effectiveness of a collision process in bringing about a physical change is measured by a parameter called the *collision cross section*. This term originates from the theory of collisions between hard spheres. When such spheres of radii r_1 and r_2 collide, they approach each other until their centers are separated by a distance $a = r_1 + r_2$. Large forces come into play at this distance and a collision is said to occur. In calculating the number of collisional encounters among particles distributed at random, it is permissible to replace the original problem by one involving particles of radius a with pointlike particles because a collision occurs when the point representing the second particle comes within the distance $a = r_1 + r_2$ of the center of the first particle. Consider now a volume V in space that contains small, hard spheres of radius a distributed in a random manner with density N, which is not too large, so that the total volume of the space occupied by the spheres themselves is small compared to V. If a beam of point-like particles is shot through the given region, and if the tiny particles are removed from the beam whenever they strike a sphere, the proportion of tiny particles lost from the beam over a distance dx is $Na^2\pi \, dx$ because the unit cross section of the beam encounters Ndx particles in the volume it sweeps through, and each sphere presents an area $a^2\pi$. Hence the beam is attenuated according to the formula $I = I_0 \exp(-Na^2\pi x)$. In a collision process in which the hard sphere approximation is no longer applicable, we still expect the number of the unaffected particles to vary with distance as $\exp(-N\sigma x)$. Here σ is a quantity with the dimension of area. It is called the *collision cross section* of the encounter. This concept is not restricted to electron—atom collisions, in fact, it is frequently used in connection with collisions involving electrons, atoms, and molecules in all combinations.

If two atoms, each of 1 Å radius, were to collide as rigid spheres, the cross section of this process would be $4\pi(\text{Å})^2$, or approximately 12.6×10^{-16} cm². This figure serves as a mark for orientation concerning the order of magnitude of collision cross sections.

As it is the case for atomic transitions caused (or accompanied) by electromagnetic radiation, not all possible transitions between atomic energy levels are equally likely to occur in the event of an electron impact. The ratios of the probabilities of these transitions are the ratios of the cross sections, each cross section referring to a particular transition. In fact, all the basic information necessary for the prediction of probabilities and process rates is contained in these cross sections. In principle, there are quantum mechanical methods by means of which the excitation cross sections of various atomic transitions may be calculated for impact by electrons with a given kinetic energy. The calculations require approximations and lead to rather complicated selection rules, whose validity depends on the energy of the impacting electrons and also on the coupling prevailing among the electrons of the atom [2]. At best, these rules provide a guide as to which transitions are favored in given circumstances. When the kinetic energy of the impacting electrons greatly exceeds the energy difference between the ground state of the atom and the state characterized by the quantum number n, it can be shown that, in a first approximation, the collision cross section for an inelastic transition from the state 0 to the state n is proportional to the square of the matrix element μ defined in Section 5 of Chapter 1. Thus, under the restrictions stated above, inelastic collision cross sections appear to follow the selection rules applicable to electric dipole transitions. It would then appear that electron-impact excitation is ruled out when a transition is forbidden by the electric dipole selection rules. This is not the case, however, because of the restricted applicability of the approximations used. In cases in which electric dipole transitions are forbidden, the next term in the electron-impact calculation may result in a significant transition probability. This is especially true when the impacting electron has a kinetic energy nearly equal the threshold energy necessary for excitation [3]. Because the electron-impact excitation cross sections are not strictly proportional to the intensities derived from the electric dipole radiation theory, it is possible to excite atoms by electron impact into states not optically connected to the ground state; that is, states from which a direct return to the ground state by electric dipole radiation is not possible.

The excitation of molecules by electron impact is too complicated for a quantitative theoretical analysis. Molecules may store energy in a variety of ways, and the response of a molecule to an approaching electron depends greatly on the structure of the molecule. Therefore the conditions for excitation vary greatly from one molecule to another. Although molecules can change their vibrational and rotational energies in the course of an electron impact, such processes have comparatively small cross sections. Hence in molecular lasers the excitation of a molecule to a high vibrational level often takes place indirectly, as, for example, by excitation of the molecule to a higher electronic level and by an accompanying change of the vibrational energy, which takes place as a result of the Franck-Condon principle. The first change may be followed by a cascade process, at the end of which the molecule may be found in a high vibrational level of the ground electronic state.

Resonant Excitation Transfer. When two atoms collide and one of them is in an excited state before collision, there exists a finite probability that the originally excited atom loses its energy while the other atom gains internal energy. This phenomenon is easier to observe when the colliding atoms belong to different species. Denoting the excited atoms by stars, the collision may be written in the form

$$A^* + B \to A + B^* \pm \Delta E. \tag{1.4}$$

Here the term ΔE indicates that the energy given up by atom A need not be exactly equal to the energy received by atom B. The energy difference, ΔE, appears as a gain or loss in the kinetic energy of translational motions.

The theory of excitation-transfer by collision assumes the temporary formation of a "molecule" whose constituents are the colliding atoms. It indicates that for the case of exact resonance, that is when $\Delta E = 0$, transfer cross sections of the order of 10^{-13} cm^2 may be obtained, which are two to three orders of magnitude larger than the physical cross sections of the atoms. For $\Delta E \neq 0$ the transfer cross section depends on this energy difference and on the velocity of approach of the particles. When collisions occur as a result of random motions in a gas mixture in thermal equilibrium, the cross section is a function of the gas temperature. Cross sections of about 10^{-16} cm^2 were calculated for energy transfer in atom—atom collisions when $\Delta E/kT \approx 1$. Thus in the low pressure discharges used in most gas lasers cross sections of the order of 10^{-16} cm^2 are obtainable for energy differences of 0.025 to 0.050 eV, or 200 to 400 cm^{-1}. The transfer cross sections decrease rapidly with increasing ΔE [2, 3]. Because the transfer process depends so much on the near coincidence of the energy differences, it is called *resonant transfer*.

Experimentally, excitation transfer phenomena were known for a long time [4], and the early search for gas laser materials concentrated heavily on systems exploiting known possibilities of excitation transfer from metastable atoms of one element to atoms of another. Metastable states are preferred to ensure ample supply of donor atoms. The visible– and the near infrared helium-neon lasers are prime examples of the success of the resonant excitation transfer method as a means of laser excitation.

Resonant excitation transfer may also take place in molecular collisions. In fact, such phenomena occur with a greater frequency than the corresponding phenomena involving atoms only. The variety of events that may occur is illustrated in the example involving the collision of an excited atom A with a diatomic molecule BC. The reaction may begin as

$$A^* + (BC) \rightarrow A + (BC)^* \pm \Delta E_1, \qquad (1.5)$$

indicating the formation of an excited molecule $(BC)^*$. This state of the molecule may not be a bonding state. If it is not, the molecule will fly apart with the reaction

$$(BC)^* \rightarrow B^* + C^* \pm \Delta E_2. \qquad (1.6)$$

Here the atoms B and C may, or may not, end up in excited states. Experimentally, one usually does not observe the products of the first reaction, and the collision seems to follow the formula

$$A^* + (BC) \rightarrow A + B^* + C^* \pm \Delta E. \qquad (1.7)$$

Naturally, it is not essential that A^* be an excited *single* atom, it could equally be an excited molecule.

For collisions involving molecules, the energy coincidence requirement, that is, the smallness of ΔE, is not nearly as stringent as it is for atom—atom collisions. The transfer cross sections in molecular collisions are comparable to those in an atom—atom collisions with much smaller energy discrepancy. There are two reasons for this greater flexibility of energy transfer over a larger gap. First, in the molecular case there is a greater freedom open to each particle. Usually, there are a number of repulsive (nonbonding) molecular states of slightly differing energy available in a given energy range. Several of these states may lead to the same levels of the dissociated molecule. This itself makes the energy-coincidence requirement less stringent. Second, when a molecule is involved, the chance of the reaction being reversed when the particles are separating is much smaller than in the case of two atoms because the large number of paths available to the particles decreases the probability that they will retrace the original path.

The dissociative excitation process formally described in the reactions (1.6) and (1.7) is the basis for laser action in neon-oxygen and argon-oxygen lasers. The transfer of excitation between molecules of different compounds is the basic mechanism for the operation of the high-power carbon dioxide lasers.

Excitation by a Chemical Reaction. Collisions between atoms and molecules may lead to effects other than transfer of excitation energy. Even when neither of the colliding particles was originally excited, a chemical reaction may take place. One of the common occurrences is shown schematically as follows:

$$A + BC \rightarrow AB + C. \qquad (1.8)$$

Chemical energy may become free in such a reaction, and, as a consequence, it is possible that the new molecule AB, or the newly freed atom C, or both, end up in an excited electronic or vibrational state. When this occurs there is a possibility of exploiting the chemical reaction for the production of population inversion in an atom or a molecule. A laser in the visible- or near infrared region has a very high energy demand on the chemical scale, because 2 eV is about the minimum energy required for a red quantum and 1 eV is equivalent to 23 kcal/mole. It must be remembered that the major portion of the chemical energy liberated is not likely to go into the excitation of the preferred atomic level. The situation is more favorable in the middle-infrared region for wavelengths beyond 2 μm.

An example for chemical excitation is furnished by the formation of HCl from H_2 and Cl_2. The complete molecules do not react with each other, therefore the reaction must be initiated by atomic chlorine which is easily produced in the necessary concentration by shining light on chlorine gas. The following sequence of reactions takes place:

$$Cl_2 \rightarrow Cl + Cl, \qquad (1.9)$$

$$Cl + H_2 \rightarrow HCl + H, \qquad (1.10)$$

$$H + Cl_2 \rightarrow (HCl)^* + Cl. \qquad (1.11)$$

The star in (1.11) indicates the formation of a HCl molecule in an excited vibrational level capable of producing stimulated radiation (see Section 9.14). Free chlorine atoms are reproduced in the last reaction so that the process continues, once triggered. It should be noted that only about 15% of the chemical energy liberated is converted into useful vibrational excitation of HCl.

The Competition of Energy Transfer Processes. We have already noted that several processes of excitation and de-excitation are operative in gases. All processes are subject to the principle of detailed balance, which relates the probability of a process to that of the inverse process. Let us consider a large reservoir R of interacting particles in thermal equilibrium with each other at a temperature T, and an experimental system E capable of interaction with the particles of R by means of a variety of physical processes. The *principle of detailed balance* asserts that each process acting separately will ultimately produce the same statistical distribution of energy in E, namely a Boltzmann distribution with a temperature T. In the situation of interest here, R may be represented by the electrons in the discharge, whose energy distribution is at least approximately that required for the validity of the argument. A similar result is obtained if we consider the interaction of the atomic system with a radiation field whose energy distribution is given by Planck's radiation law.

Since all such processes tend to produce thermal equilibrium if acting alone, the question arises of how one can ever obtain a steady-state population inversion. The answer is that the rates at which these processes proceed toward equilibrium are all different. Consequently, we may play one of the mechanisms against the others and operate the system in an intermediate steady-state condition, where no one process dominates and where no single temperature characterizes the population of all atomic levels. The situation is helped in multilevel systems if strong departures can be obtained from a Maxwellian (equilibrium) distribution of the interacting particles, or if a highly monochromatic radiation field is employed in place of blackbody radiation. There is, however, for each excitation process an inverse process which tends to produce thermal equilibrium, and the production of population inversion involves minimizing the effects of the inverse process at some stage. This may be accomplished, for example, by allowing light from some levels to escape through the walls or by allowing metastable atoms to diffuse out of the discharge. Like the thermodynamic engines, the lasers must have a sink. This sink, however, may be one or more steps removed from the lower laser level.

The following energy exchange processes are important in a discharge that takes place in a single monatomic gas:

1. Electron collision of the first kind, in which an atom gains energy from an electron.

2. Electron collision of the second kind, in which an excited atom loses energy to an electron.

3. Spontaneous emission of radiation from an excited atom.

4. Absorption of radiation by an atom.

5. Stimulated emission of radiation by an atom.

6. De-excitation on collision with the walls of the vessel.

In a system containing a mixture of several gases, including molecular gases, the following additional phenomena may affect the distribution of atomic systems:

7. Resonant excitation transfer.

8. Dissociation and recombination phenomena; chemical reactions.

The rates at which these processes take place are determined by the number of atoms available in the proper states and by the probabilities that these phenomena will occur in the unit time for an individual particle. In the case of the radiative phenomena, the probabilities are calculable from the lifetimes of the transitions and the prevailing radiation densities. In the case of the collision phenomena, the probabilities are deduced from the cross sections of the phenomena involved and the proportions of the constituents present.

As already implied, the design of a gas laser involves the selection of materials, partial pressures, discharge parameters, and the walls of the confining vessels in such a manner as to insure the accumulation of atoms or molecules in an upper energy level and the simultaneous rapid removal of atoms or molecules from a lower level so that a population inversion results. Frequently, several processes are involved simultaneously so as to reinforce each other. Although the theoretical discussion suggests that a gas laser may be invented and designed entirely on paper, using known transition rates and collision cross sections, this is in practice not the case. While the development of gas lasers was aided and guided by the principles outlined above, the actual discoveries were made by many trials. Gas lasers frequently operated before their excitation mechanisms were fully understood. Nevertheless, they were often improved significantly after their basic mechanisms were understood, and modifications were made to eliminate bottlenecks in their operation.

9.2 AMPLIFICATION AND OSCILLATION CONDITIONS IN GAS LASERS

The general conditions for gain or optical amplification were developed in Chapters 1 and 3. It was found that when extraneous processes, such as absorption and scattering by foreign material are eliminated, the amplification rate in a laser material is given by the formula

$$\alpha(\nu) = k(\nu)_0 n. \tag{2.1}$$

Here $k(\nu)_0$ is the absorption curve of the unexcited material, and n is the relative population inversion. In the notation customary for four-level lasers, using 3 for the upper and 2 for the lower laser level,

$$n = \frac{1}{N_0}\left(\frac{g_2}{g_3} N_3 - N_2\right). \tag{2.2}$$

The symbol N_0 denotes the total number of active particles (atoms) in the volume element, the g's are the statistical weights (multiplicities) of the levels involved.

Optical gain is achieved in the material when n is positive. Laser oscillations can be obtained when n is not only positive, but exceeds a minimum positive value set by the requirement that the gain in the entire laser structure must at least be equal to the loss due to all causes.

The condition for amplification, $n \geq 0$ may be put in the form

$$\frac{N_3\, g_2}{N_2\, g_3} \geq 1, \tag{2.3}$$

where the equality corresponds to the condition of zero gain.

Gas lasers have certain features which make gain calculations very difficult and which almost exclude the possibility of reliable theoretical predictions based entirely on basic transition rates and collision cross sections. While in a solid laser the atoms generally follow a well-defined cycle passing through three or four levels, the energy exchange processes of active atoms in a gas laser do not form a unique, well-defined cycle. Generally there are several paths available for a gas atom in a given state, because several excitation and de-excitation processes take place in the gas simultaneously. Population inversion, if it is attained at all, is the result of the competition of several dissimilar processes. As a consequence, the lifetime of an atom in an excited state is determined by the rates of several decay processes. Since gas lasers are usually excited by electrical discharges through them, the impacts occurring in the discharge excite the atoms into many states, and the excited atoms cascade down the energy scale over a variety of processes. As a result of these, the lower laser level will also be populated by the excitation mechanism that is used to populate the upper level. Let W_2 be the rate at which atoms are fed into level 2, the terminal laser level, as a result of the excitation process, excluding the atoms which arrive at level 2 as a result of spontaneous and stimulated radiative transitions from level 3. Let furthermore R_2^{-1} be the lifetime of atoms on level 2. The population balance– or rate equation for level 2 is then

$$\frac{dN_2}{dt} = W_2 + (A_{32} + S_{32})N_3 - R_2N_2, \tag{2.4}$$

where A_{32} and S_{32} are the spontaneous and stimulated transition rates for the laser transition. The laser is to be operated in a stationary regime, requiring that

$$W_2 + (A_{32} + S_{32})N_3 = R_2N_2. \tag{2.5}$$

Therefore, even when stimulated transitions are neglected, it follows that

$$R_2N_2 > A_{32}N_3. \tag{2.6}$$

Combining this inequality with (2.3), the condition of zero gain becomes

$$\frac{R_2}{A_{32}} \frac{g_2}{g_3} > 1. \tag{2.7}$$

Thus for zero gain one must have

$$R_2 > A_{32} \frac{g_3}{g_2}. \tag{2.8}$$

This is a necessary condition, not a sufficient one, and not an exact one because the amount by which R_2 must exceed the expression on the right side of (2.8) depends on the rates of the processes that feed the population on level 2. Nevertheless, it serves a useful purpose. The right side of the inequality depends only on atomic constants. The left side is under the control of the experimenter to some extent because he may be able to introduce or reinforce processes that tend to increase the decay rate R_2 from the terminal level.

A superficial look at (2.8) suggests that a transition with a small value of A_{32} is to be preferred for a laser. This conclusion is wrong, as one readily discovers by examining the condition for attaining laser oscillations. It was shown in Section 3.1 that in a laser of active length L and loss parameter γ, the threshold of oscillation is attained when

$$nN_0 = \left(\frac{g_2}{g_3} N_3 - N_2\right) = \frac{\gamma}{L\kappa g(0)}. \tag{2.9}$$

Here $g(0)$ is the peak of the normalized lineshape and κ is the constant introduced in Section 1.3 (3.17). In the notation used in the present section

$$\kappa = \frac{g_3}{g_2} \frac{A_{32}\lambda^2}{8\pi}. \tag{2.10}$$

Since $g(0)$ is inversely proportional to the linewidth $\Delta\nu$, the condition for the attainment of threshold becomes

$$\frac{g_2}{g_3} N_3 - N_2 = C \frac{\Delta\nu}{A_{32}}. \tag{2.11}$$

The constant C contains the design factors of the laser, the wavelength, and another constant that relates the linewidth to the peak value of the normalized lineshape. Clearly, the threshold is the easier to reach the larger A_{32} and the smaller the linewidth. Therefore, other things being equal, transitions with large matrix elements are to be preferred and so are conditions that produce a smaller linewidth. Because the factor λ^2 enters into the threshold condition in the same manner as A_{32}, other things being equal, lasers with longer wavelengths require lesser excitation than those with shorter wavelengths.

The use of the notation adapted for four-level lasers does not imply that there are usually *exactly* four levels involved in the useful operating cycle of a gas laser. In fact, in the most common gas lasers the atoms do not return from the terminal level of the laser transition *directly* to the ground state. They pass through an intermediate state, or possibly through several. In some lasers, atoms leave the terminal level (2) by giving up their energy in inelastic collisions to atoms or molecules of another gas, deliberately introduced for the purpose of making R_2 as large as possible.

9.3 SPECTRAL DISTRIBUTION OF LASER RADIATION. LAMB DIP

The objective of this section is the analysis of the distribution of laser radiation in frequency on a very fine scale. Accordingly, we confine our attention to transitions occurring between a fixed pair of energy levels in a large assembly of atomic systems. The atomic systems form a gas; they move around and collide with each other as well as with the walls in the manner described in the kinetic theory of gases. The properties of the atomic assembly (gas) affect the spectral lines, and this interaction must now be examined in some detail.

Having chosen a fixed transition for examination, we focus on a single spectral line, limiting the range of attention typically to a wavelength region between 0.1 and 1.0 Å wide. The corresponding frequency-spread in the red end of the visible spectrum varies from 6×10^9 to 6×10^{10} Hz, the wavenumber-spread from 0.2 to 2.0 cm^{-1}.

The variation of the intensity of emission and absorption within a spectral line is described by the function $k(\nu)$, introduced in Section 1.3 as an empirical function obtained from spectroscopic experience or, alternatively, by the normalized lineshape function $g(\nu,\nu_0)$, discussed in Sections 1.3 and 1.4. These functions, as was pointed out, are determined not by the energy-level structure of the atom, but by the motions and interactions of the entire atomic assembly. Although there is a finite spectral width associated with isolated atoms at rest, this intrinsic, or natural linewidth is orders of magnitudes smaller than the spread with which we are now concerned. The two main causes of line-broadening in gases are the Doppler effect and collisions. These effects were briefly described in Section 1.4, and the mathematical forms of the lineshapes resulting from each of these effects acting by itself were indicated. Generally, both effects are present simultaneously, but at higher pressures (0.1 atm) collision-broadening dominates, while at low pressures (1 torr) Doppler-broadening is the determining effect.

The lineshapes appropriate to collision-broadening and Doppler-broadening are described by (4.2) and (4.6) in Chapter 1. They not only differ in their mathematical form, they result from entirely different statistical situations. In the case of Doppler-broadening, the contribution of an atom is confined to a narrow frequency range, the location of which relative to the line center is determined by the velocity of the atom relative to the observer. The absorption curve of the gas is the result of the distribution of atoms with respect to velocity components. In the case of collision broadening, atoms cannot be identified with specific spectral regions, the broadening is spread over the entire atomic assembly. This is called *homogeneous broadening* in contrast to the Doppler-broadening, which is referred to as *inhomogeneous broadening*.

In lasers it is necessary to consider the interaction of an excited atomic assembly with a radiation field having extremely narrow frequency spread. This spread is determined by the Fabry-Perot cavity and is generally smaller than the linewidth of the collision- or Doppler-broadened line of the assembly. It is of interest to calculate what happens to an excited atomic assembly as a result of interaction with a radiation field having a narrow spectral spread and an intensity high enough to reduce substantially the number of excited atoms in the assembly. Naturally the interaction calculation must involve the distribution in frequency of the radiation field as well as the statistical properties of the assembly. First, the response of a damped oscillator to a harmonic exciting force is calculated, then statistical averaging is carried out over the oscillator assembly and over the distribution of the exciting field. The mathematical apparatus of this calculation is developed in the textbook of Sinclair and Bell [5].

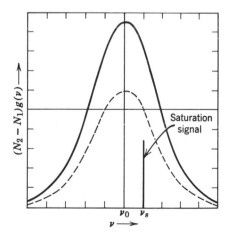

FIG. 9.4 Uniform saturation of a homogeneously broadened line by a monochromatic signal of frequency ν_s. *Solid curve:* amplification in the absence of a strong signal; *dotted line:* amplification in the presence of saturation.

The results of the calculations, which are physically quite plausible, may be summed up as follows: When an intense radiation field of narrow spectral width interacts with an excited atomic assembly having a homogeneously broadened line, the resulting emission of stimulated radiation depletes the entire assembly uniformly. This is illustrated in Fig. 9.4, p. 281. When the atomic assembly is Doppler-broadened, the radiation field (saturation signal) will act mainly on those atoms which have the proper velocity component in the direction of wave propagation. Hence the signal will feed selectively on certain atoms and not on others. Of course, the velocities of the atoms are redistributed as a result of atomic collisions. At pressures that normally prevail in gas lasers, an atom makes 10^6 to 10^7 collisions per second. When a very sharp pulse of narrow spectral content passes through the gas, a collision rate of 10^7 sec^{-1} may not be fast enough to prevent depletion among atoms falling in the favorable velocity range. A hole may thus be burned into the inhomogeneously broadened line. The distortion of the line will usually appear in the form shown in Fig 9.5, with two holes burned in the spectrum whenever the frequency of the cavity mode differs from the peak frequency of the atomic line. The two holes are produced because the light traverses the laser in two opposite directions, and stimulated emission depletes atoms traveling with equal velocity components in two opposite directions [6].

An interesting phenomenon may be observed when a short gas laser is operated in a single axial mode and this mode is shifted in frequency

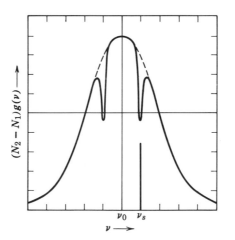

FIG. 9.5 Hole-burning in a Doppler-broadened line of a gas laser in the presence of a signal of frequency ν_s.

by varying slightly the distance between the mirrors. Such tuning may be accomplished by mounting the mirrors on rods whose lengths may be adjusted by magnetostriction. When the mode frequency is different from the peak of the atomic line, atoms traveling with a velocity component along the beam and also in the opposite direction are available for amplification. When the peak of the mode coincides with that of the atomic line, only atoms moving perpendicularly to the beam may be utilized. Then, in tuning the mode over the line, the supply of amplifying atoms is decreased when the peak ν_a of the atomic line is reached. A decrease in the output power is then observed, which may be used for the precise determination of the frequency of the atomic line.

Quantitatively, the situation is as follows: The number of atoms whose velocity has an x-component between v_x and $v_x + dv_x$ is proportional to $\exp\left(-mv_x^2/2kT\right) dv_x$. The frequency shift w, with respect to the center frequency ν_a of the atomic line, is $w = \nu - \nu_a = \nu_a v_x/c$.* Therefore, if the passage of light were restricted to the positive x-direction, the amplification as a function of the frequency (w) would be proportional to

$$P(w) = \exp\left(-\frac{\beta w^2}{\nu_a^2}\right),\tag{3.1}$$

where

$$\beta = \frac{mc^2}{2kT}.\tag{3.2}$$

We assume now that the cavity mode acts as a simple filter which accepts all frequencies within a region of width δ centered around w. Then the total amplification in this cavity mode is proportional to

$$\int_{w-\delta/2}^{w+\delta/2} P(w')\,dw'.\tag{3.3}$$

Now we take into account the fact that light actually passes both in the $+x$ and in the $-x$ direction. Then the total amplification in the above cavity mode arises from two regions symmetrically located around $w = 0$. When these regions do not overlap, the amplification is proportional to

$$F(w) = 2\int_{|w|-\delta/2}^{|w|+\delta/2} P(w')\,dw'.\tag{3.4}$$

This situation prevails when $\delta/2 < |w|$. In the opposite case, the regions of integration overlap, and the amplification is proportional to

$$F(w) = 2\int_0^t P(w')\,dw',\tag{3.5}$$

* See Section 1.4.

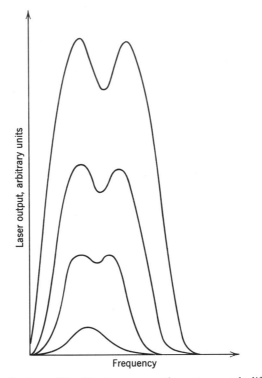

FIG. 9.6 Lamb dip in a He—Ne laser operating at several different levels of excitation. Observed near the center of the 1.15-μm line of Ne[20]. (After Szöke and Javan.)

where $t = |w| + \delta/2$. This function is an even function of w. For $w > 0$ its derivative is

$$\frac{dF(w)}{dw} = 2P\left(w + \frac{\delta}{2}\right).$$
(3.6)

This function is positive, therefore $F(w)$ is an increasing function for small positive values of w. Since $F(w)$ is an even function, it must have a local minimum at $w = 0$. Because the amplification passes through a minimum as the center of the oscillation frequency passes through ν_a, the power output of the laser also passes through a local minimum.

This variation of the laser output of a gas laser was predicted theoretically by Lamb [7] and is called the *Lamb dip;* it was observed in

He—Ne lasers under highly controlled conditions. A pure Ne isotope was used to avoid the masking of the effect by the superposition of radiation originating from different Ne isotopes, whose lines are close together [8, 9]. Figure 9.6 shows the relevant results of Szöke and Javan [9]. The curves measured at different levels of excitation demonstrate that the dip becomes more significant at high power levels, as was predicted by Lamb [7].

9.4 CLASSIFICATION AND ORGANIZATION OF THE MATERIAL

The name gas laser is applicable to a large number of devices that differ much from each other in construction, operating material, and performance. Many gas lasers are capable of emitting coherent radiation over a number of spectral lines; some function in several, widely separated spectral regions. A complete listing of spectral lines observed from gas lasers, if compiled in 1969, would contain over 2000 entries. It would be inconsistent with the general purpose of this book to attempt a complete listing of the laser lines reported. In fact, such a list would be quite meaningless without extensive notes indicating the conditions under which laser oscillations may be obtained on these lines. Nor does it seem desirable to attempt to include here every element and every compound that was ever successfully used as a laser material. Our discussion will cover all gas laser types, with detailed description confined to an important representative of each type and with emphasis on those lasers which have found frequent applications. Detailed spectroscopic data will be given only occasionally, and then primarily for orientation and illustration. The following principles will be adopted for the classification and organization of the material:

According to the chemical state of the active atomic system, the basic distinction is made between atomic, ionic, and molecular lasers. Each of these classes is further subdivided according to the specific elements, ions, and molecules involved, and in the case of molecules a distinction is made between lasers based on electronic–, vibrational–, and rotational transitions.

According to the principal excitation mechanism, it is appropriate to distinguish between lasers excited optically, by direct electron collision, by resonant excitation transfer, and by chemical reactions.

The active materials of atomic lasers are *complete atoms* of monatomic gases and vapors and complete single atoms obtained from dissociation of polyatomic molecules. The best-known of these lasers are those employing the noble gases, primarily neon. Atoms of most halogens as well as oxygen and nitrogen produce many gas laser lines. Of the metal vapors, mercury is frequently used, others, such as cesium, copper, lead, and manganese, relatively rarely. Most of the atomic lasers are excited by low-current glow discharges and are operated in a continuous regime. With very few exceptions, they are low-power, high-precision lasers.

The active materials of ionic lasers are *atoms* of the common gases *with one or more electron removed*. Coherent light is emitted as a result of an electronic transition among the ionic levels. The blue and green-emitting argon laser is the best-known example of an ionic gas laser. Similar lasers have been operated using ions of the other noble gases. The ions of halogens, mercury, carbon, oxygen, and nitrogen are also suitable laser materials. It is customary to use the following spectroscopic notation when describing ionic gas lasers: The energy levels of the complete atom are identified (when necessary) with the Roman numeral I. The numerals II, III, and IV are used to identify ions with 1, 2, and 3 electrons missing from the complete atom. Thus Hg II denotes the spectrum of once-ionized mercury.* Numerous laser lines were obtained from highly ionized elements as Ar III and IV, Xe III, Kr III, O III and IV, N III and IV. Excitation of the ions, especially the higher ones, requires considerable energy since the element must first be ionized. Such lasers are usually excited by high-current pulsed arc discharges. Ion lasers are frequently high-power devices requiring considerable engineering design. They are excellent sources of coherent radiation in the visible– and ultraviolet region of the spectrum, in contrast to the atomic lasers most of which operate in the infrared region.

Molecular lasers have been obtained from the common atmospheric gases: N_2, H_2O, CO, CO_2, and many others including HCl, HCN, and CS_2. Resonant excitation transfer is quite common as a contributing mechanism of molecular laser operation, and for this reason many molecular lasers contain more than one gas. These lasers are mostly pulse-excited, requiring short and very energetic electrical pulses for their operation. Nitrogen and carbon monoxide emit not only in the infrared region, where most molecular lasers operate, but in the visible region as well. The emission in the short wavelength region is attributed to electronic transitions in the N_2 and CO molecules. For the most part molecular lasers operate on vibration—rotational transitions. They emit on many lines in the 5– to 12-μm region. The CO_2 laser, operating at

---

* We recognize that this notation is not consistent with the notation employed for ions located in a crystal lattice, where Cr^{3+} stands for a chromium atom with 3 electrons missing. However, the spectroscopic notation is so well established historically that it is best to use it as it is.

10.6 μm, is the best-known molecular laser because of its high power capability. This laser contains additional molecular gases.

Turning to the methods of excitation, we note that optical pumping, which is universally useful for the excitation of solids and liquids, is not a practical method for the excitation of gases. Only one optically excited gas laser has been constructed and that is more of a curiosity than a practical instrument. Optical excitation played an interesting role in the early history of gas lasers, and for this reason the optically excited Cs-vapor laser will be briefly discussed in Section 9.6.

In the most popular gas laser types the basic source of energy is an electric discharge; the dominant process is resonant transfer of excitation. The prototypes of those lasers in which such transfer takes place from one atom to another are the common helium—neon lasers operating at 0.6328, 1.15, and 3.39 μm. In these lasers excitation is transferred from metastable He atoms to Ne atoms. The scheme is successful because of the chance near-coincidence of two metastable He levels with Ne levels. Such coincidences are rare in atomic spectroscopy, and for that reason the majority of atomic laser lines are obtained by electron impact without transfer of excitation. The atomic oxygen and carbon lasers are somewhat exceptional because their excitation is largely provided by resonant transfer of excitation from a metastable atom to a molecule which then disintegrates producing an oxygen or carbon atom in an excited state. It has already been noted that resonant excitation transfer is more likely when a molecule is involved than when the transfer takes place among single atoms. In fact, such transfer is most common as the dominant or contributory source of excitation in molecular lasers.

Direct electron impact, sometimes followed by a cascade of decay processes, is the source of excitation for the vast majority of atomic and ionic lasers. The lasers in pure vapors and gases are powered by such impacts alone. Ionic lasers sometimes require several electron impacts to reach the starting level.

Excitation of atoms and molecules may be the result of a chemical reaction, as is the case in HCl, or it may take place as a combined effect resulting from photoexcitation and a chemical reaction.

9.5 GAS LASER CONSTRUCTION

Although the enumeration of gas laser types in the preceding section is not complete, it contains enough variety to suggest that gas lasers come in many sizes and configurations. Statements about gas lasers must be made with reservations because exceptions may be found to just about any assertion one would care to make. It is therefore best to preface this entire Section with the warning that its purpose is merely to provide the reader with a picture of what he can expect to encounter in the vast majority of cases. The statements that will follow are to be interpreted with the reservation that they are valid frequently, but not always. Gas lasers excited by means other than an electric discharge are not included in this section.

Gas lasers operate with a much smaller gain per unit length than lasers made of ionic crystals. Consequently, they require a minimum length about an order of magnitude larger than solid lasers of the type of ruby. Ordinarily, gas lasers vary in length from 30 to 100 cm, although some as short as 10 cm or as long as 10 m have been built for special purposes. The gas is confined in a long, narrow cylindrical tube of glass or quartz with an inside diameter between 2 and 15 mm. Collision of the active particles with the walls is in most gas lasers an essential part of the cycle. For this reason it is generally impossible to scale up the gas lasers and obtain an output proportional to the cross sectional area. Consequently, the active part of a glass laser is a slender, long structure.

The first successful glass laser was provided with plane mirrors placed within the gas envelope. The orientation of these mirrors was adjustable in a mount made flexible by means of metal bellows, which can be seen in Fig. 9.7. The picture also shows the radio-frequency transformer and the leads used to couple the high-frequency electric field to the electrodes placed externally on the discharge tube.

As technology progressed, gas lasers with external spherical mirrors were developed. The selection of these mirrors is guided by the design principles explained in Section 3.9. The tubes containing the gas are terminated by optical flats oriented at Brewster's angle, to eliminate reflection in one polarization. The laser normally operates only in the polarization for which the windows are transparent. When operating in the far infrared region, the material of the flat windows is selected to avoid unwanted absorption.

The design of the laser tube can be extremely simple for the case of an atomic gas laser that operates with a glow discharge. These discharges have high impedance, they draw comparatively small currents. When high-frequency (20 to 30 MHz) current is used for excitation, it may be coupled to the discharge by means of external electrodes, as was shown in Fig. 9.7. Usually excitation by means of internal electrodes is preferred, and in that case, the laser tube may have the structure illustrated in Fig. 9.8. The electrodes are most frequently introduced in side arms, but hollow, cylindrical electrodes in the main tube are also

FIG. 9.7 Javan's helium-neon laser. (Courtesy Bell Telephone Laboratories.)

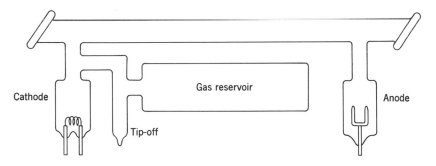

FIG. 9.8 Laser tube with internal electrodes and gas reservoir.

used. Often a gas reservoir is provided which serves to prolong the useful life of a sealed-off gas laser.

The structure of tubes used for ion– and molecular lasers is more complicated, primarily because these lasers operate with large power dissipation. It is customary to surround ion discharge tubes with a wind-

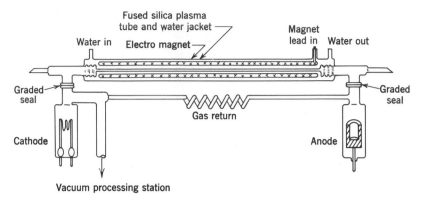

FIG. 9.9 A direct-current-excited continuous-wave gas-ion laser tube.

ing for an electromagnet which serves to stabilize the arc. Water cooling is also necessary. In dc-driven discharges, a return path is provided for the gas which tends to be driven to the cathode by cataphoresis. Because of the high power dissipation, envelopes of ion– and molecular lasers are subject to large thermal stresses. These are minimized by incorporating coils and bellows and by making the most exposed parts out of fused quartz. A laser tube designed for ion lasers is shown schematically in Fig. 9.9. When radio-frequency excitation is used, capacitative coupling is not suitable for the low-impedance ion lasers. They can be coupled to the radio-frequency field inductively, as shown in Fig. 9.10.

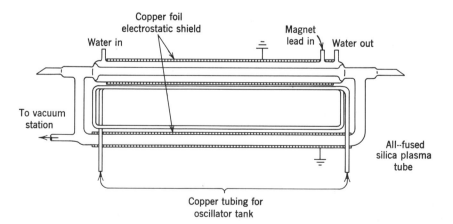

FIG. 9.10 A radio-frequency-excited continuous-wave gas-ion laser plasma tube.

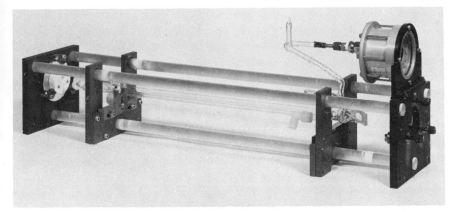

FIG. 9.11 Photograph of a small CO_2 laser. (Hughes Research Laboratories.)

The discharge in this case forms a ring which is the secondary of a radio-frequency transformer. Such a structure is only convenient in an ion laser with high gain per unit length because it is not practical to build a large ring.

The main requirement in the mechanical design of a gas laser is to provide the mirror holders with a rigid support which does not change its dimensions during operation. The laser tube is supported by members resting on the same frame, and adjustments are provided to align the tube with the optical axis of the interferometer. There is some leeway in the adjustment of the tube, but the mirrors must be adjusted with great precision and this adjustment must be held. Figure 9.11 is the photograph of a simple, sturdy laser structure developed at Hughes Research Laboratories for CO_2 lasers. The mirror holders (seen at left) are held on to the heavy vertical steel plates by spring-loaded stubs and are pushed from the other side by means of three precision screws, visible on the right end of the photograph. The steel plates are held in place by three horizontal quartz rods which also carry the two cradles that position the discharge tube. The position of the tubes can be adjusted, making use of the freedom provided by the slotted holes in the cradle mechanisms. The tube in this picture is provided with internal electrodes largely obscured by one of the quartz rods. Cooling water is fed through two plastic connectors, one of which is visible near the right end of the tube.

The mirrors used in the visible– and the near-infrared region are usually multilayer dielectric-coated concave glass lenses. The reflectivity of such a mirror is quite high (99%) for the favored laser line. Beyond

5 μm, the construction of such selectively reflecting mirrors is difficult. In the middle- and far-infrared region completely reflecting gold-coated mirrors are used, with a small hole at the center of the mirror through which the radiation is to be emitted. Good optical quality of the mirrors and their perfect alignment are prime requirements of gas laser operation.

Power supplies used for glow discharge lasers, such as the He—Ne laser, are very simple. In fact, neon-sign transformers may be used with proper ballast. It is necessary to generate much higher voltage than the voltage drop across the tube because the discharge has negative resistance characteristics, and the circuit must be stabilized by means of a ballast resistance in series with the tube. The power supply should provide between 1500 and 3000 V to allow for drop over the ballast which is often between 50 and 100 kΩ. Currents between 5 and 25 mA are usually satisfactory for most of the atomic laser types.

The power supply requirements of ion lasers are entirely different because these are low-impedance high-current devices. A typical argon ion laser operates with a discharge current density of 300 A/cm^2 so that 10 to 50 A-current is required for a discharge tube of convenient size. The voltage required is in the range of 200 to 400 V. The design of the power supply will depend on many variables. The most important factor is the nature of laser operation, whether it is to be pulsed or continuous. Beyond this question, the impedance of the discharge is to be considered, which depends on the dimensions of the laser and on the composition of the gas.

Molecular lasers operate in glow discharges. The common, small CO_2 power lasers operate with a discharge current of 5 to 30 mA and a potential drop over the discharge of 5 to 10 kV. The power supply has to provide at least 15 kV because a large ballast resistor is necessary in series with the discharge tube.

The design of ion- and molecular lasers involves many technical difficulties which result from the severe loading of the components as a result of the thermal stresses, the abrasion of the tube by fast-moving particles, sputtering at the electrodes, and hazards caused by high currents or high voltages. *Although atomic lasers are relatively harmless, ion- and molecular lasers can be very dangerous because of their electrical and fire hazards.*

9.6 OPTICALLY PUMPED GAS LASERS

Excitation of a gas laser by irradiation is a matter of historical interest. In principle, this is the simplest method for achieving inversion in a three- or four-level system. As was explained in connection with solid lasers, the success of such a scheme depends to a large extent on the availability of a radiation source capable of exciting atoms selectively either to a laser-starting level or to another level from which they spontaneously decay to a laser-starting level. The rate of the excitation process must be fast enough to compete with other processes occurring simultaneously. Thus one needs a powerful radiation source matched to the selective requirements of the active system.

It was proposed in the famous paper of Schawlow and Townes [10] in 1958 that gas lasers be excited by irradiation from a spectral lamp using a spectral line of the same gas or a coincident spectral line of another. Considerable amount of speculation and experimentation went into the creation of an alkali vapor laser excited by one of its own lines. No success was ever achieved, and it is very unlikely that such a scheme would ever produce enough inversion to make a laser of reasonable length. Excitation by a co-incident strong line of another element led to success in Cs after a tremendous amount of work, that in retrospect seems disproportionate to the result attained. The matter is interesting largely from the historical point of view because the helium-line-excited cesium laser was in a sense derived theoretically. The details of the relevant transitions were explored in advance, that is, before the cumbersome and difficult construction of a Cs-vapor laser was undertaken.

The history of this laser started in 1930 when Boeckner demonstrated that the fluorescence of cesium can be excited by irradiation with the 3888-Å helium line which coincides almost perfectly with a cesium line. In a series of experiments performed in the Soviet Union in 1957, Butayeva and Fabrikant searched for population inversion of cesium connected with transitions in the visible region. The results were inconclusive. The rather complicated structure of the fluorescence of cesium was studied in detail by Townes and his students at Columbia University [11]. They concluded that conditions were correct for obtaining laser oscillations at 3.20 and 7.18 μm. After making a series of additional measurements and overcoming serious technical difficulties connected with the confinement of the highly reactive Cs vapor at a temperature of 175°C, the irradiation with a powerful He discharge, and the alignment of the optical system, Rabinowitz, Jacobs, and Gould [12] observed laser action at the predicted frequencies. The energy levels and transitions involved in this laser action are shown in Fig. 9.12. The main problem in connection with this laser is not the availability of power from the He lamp. There are a number of processes involved which must be (and were) examined because their rates affect the population inversion on either the $8P^o_{1/2} \rightarrow 6D_{3/2}$ or the $8P^o_{1/2} \rightarrow 8S_{1/2}$ transitions. The transitions from the starting level to the $7S$ and $5D$ levels may not go too fast, and the drain-

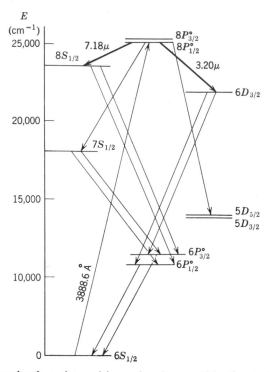

FIG. 9.12 Energy levels and transitions of cesium participating in the fluorescence cycle of the optically excited lasers.

ing of the terminal levels must proceed fast enough to drain these levels. Fortunately all these conditions are met and the Cs laser works.

Several other schemes based on the coincidence of spectral lines were examined in the early 1960's and were eventually abandoned. A concise description of these schemes is found in Patels' review article [3].

9.7 SPECTROSCOPY OF THE NOBLE GASES

The gases helium, neon, argon, krypton, and xenon are the favored materials in gas laser technology. Not only are they used as primary, active materials whose transitions produce coherent light, but they are frequently incorporated in lasers as auxiliary materials and as such participate in the excitation cycle. Although it is possible to construct and operate a laser with little knowledge of spectroscopy, it is not possible to understand the processes that take place within the laser unless one is oriented concerning the energy level structure of the elements involved. The purpose of this section is to provide an orientation concerning the noble gases, primarily concerning helium and neon and to introduce the nomenclatures of their energy levels.

The complete He atom has two electrons. Its energy level structure is illustrated in Fig. 9.13. The total angular momentum $\mathbf{J}$, the total orbital angular momentum $\mathbf{L}$, and the total spin angular momentum $\mathbf{S}$ are constants of the motion. The ground state of He is a $1s^2\ {}^1S_0$ state. Excita-

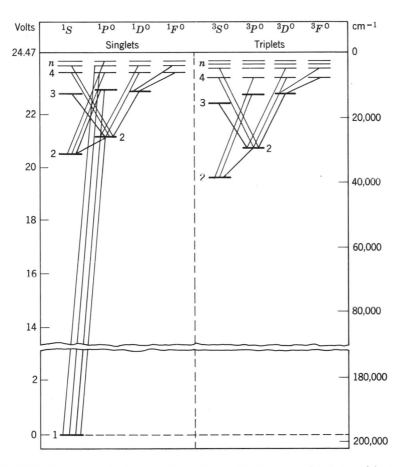

FIG. 9.13 Energy-level diagram for helium. Typical permitted transitions are shown by slant lines. The scale on the left is in electron volts measured from the ground state; that on the right is in wave numbers measured from the ionization level.

tion of one of the electrons results in terms of the type:

$$1s2s, \ 1s2p, \ 1s3s, \ 1s3p, \ 1s3d, \ \ldots$$

When the spins of the electrons are aligned opposite to each other, the states are singlets: $S = 0$. When they are aligned parallel, the states are triplets: $S = 1$. Electric dipole transitions between singlet and triplet levels are forbidden, therefore it it practical to display the singlet and triplet levels separately. This is done in Fig. 9.13, which also indicates the transitions among the lower levels that are permitted by the selection rules. There is no downward escape possible from the $2s^1S$ and $2s^3S$ states; these states are metastable. Accumulation of He atoms in these metastable states is an important factor in the mechanisms of several lasers. Although the classification and naming of He levels is quite simple, the situation is more confusing in the case of the other noble gases. Their spectra were explored and many of their energy levels were given names before the advent of quantum theory, or at any rate, before quantum theory was successfully applied to complex atoms. These old names are known as *Paschen symbols;* they are still preferred by many authors for their relative simplicity and because the authors are used to them. These symbols are just names, even though they often look like quantum mechanical symbols. Systematic description of the energy levels of rare gases is accomplished by means of the more modern *Racah symbols.* They indicate the constants of motion that identify the stationary energy levels.

The common characteristic of the electronic structure of the noble gases neon, argon, krypton, and xenon is that the highest p-shells are filled, and there are no electrons outside of these shells when the atom is in its ground state. Since the p-shells hold six electrons, the symbols of the electron configurations of these elements end with $2p^6$, $3p^6$, $4p^6$, and $5p^6$, respectively. The total angular momentum $\mathbf{J}$, the orbital angular momentum $\mathbf{L}$, and the spin angular momentum of such a closed shell configuration are all zero. When excitation takes place, one of the electrons moves out of this closed shell, leaving a core of five p-electrons behind. Thus, in the case of Ne, excited configurations of the type $2p^53s$, $2p^53p$, $2p^53d$, $2p^54s$, etc., arise.*

We shall be concerned with excited states of neon and the heavier noble gases in which only one electron is removed from the otherwise complete p-shell. The five-electron core has an orbital angular momentum and a spin angular momentum equal and opposite to the one that the missing electron had before its removal. This means that the orbital

---

* The symbols of the completed shells $1s^22s^2$, which logically precede the $2p^5$ symbol, have been omitted.

angular momentum quantum number for the core is $L = 1$ and the spin quantum number is $S = \frac{1}{2}$. These angular momenta combine either parallel or antiparallel, giving a total core angular momentum quantum number $J_c = \frac{3}{2}$ or $\frac{1}{2}$.

The outer electron of an excited noble gas atom is not coupled to the core electrons according to the rules of the $L - S$-, or Russell-Saunders, coupling discussed in Section 1.6. The interactions are such that in good approximation the orbital angular momentum $\mathbf{I}$ of the outer electron is coupled to the total angular momentum $\mathbf{J}_c$ of the core.* The resultant vector $\mathbf{K} = \mathbf{J}_c + \mathbf{I}$ is then coupled to the spin of the outer electron to give a total angular momentum, whose absolute value is $K \pm \frac{1}{2}$. Such coupling is called *pair coupling*. The terms of an atom in which pair coupling prevails are designated by Racah's symbols, which consist of the symbol of the outer electron configuration followed by $[K]$. These symbols are exemplified by the first excited states of neon: $3s[\frac{3}{2}]$ and $3s'[\frac{1}{2}]$. They are arrived at as follows: No prime is used over the letter symbol when the core orbital angular momentum and the spin are parallel, that is, when $J_c - \frac{3}{2}$. When they are opposite, $J_c = \frac{1}{2}$, a prime is used. Since the angular momentum of the $3s$ electron is 0, we must have $K = \frac{3}{2}$ in the first case and $K = \frac{1}{2}$ in the second. The total angular momentum $\mathbf{J}$ of the atom is obtained by adding (vectorially) the spin of the $3s$ electron to $\mathbf{K}$. Thus, for $3s[\frac{3}{2}]$ the possibilities are $J = 2$ or 1; for $K = \frac{1}{2}$, on the other hand, $J = 1$ or 0. These values of J are frequently written as subscripts; thus the symbols of the four lowest excited states of neon are $3s[\frac{3}{2}]_2$, $3s[\frac{3}{2}]_1$, $3s'[\frac{1}{2}]_0$, and $3s'[\frac{1}{2}]_1$. The reader can convince himself that the $3p[K]$ configurations ($J_c = \frac{3}{2}$) of Ne are possible with $K = \frac{1}{2}$, $\frac{3}{2}$, and $\frac{5}{2}$, whereas the $3p'[K]$ configurations ($J_c = \frac{1}{2}$) permit only $K = \frac{1}{2}$ and $\frac{3}{2}$. Since the spin of the outer electron permits two orientations for each orbit, we obtain a total of ten states of the $3p$ type.

Racah's notation is admittedly complicated, but it describes the physical situation, or at least, it provides a model. Unfortunately, the most commonly used notation, at least for Ne, is the Paschen notation, which is simply a system of shorthand symbols. Although the letters s, p, and d are used, we cannot safely infer that a Paschen symbol with the letter s always refers to an outer electron in an s orbit. Paschen's symbols must be treated as arbitrary names given to levels. To find out what quantum-mechanical model represents the level, we must consult a list of these symbols which relates these to the model. Moore's atomic-energy-level tables [3] contain the necessary information.

For the sake of orientation, we include a chart of the lowest excited

---

* Bold face letters are used to distinguish a vector $(\mathbf{I})$ from its absolute value (l).

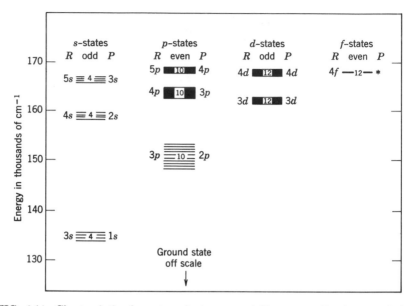

FIG. 9.14 Chart of the lowest excited states of Ne atoms. Paschen symbols on right, electron configuration of the Racah symbols on left, number of terms in center. (* indicates that some terms in this group have special Paschen symbols.)

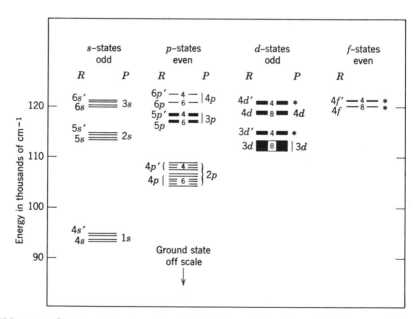

FIG. 9.15 Chart of the lowest excited states of Ar atoms. Paschen symbols on right, electron configuration of the Racah symbols on left, number of states in center. (* indicates miscellaneous or improvised symbols.)

levels of Ne (Fig. 9.14) and a similar chart for Ar, which is representative of the heavier noble gases (Fig. 9.15). The change in the angular momentum of the core from $J_c = \frac{1}{2}$ to $\frac{3}{2}$ affects the energy of the neon levels by only relatively small amounts. For this reason, the levels which differ by only the value of J_c are not shown separately on the chart of neon, and the primes of the Racah symbols are also omitted. The separation of the energy levels with differing values of J_c progressively increases with increasing atomic number. It is already conspicuous in the case of argon, and the grouping of the levels according to the value of J_c is shown on the argon chart.

Each column of the charts contains infinitely many energy levels; their Racah configuration numbers are ns, ns', np, np', etc. All unprimed sequences tend to the same limit, the ionization energy required to produce a singly ionized noble gas atom in the $^2P^o_{3/2}$ state; all primed sequences tend to another limit, which represents the ionization energy required to produce an ion in the $^2P^o_{1/2}$ state.* The latter energy is greater (in accordance with Hund's rules), and the energy difference increases with increasing atomic number.

Table 9.1, below and on page 300, contains the numerical values of 44 levels of Ne including the groups $3s$, $4s$, $5s$, $3p$, $4p$, and $3d$ (Racah). Transitions among these levels account for the majority of observed Ne laser lines. The ground level is assigned the energy 0. Frequently, tables are constructed from the ionization level E_∞ down. This system is illustrated by the scale of Fig. 9.13 on the right margin of the figure. The entries of such tables are $E_\infty - E_n$; they are called term values. For Ne, the value of E_∞ is 173,931.7 cm^{-1}. Neon levels included in our short table represent but a fraction of the Ne levels contained in Moore's tables [13]. It is interesting to note that the lowest excited level of Ne is over 134,000 cm^{-1} above ground level and to compare this value with the first excited levels of ions which play a role in solid-state lasers.

* The superscript o indicates a term of *odd* parity.

TABLE 9.1
Energy Levels of Neon (Ne I)
(From Moore's Atomic Energy Levels)

Paschen	Racah	J	Energy Level (cm^{-1})
Ground	$2p^{6\,1}S$	0	0
$1s_5$	$3s[\frac{3}{2}]^o$	2	134 043.8
$1s_4$	$3s[\frac{3}{2}]^o$	1	134 461.2
$1s_3$	$3s'[\frac{1}{2}]^o$	0	134 820.6

TABLE 9.1 (*Continued*)

Paschen	Racah	J	Energy Level (cm^{-1})
$1s_2$	$3s'[\frac{1}{2}]^\circ$	1	135 890.7
$2p_{10}$	$3p[\frac{1}{2}]$	1	148 259.7
$2p_9$	$3p[\frac{5}{2}]$	3	149 659.0
$2p_8$	$3p[\frac{5}{2}]$	2	149 826.2
$2p_7$	$3p[\frac{3}{2}]$	1	150 123.6
$2p_6$	$3p[\frac{3}{2}]$	2	150 317.8
$2p_5$	$3p'[\frac{3}{2}]$	1	150 774.1
$2p_4$	$3p'[\frac{3}{2}]$	2	150 860.5
$2p_3$	$3p'[\frac{1}{2}]$	0	150 919.4
$2p_2$	$3p'[\frac{1}{2}]$	1	151 040.4
$2p_1$	$3p'[\frac{1}{2}]$	0	152 972.7
$2s_5$	$4s[\frac{3}{2}]^\circ$	2	158 603.1
$2s_4$	$4s[\frac{3}{2}]^\circ$	1	158 798.0
$2s_3$	$4s'[\frac{1}{2}]^\circ$	0	159 381.9
$2s_2$	$4s'[\frac{1}{2}]^\circ$	1	159 536.6
$3d_6$	$3d[\frac{1}{2}]^\circ$	0	161 511.6
$3d_5$	$3d[\frac{1}{2}]^\circ$	1	161 526.1
$3d_4'$	$3d[\frac{7}{2}]^\circ$	4	161 592.3
$3d_4$	$3d[\frac{7}{2}]^\circ$	3	161 594.1
$3d_3$	$3d[\frac{3}{2}]^\circ$	2	161 609.2
$3d_2$	$3d[\frac{3}{2}]^\circ$	1	161 638.6
$3d_1''$	$3d[\frac{5}{2}]^\circ$	2	161 701.6
$3d_1'$	$3d[\frac{5}{2}]^\circ$	3	161 703.4
$3s_1''''$	$3d'[\frac{5}{2}]^\circ$	2	162 410.6
$3s_1'''$	$3d'[\frac{5}{2}]^\circ$	3	162 412.1
$3s_1''$	$3d'[\frac{3}{2}]^\circ$	2	162 421.9
$3s_1'$	$3d'[\frac{3}{2}]^\circ$	1	162 437.6
$3p_{10}$	$4p[\frac{1}{2}]$	1	162 519.9
$3p_9$	$4p[\frac{5}{2}]$	3	162 832.7
$3p_8$	$4p[\frac{5}{2}]$	2	162 901.1
$3p_7$	$4p[\frac{3}{2}]$	1	163 014.6
$3p_6$	$4p[\frac{3}{2}]$	2	163 040.3
$3p_3$	$4p[\frac{1}{2}]$	0	163 403.3
$3p_5$	$4p'[\frac{3}{2}]$	1	163 659.2
$3p_2$	$4p'[\frac{1}{2}]$	1	163 709.7
$3p_4$	$4p'[\frac{3}{2}]$	2	163 710.6
$3p_1$	$4p'[\frac{1}{2}]$	0	164 287.9
$3s_5$	$5s[\frac{3}{2}]^\circ$	2	165 830.1
$3s_4$	$5s[\frac{3}{2}]^\circ$	1	165 914.8
$3s_3$	$5s'[\frac{1}{2}]^\circ$	0	166 608.3
$3s_2$	$5s'[\frac{1}{2}]^\circ$	1	166 658.5

Some conclusions can be drawn about the probabilities of radiative transitions between the levels listed in the tables from the selection rules and the quantum numbers listed in the tables. The actual calculation of the transition rates of permitted transitions is a very complicated task. It requires an approximate knowledge of the wave functions characterizing the states. The calculation of the transition rates, line strengths, and related properties of noble gases is described in the literature [14, 15]. Closely related to the transition rates are the decay rates of the states. These rates can be determined experimentally under conditions similar to those found in the laser. Although they represent engineering-type data, they have a certain physical meaning. The decay rate of any one state represents the sum of transition rates, radiative and otherwise, to all lower levels [16].

The spectrum of an ion is quite different from the spectrum of the neutral atom from which the ion derives. Similarities exist however among the spectra of atoms and ions which contain the same number of electrons; such structures are called *isoelectronic*. Oxygen atom, singly ionized fluorine, and doubly ionized neon, for example, form an *isoelectronic sequence*. It is customary in spectroscopy to distinguish between the spectra of an atom and its various ions in the following manner: The spectrum of the neutral atom is provided with a Roman numeral I, that of the singly ionized atom by II, and so on. Thus we have so far discussed the energy-level structures of Ne I, Ar I, etc.

The principal features of the energy-level structure of singly ionized noble gases are illustrated in the example of Ne II. The ground state configuration of Ne II is $1s^2 2s^2 2p^5$. Excited configurations are formed by promoting one of the $2p$ electrons to a higher orbit, say $3s$, $3p$, $3d$, etc., or, exceptionally, by promoting one of the $2s$ electrons to obtain the configuration $1s^2 2s 2p^6$. Leaving aside the exceptional situation, Ne II will consists of a core with a $1s^2 2s^2 2p^4$ configuration and an external electron which may be located in a variety of orbits with a principal quantum number at least 3. In the case of Ar II, the core has a $1s^2 2s^2 2p^6 3s^2 3p^4$ configuration and the external electron may be in either a $3d$ orbit or in any orbit whose principal quantum number is at least 4. The halogens are isoelectronic with the singly ionized noble gases; therefore, what was said about Ne II and Ar II also applies to F I and Cl I respectively.

In these atoms two electrons are missing from the core so that the spin of the core is 0 or 1. When this spin is combined with the spin of the outer electron, the resultant spin coordinate S has the value $\frac{1}{2}$ or $\frac{3}{2}$. The energy-level scheme will therefore consist of doublets $S = \frac{1}{2}$ and of quartets $S = \frac{3}{2}$. (Transitions between terms of different spin are not completely absent because the Russell-Saunders scheme is not completely applicable.)

Removal of one of the p electrons may leave the core itself in several different states, which are either singlets or triplets. Application of Pauli's principle leads to the conclusion that the possible terms are ^3P, ^1D, and ^1S. The lowest energy is associated with the ^3P core. A term of Ne II, with a lowest-energy core and with the outside electron in the lowest s orbit, is $3s\ ^4P$. Another, with spin of the outside electron reversed, is $3s\ ^2P$. The first term permits the J values $\frac{5}{2}$, $\frac{3}{2}$, and $\frac{1}{2}$, the second only $\frac{3}{2}$ and $\frac{1}{2}$. When the core is in its next lowest state (^1D) the electronic symbol is provided with a prime; thus a term designated by $3s'\ ^2D$ may be obtained. Double primes indicate a ^1S core.

In attempting to organize the energy levels of atoms, we must keep in mind that the Russell-Saunders coupling scheme, as well as other coupling schemes, represents only approximations which do not have universal validity. With increasing atomic number, forces responsible for Russell-Saunders coupling decrease in comparison with other forces. Even within one element the ratios of these forces are different for electrons that pass near the nucleus and for those whose orbits are farther removed. For heavier elements and for large quantum numbers n and l of the outer electron, the coupling approaches pair coupling.

Spectral lines, whether derived from spontaneous or from stimulated emission, can be given a transition assignment among the terms tabulated in the atomic energy tables insofar as the tables are complete. With the extension of stimulated emission spectroscopy into the infrared region, lines are occasionally observed which result from transitions among levels not yet tabulated.

9.8   HELIUM—NEON LASERS

The neon lasers whose operation is based on resonant transfer of excitation from helium are the best known gas lasers. They are the first type discovered and they are the lasers most widely sold and used.

Helium—neon lasers operate in three distinct spectral ranges: in the red at 6328 Å, in the near-infrared around 1.15 μm, and further in the infrared at 3.39 μm. The origin and interrelation of these laser lines is explained with the aid of the partial energy-level diagram in Fig. 9.16. Several groups of energy levels of neon are shown in this diagram together with two metastable helium levels: 2^1S and 2^3S. The groups of neon levels are indicated by their Paschen symbols; the levels themselves can be found in Table 9.1. The closeness of the $2s$ and $3s$ groups to He levels is apparent from the figure. Quantitatively the situation is as follows: The 2^3S helium level is 159,850 cm^{-1} above the ground;

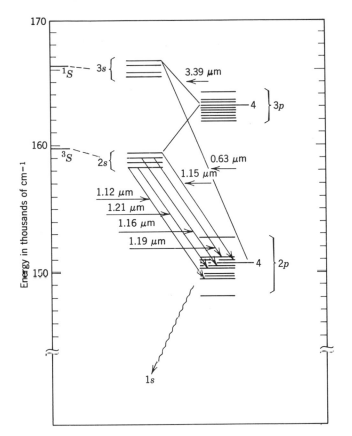

FIG. 9.16 Helium—neon energy levels and strongest laser transitions.

the 2s group of neon levels extends from 158,600 to 159,540 cm⁻¹. The 2¹S helium level is at 166,272 cm⁻¹, whereas the 3s neon group ranges from 165,830 to 166,660 cm⁻¹. The principal laser lines that arise from population inversion caused by excitation transfer from helium to neon are also shown in Fig. 9.16. Their wavelengths and wavenumbers are listed in Table 9.2, with the transitions indicated by both the Paschen and the Racah symbols.

Helium—neon lasers were first discovered in 1960. Javan, Bennett, and Herriott [17] observed stimulated emission on five nearby lines in the near-infrared region. These are lines 2 to 6 in Table 9.2. The strongest oscillation occurs in line 3 at 1.1523 μm. The radiation arises from stimulated transitions from the 2s group of levels whose population is enhanced by transfer of excitation from 2^3S levels of He. For some time after their discovery these lasers were the only gas lasers in existence. There are 30 transitions permitted by the selection rules between the four $2s$ and the ten $2p$ levels. All but eight of these were subsequently observed in stimulated emission, but most of them can only be seen when the five easily-oscillating lines are suppressed. In wavelength, these $2s \rightarrow 2p$ lines range from 0.89 to 1.72 μm.

TABLE 9.2
SELECTED LASER LINES OF NEON

Line Number	λ_{air}, (μm)	σ, (cm^{-1})	Transition	
			Paschen	Racah
1	0.6328	15798.0	$3s_2$–$2p_4$	$5s'[\frac{1}{2}]_1^o$–$3p'[\frac{3}{2}]_2$
2	1.1177	8944.07	$2s_5$–$2p_9$	$4s[\frac{3}{2}]_2^o$–$3p[\frac{3}{2}]_3$
3	1.1523	8676.10	$2s_2$–$2p_4$	$4s'[\frac{1}{2}]_1^o$–$3p'[\frac{3}{2}]_2$
4	1.1614	8607.87	$2s_3$–$2p_5$	$4s'[\frac{1}{2}]_0^o$–$3p'[\frac{3}{2}]_1$
5	1.1985	8341.53	$2s_3$–$2p_2$	$4s'[\frac{1}{2}]_1^o$–$3p'[\frac{1}{2}]_1$
6	1.2066	8285.25	$2s_5$–$2p_6$	$4s[\frac{3}{2}]_2^o$–$3p[\frac{3}{2}]_2$
7	1.5231	6563.87	$2s_2$–$2p_1$	$4s'[\frac{1}{2}]_1^o$–$3p'[\frac{1}{2}]_0$
8	3.3913	2947.90	$3s_2$–$3p_4$	$5s'[\frac{1}{2}]_1^o$–$4p'[\frac{3}{2}]_2$

White and Rigden [18] discovered the visible He—Ne laser (line 1) in 1962. The radiation arises from the $3s_2 \rightarrow 2p_4$ transition; the population of the upper level is enhanced by the transfer of excitation from the 2^1S state of He. This laser is the most convenient one to use for demonstration and alignment purposes; many companies market such lasers for general use.

It is somewhat inaccurate to refer to the 6328 Å laser as *the* visible Ne laser since laser oscillations can be produced in other visible Ne lines. Once the $3s_2$ level of Ne is heavily populated, one can force laser transitions into a number of the $2p$ levels other than $2p_4$. The gain in these transitions is less than that in the $3s_2 \rightarrow 2p_4$ transition; therefore longer discharge tubes are used, with dispersive prisms incorporated into the path of the beam within the mirrors to separate the desired wavelength and enhance its oscillations, suppressing all others [19]. These unusual visible Ne laser lines extend in wavelength from 0.59 to 0.73 μm.

Shortly after the discovery of the visible He—Ne laser, Bloom, Bell, and Rempel observed that infrared radiation of over 3-μm wavelength frequently accompanies the emission of the visible line [20]. The emission of this infrared radiation interferes with the operation of the visible laser.

The transition responsible for the infrared radiation originates at the $3s_2$ level, the starting level of the visible radiation. It terminates at the $3p_4$ level, yielding a radiation of 3.3913 μm in wavelength (line 8). The operation of this 3.39-μm laser not only depletes the $3s_2$ level—to the detriment of the population inversion required for the visible laser—it overpopulates the $3p_4$ level and thereby offers the possibility of enhancing laser action originating at the latter level. Such multiple, or cascade, lasers were actually observed with several alternative branches from the initial $3s$ group to the terminal $2p$ group. In order to stabilize He—Ne lasers designed for operation at 6328 Å it is customary to inhibit oscillations at 3.39 μm by means of auxiliary devices. This will be discussed later in some detail.

It is very easy to attribute to He a role that it does not entirely deserve, and this was frequently done in the early literature. It should be noted that several He—Ne lines are obtainable in pure neon, and many Ne lines can be produced in stimulated emission whose upper level is not connected with an excited state of He. Therefore we ought to conclude that the transfer of excitation from He substantially enhances population inversion in certain levels of Ne without being the exclusive cause of such inversion.

Helium—neon lasers are built for a variety of purposes and their designs vary accordingly. Commercial lasers are usually packaged with their power supplies. They are designed to be safe and portable. Laboratory lasers may look like the structure in Fig. 9.17, that clearly shows the dangerous high voltage leads and the precision adjustments for the orientation of the mirrors.

Considerable work has been done to determine experimentally the optimal characteristics of various He—Ne lasers and their radiative outputs as functions of the design parameters. The reader interested in such matter may find a wealth of information in books and review articles devoted entirely to gas lasers [5, 21, 22] and in specialized articles dealing with design optimization problems [23—26]. The best operation of the 6328-Å and the 3.39-μm lasers is attained for a 5:1 ratio of He pressure to Ne pressure. For optimum laser gain, the product of the total gas pressure p and the tube diameter D should lie between 2.9 and 3.6 torr mm. The tube diameter is usually chosen between 1 and 10 mm. In this range, the maximum gain per unit length is inversely

FIG. 9.17 A short laboratory-type He—Ne laser. (Hughes Research Laboratories.)

proportional to the tube diameter. It is to be noted that the discharge current is not considered a constant. While determining the maximum gain for a laser tube of a given diameter, the current is adjusted to optimize the gain. Another matter of importance is that the conditions for maximal gain are not the same as those for maximal power output. The power output is approximately proportional to gain times the volume of the laser. For the near-infrared neon lasers the optimal gas composition is 1 torr He and 0.1 torr Ne. Tube diameters of 5 to 8 mm are most suitable for these lasers, whose excitation is derived from the 2^3S helium levels. The neon lasers discussed earlier are excited from the 2^1S helium levels.

The greatest practical interest is in the visible neon laser. It was already noted that the $3s_2$ neon level is the common starting level of the 6328-Å and of the 3.39-μm lasers. The $3s_2 \rightarrow 3p_4$ transitions that result in the emission of the 3.39-μm line deplete the population of the $3s_2$ level, thus decreasing the gain available for laser operation in the 6328-Å line.

The ratio of the wavelengths of these lines is 5.36. When the same number of transitions take place from $3s_2$ to $3p_4$ and from $3s_2$ to $2p_4$

levels, the energy of the visible radiation is 5.36 times greater than that of the infrared radiation, thanks to the $h\nu$ law. But this is not the essential point! The visible laser is at a serious disadvantage as far as amplification or gain per unit length is concerned. In a moderately long laser the advantage of the infrared line is compounded exponentially so that, in the absence of inhibiting factors, the infrared radiation takes over because its density builds up much faster than that of the visible radiation. It was shown in Sections 1.3 and 1.4 that, other things being equal, the peak amplification, or gain per unit length, is inversely proportional to the linewidth. In a gas laser, the linewidth is determined largely by the Doppler effect, which produces a broadening in frequency proportional to ν [see (4.4) in Chapter 1]. Measurements of the power outputs of short lasers operating both in the visible and in the infrared region at 3.39 μm indicate that the integrated amplification over these two lines is nearly equal. Then the ratio of the peak amplifications is

$$\frac{\alpha_1}{\alpha_2} = \frac{\Delta\nu_2}{\Delta\nu_1} = \frac{\nu_2}{\nu_1} = \frac{\lambda_1}{\lambda_2}.$$

This ratio is 5.4:1 in favor of the infrared line. Because of this advantage, lasers intended to operate at 6328 Å include devices for the suppression of the 3.39-μm radiation.

The less sophisticated suppression devices reduce the proportion of infrared radiation returned by the mirrors to the laser. The reduction may be accomplished by inserting an absorption cell between the laser windows and the mirrors and filling it with a substance transparent in the visible, but highly absorbent in the infrared region. Methane at atmospheric pressure is suitable for this purpose. It is used in laboratory-type He—Ne lasers, but not in commercial equipment because of the difficulty of maintaining the methane gas in a sealed enclosure over a long period of time. Commercial lasers frequently contain a dispersive prism between one of the mirrors and a window in the manner illustrated in Fig. 9.18. The mirror is adjusted so as to return only the visible

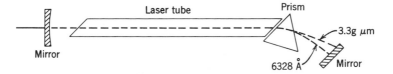

FIG. 9.18 Suppression of the infrared radiation by means of a dispersive prism. The divergence of the rays is exaggerated.

ray in the proper direction. Both methods have the disadvantage of not preventing the growth of the 3.39-μm radiation within the laser even though they prevent oscillations. When the tube is long, the 3.39-μm radiation may grow from noise to a significant level in a single passage and may therefore deplete the excitation of Ne near both ends of the tube.

A more elegant method destroys the intrinsic advantage of the 3.39-μm line by broadening its linewidth to match that of the 6328-Å line. This is accomplished by creating an inhomogeneous magnetic field in the plasma by means of small ceramic magnets placed around the tube. The magnetic field produces a Zeeman splitting of the lines that is independent of the frequency. In an inhomogenous field, the line is "smeared out." The smearing is the same for the visible– and for the infrared line. When properly handled, this technique does not broaden the visible line appreciably, but it widens the originally narrow infrared line. This method is frequently used in commercial He—Ne lasers.

The visible He—Ne laser is widely used as a tool for alignment and as a source of coherent radiation in holography. In these applications the extreme sharpness of the spectrum and the frequency stability of the laser are not of great concern. The laser is generally operated at a high level of excitation to obtain a sizeable output. Under these conditions, several axial modes are simultaneously oscillating, and the spectrum of the laser has the comblike structure shown in Fig. 3.6. Sometimes off-axis modes are also present and contribute additional frequencies. In interferometric applications of the laser, monochromaticity and extreme frequency stability are very important. For such a purpose, He—Ne lasers are operated in a single mode and are provided with special devices to keep their frequency constant in the presence of environmental fluctuations. These frequency stabilization techniques are discussed in Section 9.15.

9.9 A SURVEY OF ATOMIC NOBLE GAS LASERS

Ordinary (Steady) Lasers. The helium—neon lasers described in Section 9.8 are the best-known and the practically most important representatives of a large group of lasers whose working elements are complete atoms of noble gases. Stimulated emission has been demonstrated in such gases over several hundred spectral line, extending from 0.6 μm to beyond 130 μm.

The vast majority of these atomic noble gas lasers is similar in appearance and operation to the He—Ne lasers described in Section 9.8. The gases are confined in tubes of the type shown in Fig. 9.7 or 9.8. Most such lasers can be operated with tubes of a few millimeter in diameter and about 2 m in length. Longer tubes are necessary to operate in the weaker spectral lines, and occasionally stronger lines must be suppressed to obtain oscillations in the weaker ones. The favorable pressure range is between 0.01 and 1.0 torr, varying somewhat from gas to gas and from line to line.

Excitation is accomplished by means of a steady glow discharge which may be maintained by alternating or direct current. The optimum discharge current varies between 5 and 75 mA. Gain and laser output reach relatively flat maxima as functions of the discharge current within the operating range.

The mechanism by which population inversion is established in the discharge varies from one laser to another. A few noble gas lasers involve the transfer of excitation from one element to another in the manner of the He—Ne lasers. The basic source of the excitation energy is always the kinetic energy of the electrons in the discharge and this, in turn, is derived from the work done by the impressed electric field. The uncharged atoms of the gas move in random directions, and quite slowly compared to the electrons. The latter establish a state of near thermal equilibrium with one another at a temperature equivalent to a few thousand degrees, while the gas remains at a much lower temperature. The excited levels of the atoms involved are usually populated by excitation in a single electron collision from the ground state or by decay from higher atomic levels. A variety of energy-exchange phenomena take place in the plasma involving the emission and absorption of radiation, and these phenomena proceed at different rates. The result of the competition of these processes is the evolution of a steady, nonequilibrium distribution of atoms among a large number of energy levels. This distribution may contain inversions that can be exploited for laser action.

These ordinary noble gas lasers are operated in a steady condition, that is, with the discharge current kept constant. With a few exceptions, they have a small gain of the order of 0.01 cm^{-1} or less, even under optimum conditions. Their output power is quite small compared to other lasers. The He—Ne lasers are among the most powerful lasers of this type. Their typical output is 1 mW. Most ordinary atomic gas lasers have outputs one or two orders of magnitude less than this value.

Exceptional Lasers. There are atomic noble gas lasers that resemble the ionic lasers in their operating characteristics. Certain far-infrared lasers in He, for example, can only be obtained in a high-current discharge. A number of lines in other noble gases either appear only in a pulsed, high-current discharge or appear in such discharge with a gain and power output orders of magnitude higher than obtainable in a steady glow discharge. The nature of population distribution in these gases is such that, for certain transitions, inversion exists only for a short period of time (microseconds) after the onset of the discharge. The upper laser level fills up faster than the lower level, but eventually the lower level is also filled and laser oscillations cannot continue. This situation usually arises when the lower level drains slowly. (It has a long lifetime.) In an intermediate case some inversion may remain, but its steady state value is much smaller than the maximum reached shortly after the onset of the discharge. In this case, laser action can be achieved in steady operation at a lower level than is possible with pulsed excitation.

Excitation of pulsed lasers is accomplished usually by charging capacitors from a power supply to a potential of tens of kilovolts, then discharging them through the gas in short pulses, with peak currents reaching several hundred amperes. The duration of each light pulse seldom exceeds one microsecond. Occasionally, very high gains are achieved in these pulsed lasers. In fact, the gain may be so high that stimulated emission is observed in a 50-to-100-cm tube *without mirrors*, indicating the buildup of a pulse in a single passage through the tube. Such emission is called *superradiant emission*.

Under exceptional atomic gas laser types we include all high-current types, those that are pulsed as well as those excited by a continuous arc discharge.

Both the ordinary and the exceptional lasers will now be discussed briefly, element by element. A detailed tabulation of all observed laser-lines will not be given here. Such tables, complete to include lines discovered before 1965, are found in another work of the author [27] and also in the review article of Patel [3].

Helium. Laser oscillations may be obtained from this element in two widely separated spectral regions. The near-infrared lasers at 1.9543 μm and at 2.0603 μm (measured in air) are of the ordinary atomic gas laser type. They were discovered early; the first one in 1962 by a research team at Bell Telephone Laboratories [28], the second one in 1963 by a group in France [29]. The exceptional far-infrared lasers of He at 95.8 μm and at 216 μm were discovered in 1969 by Levine and Javan at the Massachusetts Institute of Technology [30]. Both types of He lasers have unique and interesting features. They will be described in detail because their operation is reasonably well understood and because the relatively simple structure of helium is a good vehicle

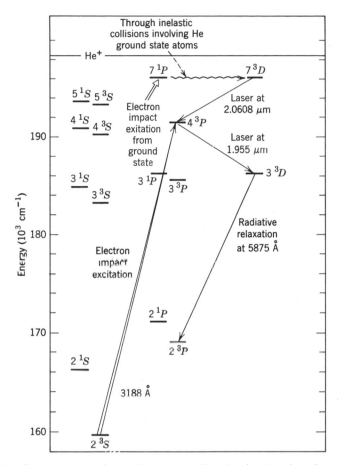

FIG. 9.19 Partial energy-level diagram of He showing levels relevant to laser action in He the near-infrared region.

for an introduction into the procedures and intricacies of the analysis of laser action.

The near-infrared lasers are excitable in a typical glow discharge. The energy levels relevant to their operation are shown in Fig. 9.19. The terminal level of the 2.06-μm laser is the 4^3P level which is also the starting level of the 1.95-μm laser. These two transitions lead to laser oscillations under widely different conditions because the first line is favored when the population of the 4^3P level is low, the second when it is high. The first line is obtained when the gas pressure is high (optimum 8 torr), the second line when it is low (optimum 0.3 torr).

The transition $7^3D \rightarrow 4^3P$ lies high above the ground level of He, and it is surprising that a population inversion should occur between these levels. First, it is necessary to explain how the 7^3D level is populated to a singificant extent, since a direct transition from the ground 1^1S level to this level is forbidden by the selection rules. The primary source of excitation of the 7^3D level is transfer by collision with He atoms excited to the 7^1P level, which is very close in energy and which is connected to the ground state by an electric dipole transition. The reactions that lead to populating the 7^3D level may be written as follows

$$\text{He} + (e + \text{KE}) \rightarrow \text{He} \ (7^1P) + e,$$
$$\text{He} \ (7^1P) + \text{He} \rightarrow \text{He} \ (7^3D) + \text{He}.$$

Here He by itself stands for He in the ground state. The 4^3P state is not fed by such collisions. It receives excitation largely from the 2^3S level, which is metastable, and which is connected to the 4^3P level by a permitted transition. It has been found that the lifetime of the metastable 2^3S level is decreased as the pressure is increased or as impurities are added to the He gas because collisions tend to depopulate that level. It is now clear that the operation of the 2.06-μm laser is aided by the increase of pressure in two ways: The increase of pressure enhances the number of collisions among He atoms, thereby increasing the rate of transfer of excitation from the 7^1P to the 7^3D level. At the same time, the more frequent collisions reduce the supply of atoms, that reach the 4^3P state via the 2^3S state.

The 1.95-μm He laser operates best at 0.3 torr pressure. The presence of other gases as impurities is detrimental in this laser because they tend to destroy the excitation of atoms in the 2^3S metastable state.

The far-infrared laser lines of neon were obtained in a high-current (steady) discharge with the laser power increasing with discharge current up to the maximum available current of 10 A. The continuous (steady) optical output was as high as 1.5 mW in the 216-μm transition and 0.1 mW in the 95.8-μm transition. These lasers were obtained in relatively large structures, the tube having 4 m length and 6 cm inner diameter; gas pressure: 0.1 torr.

The identification of these laser transitions is as follows:

95.8 μm: 3^1P to 3^1D,
216.3 μm: 4^1P to 4^1D.

The frequency separation of these levels is so small that it was not practical to show them in Fig. 9.19, whose frequency scale is adapted to the display of the near-infrared transitions.

An interesting and almost unique aspect of these lasers is the fact that many of their features are predictable from theoretical calculations, starting from first principles in quantum mechanics. (Interestingly, they were not discovered until rather late in the laser game!) The attainment of population inversion in both lasers depends on the trapping of radiation in the gas because the levels 3^1P and 4^1P are connected to the 1^1S_0 ground state by electric dipole radiation. The lifetimes of these states in isolated atoms are very short: 1.7 and 3.9 nsec, respectively. On the other hand the terminal laser states 3^1D and 4^1D are not connected to the ground state by electric dipole radiation, therefore their lifetimes are longer. On the basis of this information alone, quite the opposite of population inversion could be expected. The strong re-absorption of the radiation emitted in the $3^1P \rightarrow 1^1S$ and $4^1P \rightarrow 1^1S$ transitions acts to extend the effective lifetime of these upper levels 73.3 and 116 nsec, respectively. These trapped radiative lifetime now exceed the lifetimes of the terminal D states and population inversion can be established. The calculations of Levine and Javan [30] show that direct excitation through electron impact with ground-state atoms accounted for 90% of the excitation of the P levels; the rest may be attributed to cascade decay processes from higher levels. The nearly linear dependence of the gain on the discharge current observed for moderate current levels is behavior to be expected from direct excitation by electron collision with ground-state atoms.

Neon. Over 150 spectral lines of Ne I were observed in stimulated emission. With the exception of the lasers discussed in Section 9.8, the population inversion of these lasers is not produced by transfer of excitation from atoms of another element. Laser oscillations in pure neon are usually obtained at a gas pressure of 0.2 torr and below. Many transitions that result in laser action produce a very small gain, therefore these neon lasers usually require tubes several meters long. Ordinary Ne lasers extend over the spectral range from 0.59 μm to 133 μm, or 0.13 mm. The origin of most of the lines below 20 μm is in transitions illustrated in Fig. 9.20. This figure is a topological diagram of the lower neon levels. The energy level scale is grossly distorted, as can be seen by comparing this Figure with 9.14, which was drawn to scale. Each box in Fig. 9.20 represents a group of levels with the common Racah symbols indicated.* The solid straight lines connecting the boxes repre-

---

* In analytical work Racah symbols are preferred because they indicate the quantum numbers of the states, while Paschen symbols are only arbitrary names.

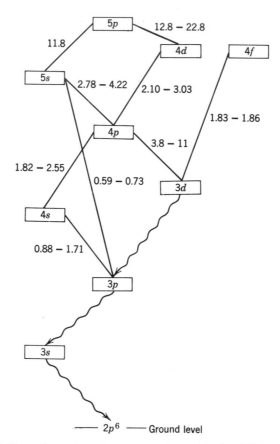

FIG. 9.20 Ordinary laser transitions among lower levels of Ne I. Each box represents groups of energy levels with common Racah symbols as indicated. *Solid lines:* laser transitions; *wavy lines:* transitions not involving laser action. The numbers adjacent to transitions indicate their wavelength range in μm. (The energy differences are not drawn to scale.)

sent groups of observed laser lines with their wavelength range indicated. The condensed nature of this diagram is illustrated by the fact that the line connecting the 5s group with the 3p group stands for all the laser transitions originating at one of the 5s Racah levels (3s in Paschen notation) and terminating at one of the 3p Racah levels (2p in Paschen notation). It so happens that only eight such lines were observed. All their transitions start at the $5s'[\tfrac{1}{2}]°$ Racah level ($3s_2$ in Paschen notation). Their wavelengths and their terminal levels are listed below.

Line	Wavelength in air (μm)	Terminal level (Racah)	Paschen
1	0.5939	$3p[\frac{5}{2}]_2$	$2p_8$
2	0.6046	$3p[\frac{3}{2}]_1$	$2p_7$
3	0.6118	$3p[\frac{3}{2}]_2$	$2p_6$
4	0.6294	$3p'[\frac{3}{2}]_1$	$2p_5$
5	0.6328	$3p'[\frac{3}{2}]_2$	$2p_4$
6	0.6352	$3p[\frac{1}{2}]_0$	$2p_3$
7	0.6401	$3p'[\frac{1}{2}]_1$	$2p_2$
8	0.7305	$3p'[\frac{1}{2}]_0$	$2p_1$

The solid line connecting the 4s group with the 3p group is representative of 22 laser lines that have been observed on transitions between the four 4s levels and the ten levels in the 3p group. The wavelength range of these lasers is from 8865 Å to 1.7162 μm.

The wavy lines in Fig. 9.20 indicate paths of return (de-excitation) by processes other than stimulated emission. Such processes are only indicated in the absence of known laser transitions.

The details of the mechanisms by which certain groups of levels are preferentially populated were the subject of many investigations [31—33], but they are not entirely clarified. When He is present in the discharge, the excitation of the 5s and 4s levels (3s and 2s Paschen!) is obtained largely by transfer in collisions, as described in Section 9.8. Nevertheless lasers originating at 5s levels were also observed in pure neon, therefore these levels may be populated by other excitation processes. The situation concerning the 4s levels is not clear. Some levels, that are terminal levels of one laser, are starting levels of another. Several such cascade lasers are known in which the operation of one laser enhances or inhibits the operation of another [34—37].

Ordinary neon laser lines cover the 1.0- to 3.4-μm range rather densely. Over 50 lines are known in this region. Many lines are also found between 7 and 8 μm, and the list of lines extends far beyond the region plotted in Fig. 9.20.

Fairly complete tables of the ordinary laser lines of noble gases may be found in many review articles and general texts [3, 27]. There are, however, many interesting noble gas lasers of the exceptional type which have been discovered later than the ordinary types and which have not yet been assimilated in the textbook literature. Some attention to these may now be in order. These exceptional lasers include lines that cannot be produced at all in a steady discharge as well as lines that can be so obtained with a moderate gain and output, but whose gain in pulse excitation becomes so high that superradiant emission results in a relatively short (1 m) tube. An exceptional laser of the first type is one whose terminal level is the lowest excited level of a noble gas. The lowest excited group of levels in Ne is the $3s$ group ($1s$ Paschen). It contains four levels which lie between 134,000 and 136,000 cm^{-1} above the ground level of Ne (see Table 9.1). Three of these levels are connected to the ground level by electric dipole transitions. Atoms in these states return spontaneously to ground level with a lifetime of a few nanoseconds and emit ultraviolet radiation around 750 Å. Transition between the last, and the lowest, of the $3s$ levels and the ground level is forbidden by the selection rules because it involves a change $\Delta J = 2$ in the total angular momentum. Therefore this state, $3s[\frac{3}{2}]_2^o$, is a metastable state.

When a Ne laser is put into operation, the $3s$ levels fill up rather quickly and drain rather *slowly* to the ground, even though the lifetimes of *isolated* atoms in three of these levels are short. The reason for the glut in the $3s$ group of levels is the trapping of the ultraviolet radiation in the discharge. This trapping prolongs the effective lifetime of the three levels connected to the ground state by more than one order of magnitude. This prolongation of the lifetime of $3s$ levels not only prevents laser action under ordinary circumstances, with these levels acting as terminal laser levels, but it inhibits laser action terminating at the $3p$ levels as well, because the latter levels must drain through the $3s$ levels. The main reason for using laser tubes of small diameter is to reduce the lifetimes of the $3s$ levels. At the walls the excitation of these levels may be lost by means of collisions and also by the escape of the ultraviolet radiation to the wall of the discharge tube.

Before the electric discharge is turned on essentially all Ne atoms are at the ground level. Then, with the discharge turned on, different levels fill up at different rates. The three $3s$ levels directly connected to the ground level will fill up rather fast, certainly faster than the $3p$ levels above them. The singular $3s[\frac{3}{2}]_2^o$ ($1s_5$ Paschen) level *will not fill up so fast* because, as we already noted, the probability of excitation by electron collision is in the first approximation proportional to the electric dipole transition probability. Atoms will be raised into this level by electron collisions, but not as fast as into their sister levels. The $3s[\frac{3}{2}]_2^o$ level will also be populated by exchange in collision with the other $3s$ levels which are close in energy. In either case the population of this level will grow at the slowest rate of the $3s$ levels. It is possible to establish a population inversion with this level as the terminal level, but only by acting very quickly, and the inversion will last only for a very short time.

The following Ne laser lines were produced with pulse excitation [38–40]*:

Wavelength	Transition Racah	Paschen
5944.8 Å	$3p'[\tfrac{3}{2}]_2-3s[\tfrac{3}{2}]_2^\circ$	$2p_4-1s_5$
6143.1 Å	$3p[\tfrac{3}{2}]_2-3s[\tfrac{3}{2}]_2^\circ$	$2p_6-1s_5$.

These exceptional laser lines only appear in pulsed discharges of very high peak-current density (1000 A/cm^2) and short risetime (25 nsec). The light pulses are self-terminating with lifetimes of 2 to 5 nsec. The gain is so large that stimulated emission may be observed in a 70 cm tube without mirrors. A peak power of 40 W was measured at 6143 Å, and 15 W at 5945 Å [40].

Some lines of Javan's He—Ne laser, the $2s$—$2p$ series, become super-radiant in pure Ne when very energetic pulse excitation is used. Andrade, Gallardo, and Bockasten [41] obtained such emission in Ne at 50, mtorr pressure with short direct current pulses of 500-A peak discharge current in the following lines

Wavelength	Transition Racah	Paschen
1.1143 μm	$4s[\tfrac{3}{2}]_1^\circ-3p[\tfrac{5}{2}]_2$	$2s_4-2p_8$
1.1523 μm	$4s'[\tfrac{1}{2}]_1^\circ-3p'[\tfrac{3}{2}]_2$	$2s_2-2p_4$
1.1767 μm	$4s'[\tfrac{1}{2}]_1^\circ-3p'[\tfrac{1}{2}]_1$	$2s_2-2p_2$.

These lines have long been known as ordinary laser lines [21].

Argon. The situation in Ar I is similar to Ne I, but not as interesting or practically important. Ordinary laser lines extend from 1.62 μm to 26.9 μm. Most transitions responsible for these lines are represented schematically in Fig. 9.21, whose structure is similar to Fig. 9.20 for Ne. Only

---

* One research group [39] also reported a weaker Ne I laser line at 5400.6 Å, attributed to a $3p'[\tfrac{1}{2}]^\circ > 3s[\tfrac{3}{2}]_1^\circ$ transition, but this was not observed by the others.

two transitions produce significant gain:

Wavelength	Transition (Racah)
1.6941 μm	$3d[\tfrac{3}{2}]^o_2 - 4p[\tfrac{3}{2}]_2$
2.0616 μm	$3d[\tfrac{3}{2}]^o_2 - 4p'[\tfrac{3}{2}]_2$.

Other Ar I lines are tabulated in references [3, 27, 28, 31–33].

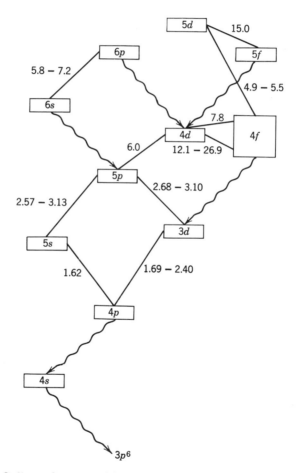

FIG. 9.21 Ordinary laser transitions in Ar I. Each box represents groups of energy levels with common Racah symbols as indicated. *Solid lines:* laser transitions; *wavy lines:* transitions not involving laser action. The numbers adjacent to transitions indicate their wavelength range in μm. (The energy differences are not drawn to scale.)

Exceptional Ar I lasers are also known. One such laser is the precise counterpart of the 6143-Å neon laser and is obtained under identical conditions [40]. Its data are as follows:

$$\lambda_{air} = 7067.3 \text{ Å, transition } 4p'[\tfrac{3}{2}]_2 \rightarrow 4s[\tfrac{3}{2}]^o_2.$$

The following exceptional laser lines of Ar I were observed by Bockasten and co-workers [41, 42] in pulsed, high current, direct current discharges:

Line	Wavelength	Transition (Racah)
1	1.2140 μm	$3d'[\tfrac{3}{2}]^o_1 - 4p'[\tfrac{3}{2}]_1$
2	1.2403 μm	$3d[\tfrac{3}{2}]^o_1 - 4p[\tfrac{3}{2}]_1$
3	1.2702 μm	$3d'[\tfrac{3}{2}]^o_1 - 4p'[\tfrac{1}{2}]_1$
4	1.4094 μm	$3d[\tfrac{3}{2}]^o_1 - 4p[\tfrac{1}{2}]_0$
5	1.6940 μm	$3d[\tfrac{3}{2}]^o_2 - 4p[\tfrac{3}{2}]_2$
6	2.3133 μm	$3d[\tfrac{1}{2}]^o_1 - 4p'[\tfrac{1}{2}]_1$

Lines 5 and 6 are also known in ordinary Ar lasers. They are strong superradiant lines in pulse excitation.

Krypton. About 30 lines of Kr I are observed in ordinary lasers. They range in wavelength from 1.68 to 7.06 μm. The two strongest lines among these are at 2.1165 and 2.1902 μm. For tabulation of these lines see [3, 27, 28, 31—33]. A list of exceptional Kr I lines is as follows [41]:

Line	Wavelength	Transition (Racah)
1	0.8104 μm	$5p[\tfrac{5}{2}]_2 - 5s[\tfrac{3}{2}]^o_2$
2	1.1458 μm	$6s[\tfrac{3}{2}]^o_1 - 5p[\tfrac{1}{2}]_1$
3	1.3177 μm	$6s[\tfrac{3}{2}]^o_1 - 5p[\tfrac{5}{2}]_2$
4	1.3623 μm	$4d[\tfrac{3}{2}]^o_1 - 5p[\tfrac{5}{2}]_2$
5	1.4427 μm	$6s[\tfrac{3}{2}]^o_1 - 5p[\tfrac{3}{2}]_1$
6	1.4765 μm	$6s[\tfrac{3}{2}]^o_1 - 5p[\tfrac{3}{2}]_2$
7	1.6853 μm	$4d[\tfrac{7}{2}]^o_3 - 5p[\tfrac{5}{2}]_3$
8	1.6897 μm	$4d[\tfrac{1}{2}]^o_1 - 5p[\tfrac{1}{2}]_1$
9	2.1902 μm	$4d[\tfrac{3}{2}]^o_2 - 5p[\tfrac{3}{2}]_2$
10	2.5234 μm	$4d[\tfrac{1}{2}]^o_1 - 5p[\tfrac{3}{2}]_2$
11	2.8613 μm	$6p[\tfrac{5}{2}]_2 - 6s[\tfrac{3}{2}]^o_2$

Line 1 is the analog of the 5945-Å Ne I line. It terminates on a metastable level and can be obtained only in a pulsed discharge. In such a discharge, Ericsson and Lidholt [40] obtained a 1000 W output on this line! All lines in the above table produce superradiant emission [41]. The last four of the lines are included among the ordinary laser lines of Kr I.

Xenon. Current lists of laser lines of Xe I contain 25 entries ranging from 2.02 μm to 18.5 μm [3, 27]. The oscillations are usually obtained in a mixture of Xe and He. There is no transfer of excitation from He to Xe; the improvement caused by the presence of He is due solely to the increase of electron density in the discharge. In contrast to Ar and Kr, the Xe I lasers are technically important because some of their lines have exceptionally high gain. The strongest Xe I lines are as follows:

Line	Wavelength	Transition (Racah)
1	2.0262 μm	$5d[\frac{3}{2}]_1^0-6p[\frac{3}{2}]_1$
2	2.6269 μm	$5d[\frac{5}{2}]_1^0-6p[\frac{5}{2}]_2$
3	2.6511 μm	$5d[\frac{3}{2}]_1^0-6p[\frac{1}{2}]_0$
4	3.1069 μm	$5d[\frac{5}{2}]_3^0-6p[\frac{3}{2}]_2$
5	3.3667 μm	$5d[\frac{5}{2}]_2^0-6p[\frac{3}{2}]_1$
6	3.5070 μm	$5d[\frac{7}{2}]_3^0-6p[\frac{5}{2}]_2$
7	4.5381 μm	$5d[\frac{3}{2}]_2^0-6p[\frac{5}{2}]_2$
8	5.5739 μm	$5d[\frac{7}{2}]_4^0-6p[\frac{5}{2}]_3$

Lines 6 and 8 are the strongest and have the highest gain. In a laser with 0.015 torr Xe and 1.5 torr He a gain of 0.13 cm^{-1} was measured on line 6 [43]. Such large a gain is possible because of the narrow linewidth of the transition. This, in turn, is the consequence of the fact that Doppler-broadening is small both because of the large mass of xenon and because of the low frequency of the transition.

The gain of the 3.6788-μm line of Xe I (not listed above) can be increased significantly by the addition of Kr to the discharge. This result is interpreted as indicating that an excitation transfer takes place from metastable 5s levels of Kr to the $5d[\frac{1}{2}]_1^0$ level of Xe I, which is the starting level of this laser transition. Thus it seems that the gases Kr and Xe form a laser analogous in its operating principle to the He—Ne laser. In a Kr—Xe laser with 0.05 torr Kr and 0.005 torr Xe, a gain of 5.4 dB/m ($\alpha = 0.012$ cm^{-1}) was measured [44].

Lines 1 and 3 in the table of ordinary Xe I lines become superradiant when operated in a suitable pulsed discharge [41]. In fact line 1 is among the strongest superradiant laser lines. Additional exceptional laser lines of Xe I are as follows:

Line	Wavelength	Transition (Racah)	Reference
1	$0.8409\ \mu$m	$6p[\frac{3}{2}]_1 - 6s[\frac{3}{2}]_2^o$	[41, 45]
2	$0.9045\ \mu$m	$6p[\frac{5}{2}]_2 - 6s[\frac{3}{2}]_2^o$	[41, 46]
3	$0.9800\ \mu$m	$6p[\frac{1}{2}]_1 - 6s[\frac{3}{2}]_2^o$	[46]
4	$1.3656\ \mu$m	$7s[\frac{3}{2}]_1^o - 6p[\frac{5}{2}]_2$	[41]
5	$1.6052\ \mu$m	$7s[\frac{3}{2}]_1^o - 6p[\frac{3}{2}]_2$	[41]
6	$1.7325\ \mu$m	$5d[\frac{3}{2}]_1^o - 6p[\frac{5}{2}]_2$	[41]

The lines 1, 2, and 3 of this table are the self-terminating lines which were already encountered in connection with Ne-, Ar-, and Kr- lasers. All these lines are superradiant under conditions described in connection with extraordinary Ne lines.

9. 0 ION GAS LASERS

The General Nature of Ionic Lasers. At first glance the distinction between atomic and ionic gas lasers does not seem to be a profound one. A laser is ionic when the transition involved in stimulated emission of radiation occurs between two levels of an ion instead of a complete neutral atom. In this sense, the ruby and the rare-earth lasers are ionic lasers, but among solid lasers this fact does not seem to be of great significance. When attention is focused on gas lasers excited by means of an electric discharge, it becomes of importance to distinguish between atomic and ionic lasers because the design and operating characteristics of these laser types are entirely different.

Although there are noteworthy differences in the physical processes involved in atomic and ionic gas lasers, the most conspicuous differences are of practical engineering nature and they will be introduced first: Ion lasers operate with considerable dissipation of power, and their peak outputs are usually orders of magnitude higher than those of atomic gas lasers. The delivery of power into the discharge tube and the removal of heat generated there create engineering problems not ordinarily encountered in connection with atomic gas lasers. Structural parts of an ion laser are subjected to more severe stresses than those of an atomic laser.

From the viewpoint of technology a distinction must be made between the pulsed ion lasers and those operated in the continuous regime because the operating characteristics of such lasers are different, even when the physical processes immediately responsible for population inversion are quite similar, which is not always the case. There are ion lasers as well as atomic lasers which operate only in a pulsed regime because they are based on a transient population inversion. Frequently, however, ionic lasers are operated in pulsed discharges even when a steady-state population inversion could be attained provided the conditions in the discharge were maintained. In such instances pulsed operation is undertaken for practical reasons, that is, to keep the average power dissipation in the tube within tolerable limits. The high-voltage, pulsed discharges which produce the extraordinary atomic laser lines also produce the pulsed ionic lines.

Ion lasers are suitable as sources of radiation for many applications because they are capable of delivering significant optical power (several watts) in spectral lines located in the visible– and near-visible range. In contrast to the atomic lasers which operate mostly in the infrared—the 6328-Å neon line being at the short end of their range—the ionic gas lasers lie mostly in the 2600-to-7000-Å range. Sources of strong, coherent radiation in this range are of practical importance, not because of the possibility of direct visual observation, which is a risky procedure at best, but because of the relatively large energy carried by each photon in this wavelength range. Such photon energy makes photoelectric and photochemical detection very convenient, and it enables one to perform experiments that are very difficult to perform with infrared photons.

The ions of the noble gases Ne, Ar, Kr, and Xe are the most prevalent ion laser materials. They are the best-explored elements for such applications and their lines account for about one half of all ionic laser lines known in 1969. Accordingly the discussion here is slanted in favor of ionic lasers with a noble gas as working material. Special attention is given to lasers of Ar II which are used in many applications.

Excitation of Ionic Lasers. The energy levels of the second spectrum lie above those of the neutral atom. This matter was briefly discussed in Section 9.7 where it was stated that the ground level of Ne II lies almost 174,000 cm^{-1} (21.56 eV) above the ground level of Ne I. An additional energy of 19.3 eV is required to produce Ne III in its ground state. Although the ionization energies of the other gases are lower, they are all considerable on the atomic scale. This energy must be invested in each excitation process in addition to the energy that may be partially recovered in the downward transition among the ionic levels. Clearly, the ionic lasers require more energetic excitation techniques than ordinary gas lasers.

Several methods of excitation are available for ionic lasers. All involve the delivery of high peak power (kilowatts) into the gas discharge. The method that is the least demanding technically provides intermittent laser operation in pulses of approximately 1-μsec duration. It is accomplished by charging capacitors in the range of 1 to 25 μF from a high voltage source capable of providing up to 10 kV and then discharging them through the gas tube. The discharge is usually initiated by means of a spark coil. This type of excitation is suitable for exciting ionic lines of noble gases with gas pressure in the discharge tube varying between 1 and 200 mtorr, lower pressures being preferred for the heavier gases. The pulse repetition rate may be quite high; values as high as 2000 pulses per second were reported in the literature, but it is usually held down to a lower value to be consistent with the safe thermal loading of the laser. The discharge tubes used in such systems are 2 to 3 mm in diameter and vary in length from 25 to 200 cm, the longer tubes requiring a higher voltage.

Such intermittent dc-excited lasers were used extensively in the early investigations of ion lasers. The original discoveries of ionic lasers by Bell [47] and Bridges [48] were made with this type of excitation, and this system was used in the fundamental spectroscopic work of Bridges and Chester [49].

An entirely different type of dc discharge is used to excite continuously operating ion lasers. Such lasers are operated in low-voltage high-current discharges with the voltage drop between the electrodes varying between 200 and 300 V, and the current between 5 and 50 A. The optimal gas pressure for such a laser is around 0.5 torr, considerably higher than the pressure in a pulsed laser. The optimal tube bore is again around 2 to 3 mm.

Continuously operating ion lasers are heavy-duty instruments provided with water cooling as shown in Fig. 9.9. According to W. B. Bridges, the very first continuous-wave ion laser was not water cooled, but operated with incandescent quartz walls. "The need for cooling was painfully evident after a few minutes of operation!" The high current densities required (over 100 A/cm^2) place severe requirements on the discharge tube structure. When the discharge is confined by the bore of the quartz tube, this bore is rapidly eroded. The confinement of the discharge is sometimes accomplished by a sequence of disks, made of highly refractory materials with good thermal conductivity such as molybdenum or tungsten. Holes are drilled through the centers of these discs and the disks are held in alignment by insulating spacers in a larger envelope.

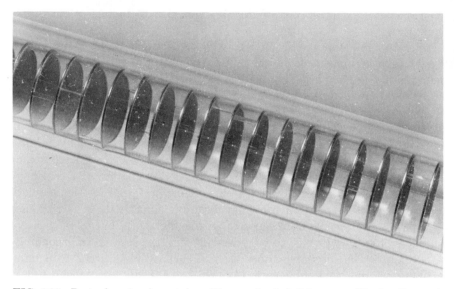

FIG. 9.22 Part of an ion laser tube with a perforated disk array. (Hughes Research Laboratories.)

Such an assembly is shown in Fig. 9.22. The discs are subjected to a large thermal load at the holes where the discharge passes through. They conduct the heat toward the perimeter of the structure. An axial magnetic field of the order of 1000 G is also used in continuous ion lasers to confine the discharge to a narrow filament near the tube axis.

Smaller continuous ion lasers may be powered by alternating current discharges coupled inductively. The configuration of such a discharge was shown in Fig. 9.10. The inventors of this device: Goldsborough, Hodges, and Bell [50] operated an Ar II laser with an optical output of 1.4 W with a radio-frequency input of 1.4 kW. The optimum discharge pressure in such a system is about the same as in dc-excited continuous lasers. The excitation of the plasma by an inductively coupled radio-frequency field provides a number of technical advantages. The most important ones are a substantial reduction of capillary bore erosion and the avoidance of all problems associated with the deterioration of electrodes that are unavoidable in the case of direct current discharges.

Turning from the excitation technology to the excitation process itself, we note that the detailed mechanism by which a population inversion is established for a pair of ionic levels probably varies greatly from one laser to the other. Very little is known about the multiplicity of processes that take place in an ionic plasma. There is reason to believe that the excitation of a Ne atom, for example, from the ground state of the neutral atom to the starting level of an ionic laser transition takes place in more than one step. It requires at least two electron collisions. This conclusion is deduced from a variety of observations, most convincing being those which indicate that the *spontaneous* emission from the upper laser levels of Ar II is proportional to the square of the discharge current [51]. Since the rate of exciting collisions is proportional to the first power of the current, the above observation indicates that two collisions are involved in the excitation of Ar II lasers. It is likely that three- and four-step processes are responsible for the excitation of lasers whose working substances are more highly ionized (Ar III, Ar IV). Because more than one collision is required for excitation, population inversion will not be achieved unless the current density in the discharge is much higher than is necessary for atomic lasers. Confinement of the discharge to a narrow filament by means of an axial magnetic field substantially enhances ion laser operation. This enhancement of the laser output is clearly seen in Fig. 9.23, which illustrates the performance of a relatively small, radio-frequency-powered Ar II laser [50]. The figure also shows the steep rate of growth of laser output with discharge current.

The steep growth of laser output with discharge current density is a characteristic of all ion lasers. It is shown again on the curve of Fig. 9.24, obtained on a small, dc-excited continuous argon ion laser. A 500-fold increase in the intensity of the 4880-Å Ar II line is observed as the discharge current is varied from 1.6 to 6.0 A [52]. Similar curves are obtained on larger lasers. Quantitative measurements on many Ar II lasers indicate that the optical power output of these lasers increases proportionally to I^6 near the threshold, and this rate drops to I^4 for higher values of the current. Nevertheless, all published performance figures indicate that the output is still sharply rising with the current when limits of the capability of the laser system are reached. This situation found in ionic lasers is in sharp contrast with the performance curves of the ordinary atomic gas lasers. The outputs of the latter come to a flat peak within the operating range so that a further increase of the discharge current results in the decrease of the laser output. The rising output curves of the ion lasers indicate that the possible limits of this type of optical power generation have not been reached.

Spectroscopy of Ion Laser Lines. Ion lasers were discovered empirically by producing a discharge in a gas or vapor placed in a configuration known to be generally favorable for stimulated emission. Once stimulated emission was observed, the wavelength of the emitted radiation was

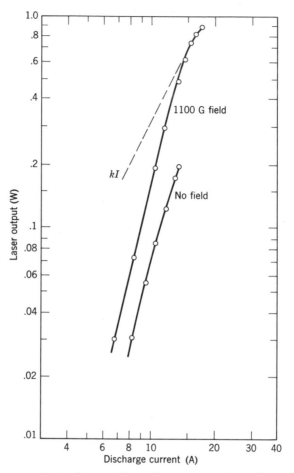

FIG. 9.23 Power output of a small continuous-wave Ar II laser with radio-frequency (8.2 MHz) excitation. Tube length 21 cm, bore 2.5 mm, p = 0.8 torr. Dotted line indicates what would be expected if output were proportional to I. (After Goldsborough, Hodges, and Bell [50].)

measured, and the optimal conditions for obtaining such radiation were explored by adjusting the parameters at the disposal of the experimenter. Whenever it was possible, gain and power output measurements were also made, and circumstances that enhanced or inhibited stimulated emission were recorded.

The first step in the theoretical analysis of laser operation is the identification of the transition that produces the stimulated radiation observed. This process is called classification. In the case of ion lasers this in itself is often quite difficult and has not been completed in many cases, in spite of the superior care and professional competence that generally distinguishes research in the field of ion lasers, in contrast to ordinary atomic lasers. The energy-level structure of the most important ions— noble gases and halogens—is more complicated and less known than the energy-level structure of the complete atoms. Some of the reasons for the complicated nature of the energy-level structure of ionized noble gases were briefly introduced in Section 9.7. We have seen that when an Ar atom is singly ionized, a core of five p electrons remains in the third shell. Excitation of this ion takes one of these five electrons to a higher orbit,* leaving a core in a $3s^23p^4$ configuration with an external electron in a higher orbit, for example, $4s$, $5s$, $4p$, $5p$, $4d$, etc. The core may be

* An exceptional case may occur (see Section 9.7).

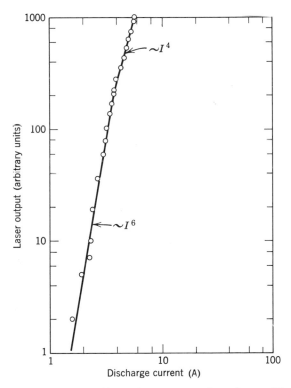

FIG. 9.24 The output of a small continuous-wave Ar II laser with direct-current excitation as a function of the discharge current. Tube length 25 cm, bore 2 mm, $p = 0.45$ torr. (After Gordon, Labuda, and Bridges [52].)

left in a ^3P, ^1D, or ^1S configuration, and the complete system of ionic configurations encompasses the combinations of each of these cores with the possible configurations of the external electron. A similar situation prevails in other noble gases and the complexities increase as we progress from Ar II to Ar III. It is not implied that the higher ionic spectra of *all* elements become progressively more complicated, but this is true of the elements in the last two columns of the periodic table, and these are the most important laser materials.

The tabulated atomic and ionic energy levels are calculated from observations of ordinary spectroscopy [13]. While the tables are reasonably complete for neutral atoms, there are more and more gaps as one proceeds to higher-order spectra. Ionic laser lines were occasionally observed to involve transitions among levels not yet tabulated. In other instances a definite classification cannot be made with certainty because, within the accuracy of the measurements, transitions between several pairs of levels could produce the observed line. The selection rules greatly aid in the classification of observed laser lines, but they must be applied with discretion because the coupling schemes that lead to the designations used for the levels are idealizations that are only approximately valid.

The most comprehensive catalog of ionic laser lines, compiled by Bridges and Chester [53], was published in 1965. It contains 230 laser lines with 186 classified.

When the laser lines are classified, the question arises what types of transitions are favored and by what detailed processes does population inversion arise in the favored pairs of levels. These questions so far remain largely unanswered although progress has been made in the case of Ar II [49, 54]. It has been found that laser transitions seldom occur with a change in core configuration. Most laser transitions occur among levels with ^3P cores, fewer among levels with ^1D cores, and none were observed involving levels with ^1S cores. Many laser lines originate at one of the $4p$ levels and terminate at lower lying s and d levels. It has been proposed that the overpopulation of the $4p$ levels is the result of many cascade-type transitions from still higher levels and that the favorable low population of the $4s$ and $3d$ levels is the consequence of the fact that these levels drain rapidly to the ground state of the ion.

The detailed study of a single ion-laser line discloses features somewhat different from those observed in neutral atomic lasers. The gain in the atomic laser line is symmetrical about the center of the line. This is not so in dc-excited ionic lasers because the ions acquire a significant velocity in the electric field, and there is a Doppler shift associated with this ion-drift velocity. The drifting of ions from the anode toward the cathode causes not only asymmetry but also instability in the mode structure of ion lasers [55].

A Survey of Accomplishments. Although the best-known ion gas lasers operate with noble gases, these are not the only ion laser materials. In fact laser action in gaseous ions was first observed by Bell [47] in Hg II. The discovery of the mercury ion laser at the end of 1963 demonstrated immediately three important characteristics of ion lasers: high power, high gain, and shorter wavelength than normal atomic lasers. Within one year of Bell's discovery, about 200 ionic laser lines were discovered and by 1968 the number of such lines grew to over 400. The largest number of ionic laser lines belongs to the noble gases, a fair number to the halogens F, Cl, I, Br, as well as to the atmospheric gases O and N. A number of lines are known in Hg; these will be

TABLE 9.3
A List of the Strongest Ionic Laser Lines

Wavelength in Air (Å)	Element	Wavelength in Air (Å)	Element
2358	Ne IV	4765	Ar II
2625	Ar IV	4825	Kr II
2984	O III	4880	Ar II
3324	Ne II	4965	Ar II
3378	Ne II	5017	Ar II
3393	Ne II	5145	Ar II
3507	Kr III	5218	Cl II
3511	Ar III	5408	I II
3638	Ar III	5419	Xe II
4067	Kr III	5592	O III
4131	Kr III	5677	Hg II
4347	O II	5680	N II
4351	O II	5682	Kr II
4415	O II	5760	I II
4417	O II	5971	Xe II
4579	Ar II	6127	I II
4603	Xe II	6150	Hg II
4619	Kr II	6471	Kr II
4658	Ar II	6765	Kr II
4680	Kr II	6871	Kr II
4762	Kr II		

discussed briefly in Section 9.12. Elements that have to be vaporized, such as B, C, S, Si, Mn, Cu, Zn, Ge, As, Cd, In, Sn, and Pb have also been used as ion laser materials. The basic source of spectroscopic data for ion lasers is the review article of Bridges and Chester [53]. It was compiled in early 1965. The material discovered after this work was completed, is scattered in many short articles. References [40] and [56—60] contain a fair number of the newer lines. Other ionic lasers are described in Section 9.12 and further references to original publications are given there. Some of the strongest ionic laser lines are listed in Table 9.3 in the order of their wavelengths. This table contains fewer than 10% of the ionic laser lines known in 1969. It should be regarded as a sampling rather than a complete listing.

9.11 ATOM—MOLECULE COLLISION LASERS

Resonant excitation transfer in an electric discharge may be employed as a means of excitation for atomic lasers in a situation where the atom to be excited is originally part of a molecule. In such a case the active element of the laser is formed in the discharge as a result of molecular dissociation. An essential attribute of a dissociation process that leads to laser action is that one of the elements formed, as the result of the dissociation, is not formed in its ground state but in one of its excited states. This principle of dissociative excitation by collision between molecules and excited (metastable) atoms was already discussed in Section 9.1. The best-explored systems in which this process leads to population inversion and laser action are $Ne-O_2$ and $Ar-O_2$. The processes are different for the two noble gases; both are illustrated in Fig. 9.25. The energy levels of oxygen are shown in this figure as reckoned from the lowest bound state of the O_2 molecule, which is 5.080 eV below the ground state of the two O atoms removed from each other. The lowest excited $(1s)$ states* of Ne and Ar are also shown on the diagram; they are indicated as Ne* and Ar*. Transfer from the Ne* levels leads directly to an unstable molecular level, from which the molecule dissociates with one of the oxygen atoms in the starting level $(3p^3P_2)$ of the laser. In the case of Ar the transfer takes place to two lower-lying levels: $2p^1D_2$ and $2p^1S_0$. These are metastable; radiative transition from these levels to the triplet ground level is forbidden by the selection rules. Because of their metastable nature, these levels become heavily populated; they

---

* The Racah symbols of these states are $3s$ for Ne and $4s$ for Ar. The reason for these levels being heavily populated in a steady discharge is explained in Section 9.9.

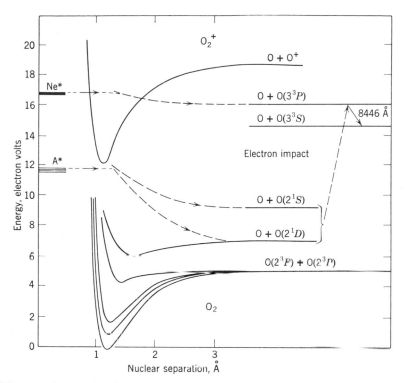

FIG. 9.25 Energy levels of oxygen, neon, and argon participating in the Ne-O_2 and Ar-O_2 lasers. Curves represent energies of a common electronic state.

become platforms from which the $3p^3P_2$ level can be reached by electron impact. The selection rules indicate that population inversion between the $3p^3P_2$ and $3s^3S_1^o$ levels cannot be established by electron collision alone, that is, without the assistance of a noble gas, because the transition from the ground $2p^3P_2$ level to the $3s^3S_1^o$ level is more likely than a transition to the $3p^3P_2$ level. As indicated on the figure, the laser transition is $3p^3P_2$–$3s^3S_1^o$, with a wavelength of 8446 Å.

Bennett, Faust, McFarlane, and Patel [61], who discovered this laser and analyzed the processes involved, obtained stimulated emission of the 8446-Å line from a radio-frequency discharge in a tube 2 m long and 7 mm in diameter. The composition of the gas was O_2:0.014 torr, Ne:0.35 torr; or alternatively O_2:0.036 torr, Ar:1.3 torr. The output power was about 2 mW in each case. The same radiation is also obtainable upon dissociation of CO, CO_2, NO, and N_2O in the presence of He or Ne.

The case of CO + He is particularly interesting. Excitation transfer takes place from the 2^3S level of He, the same level that produces the near-infrared (1.1 μm) Ne lasers. Both oxygen and carbon are left in an excited state in the dissociation reaction:

$$CO + He^*(2^3S) \rightarrow O^*(2p^1S_0) + C^*(2^5S_2) + He.$$

The excited oxygen atoms are metastable, they are in one of the states already encountered in connection with the O_2—Ar laser. The electron collision reaction

$$O^*(2p^1S_0) + (e + KE) \rightarrow O^*(3p^3P) + e$$

then leads to the population of the upper laser level of oxygen.

The excited carbon level $C^*(2^5S_2)$ is not an upper laser level of carbon, but it is possible to reach an upper laser level from this level by means of an electron collision. Alternatively, the upper levels of the carbon laser may be reached by transfer of excitation from $He^*(2^3S)$ levels to carbon atoms in their ground state [3].

Two carbon lines were observed; one at 1.069 μm, the other at 1.454 μm. Two lines of N I may be generated in the mixture that produces the oxygen line from NO and N_2O. Similarly two lines of S I were obtained by an analogous process in mixtures of SF_6 and He or Ne [62]. The laser lines of the elements C, N, and S lie in the near-infrared region between 1.0 and 1.5 μm. They were apparently not measured accurately because the data published are not entirely consistent with results obtained from ordinary spectroscopy. Some of the classifications suggested for these lines by their discoverers are open to suspicion.

9.12 LASERS OF HYDROGEN, HALOGENS, AND MERCURY

Hydrogen. A single atomic laser line of this element is known. It was discovered accidentally by Bockasten and co-workers [63] while experimenting with pulsed discharges in a He—H mixture. The oscillations occur on the first line of the Paschen series, the $n = 4$ to $n = 3$ transition. Its wavelength in air is 1.8751 μm. The optimum pressure for obtaining this laser is around 3.5 torr He and 10 mtorr H. A very powerful, short exciting pulse is recommended.

Halogens. Lasers operating in the first spectra of chlorine, bromine, and iodine are obtainable in low-current discharges similar to those that produce atomic noble-gas lasers. Although excitation may be introduced with direct current, radio-frequency excitation is preferred in the case of halogens because these gases attack most electrodes. The discharges operate better when a noble gas (usually He) is also present in the discharge. The discharge tube may be filled with HCl, HI, or HBr, and the active atoms are formed in the discharge upon dissociation of the molecules. The excitation of these lasers is accomplished by electron collisions. (Halogen lasers whose source of excitation is molecular dissociation will be discussed in Section 9.14.) Laser lines in complete halogen atoms were observed in the wavelength range between 1.3 and 3.4 μm. Thirteen identified lines are listed in Table 9.4.

Ionic halogen lasers are excited by high-current discharges. These may be pulsed or continuously operating, as is the case with noble-gas ion lasers. The difficulties encountered in connection with the construction and operation of such lasers were discussed in Section 9.10. They are present in ionic halogen lasers; in addition, the corrosive nature of the halogens, especially F and Cl, aggravates all technical problems normally encountered in laser construction.

Many laser lines are obtainable in ionized halogens, but the classification of quite a few of these lines is uncertain. Ten laser lines attributed to various ions of F were observed in the wavelength range 2700 to 6400 Å [68]. The laser lines of Cl II extend from 4100 to 6100 Å [60].

TABLE 9.4

HALOGEN LASERS

(Complete atoms only)

Line Number	Wavelength in Air (μm)	Element	Transition	References
1	1.3863	Cl	$3d\ ^2D_{5/2}-4p\ ^4D^o_{5/2}$	[64]
2	1.3893	Cl	$3d\ ^4P_{3/2}-4p\ ^4D^o_{3/2}$	[64]
3	1.4542	I	$(^3P_1)6d[2]_{3/2}-(^3P_1)7p[1]^o_{3/2}$	[64]
4	1.9755	Cl	$3d\ ^4D_{7/2}-4p\ ^4P^o_{5/2}$	[65, 66]
5	2.0199	Cl	$3d\ ^4D_{5/2}-4p\ ^4P^o_{3/2}$	[65, 66]
6	2.2854	Br	$(^3P_2)4d[3]_{7/2}-(^3P_2)5p[2]^o_{5/2}$	[67]
7	2.3511	Br	$(^3P_2)4d[3]_{5/2}-(^3P_2)5p[2]^o_{3/2}$	[67]
8	2.4470	Cl	$3d\ ^4D_{7/2}-4p\ ^4D^o_{7/2}$	[64]
9	2.5986	I	$(^3P_1)5d[2]_{3/2}-(^3P_1)6p[1]^o_{3/2}$	[64]
10	2.7571	I	$(^1D_2)5d[2]_{5/2}-(^1D_2)6p[1]^o_{3/2}$	[67]
11	2.8375	Br	$(^3P_2)4d[3]_{7/2}-(^3P_2)5p[3]^o_{5/2}$	[67]
12	3.2359	I	$(^3P_2)5d[2]_{5/2}-(^3P_2)6p[1]^o_{3/2}$	[66, 67, 68]
13	3.3405	I	$(^3P_2)5d[4]_{5/2}-(^3P_2)6p[3]^o_{5/2}$	[66, 67, 68]

Many lines of Cl III lie between 3190 and 3750 Å [68]. Twelve Br II laser lines were found in the 5180– to 6200-Å range [56, 69]. Iodine ion lasers also operate in the visible spectrum from 5400 to 7032 Å [70].

Mercury. Laser oscillations in mercury lines are usually obtained by means of electric discharges through gas mixtures containing a noble gas in addition to Hg vapor. Between 0.5 and 5 torr of He or Ar is used. The vapor pressure of Hg is regulated by controlling the temperature of the discharge tube or that of a side arm containing liquid Hg. Experiments are usually carried out in the 45 to 85°C temperature range. Mercury lasers are excited by short dc pulses obtained from capacitors charged to at least 10 kV. Both the atomic and the ionic lines may be obtained in the same experimental apparatus, but the ionic lines are favored in discharges with lower pressures. Hollow cathode discharge tubes are preferred for Hg-ion lasers. These consist of three metal tubes aligned along their common axis joined by insulating sections. The central metal tube is used as a cathode, the two outside tubes serve as anodes.

Historically, the discovery of the Hg II laser by Bell [47] preceded all other ion lasers. Although laser lines in Hg have been known since

TABLE 9.5
LASER LINES OF Hg I
(After Bockasten et al. [71])

Line Number	Wavelength in Air (μm)	Transition
1	1.1177	$7p\ {}^1P^o_1 - 7s\ {}^3S_1$
2	1.3675	$7p\ {}^3P^o_1 - 7s\ {}^3S_1$
3	1.5295	$6p'\ {}^3P^o_2 - 7s\ {}^3S_1$
4	1.6920	$5f\ {}^1F^o_3 - 6d\ {}^1D_2$
5	1.6942	$5f\ {}^3F^o_2 - 6d\ {}^3D_1$
6	1.7073	$5f\ {}^3F^o_4 - 6d\ {}^3D_3$
7	1.7110	$5f\ {}^3F^o_3 - 6d\ {}^3D_2$
8	1.7330	$7d\ {}^1D_2 - 7p\ {}^1P^o_1$
9	1.8130	$6p'\ {}^3F^o_4 - 6d\ {}^3D_3$
10	3.93	or $\begin{cases} 6d\ {}^3D_3 - 6p'\ {}^3P^o_2 \\ 5g\ G - 5f\ F \end{cases}$
11	5.88	$6p'\ {}^1P^o_1 - 7d\ {}^3D_2$
12	6.49	or $\begin{cases} 9s\ {}^1S_0 - 8p\ {}^1P^o_1 \\ 11p\ {}^3P^o_1 - 10s\ {}^3S_1 \end{cases}$

TABLE 9.6

Laser Lines in Ionized Mercury

Spectrum	Line Number	Wavelength in Air (Å)	Transition	References
II	1	5678	$5f\ {}^2F^o_{7/2}-6d\ {}^2D_{5/2}$	[47, 72, 73]
	2	6150	$7p\ {}^2P^o_{3/2}-7s\ {}^2S_{1/2}$	[47, 72, 73]
	3	7346	$7d\ {}^2D_{5/2}-7p\ {}^2P^o_{3/2}$	[47, 72]
	4	8547	$5g\ {}^2G_{7/2}-6p\ {}^2F_{5/2}$[a]	[47]
	5	9396	$10s\ {}^2S_{1/2}-8p\ {}^2P^o_{3/2}$	[72]
	6	10583	$8s\ {}^2S_{1/2}-7p\ {}^2P^o_{3/2}$	[47, 72]
	7	15554	$7p\ {}^2P^o_{3/2}-6d\ {}^2D_{5/2}$	[72]
III	1	4797	$5d\ {}^8 6s^2(J=2)-5d\ {}^9 6p_{1/2}(J=3)$	[73]

[a] Also known as Paschen's "C" level.

1963, and a number of articles were published about them, there is a certain amount of confusion in the literature concerning the origin of these lines. Tables 9.5 and 9.6 are provided in this section in order to dissipate some of the confusion and to collate the information about all confirmed and identified mercury lines. Table 9.5 is a condensation of material contained in a critical article that contains references to the original publications concerning Hg I lasers [71]. Table 9.6 is prepared largely on the basis of the article of Bloom, Bell, and Lopez [72]. It contains references to other publications. Figures 9.26 and 9.27 show the term schemes of Hg I and Hg II together with the laser lines listed in the tables.

The neutral Hg atom has a completed $5d$ shell plus two electrons in the $6s^2$ configuration when the atom is in its ground state. On excitation either one of the $6s$ electrons or one of the $5d$ electrons is raised to a higher level. The states resulting from the $5d^{10}6s$ core configuration are more common. They are displayed on the left side of Fig. 9.26. They are listed under the heading "Limit Hg II ($^2S_{1/2}$)" because this ion results when the excited electron is completely removed. The states resulting from the $5d^3 6s^2$ configuration are designated with primes; they are displayed on the right side of the figure. An excited state of Hg I has two unpaired electrons; the states are singlets or triplets.

The structure of the Hg II ion is simpler. It has only one extra electron outside of a closed shell. In principle, ionization can be accomplished by the removal of a $5d$ electron, that is, as the limit of the primed states in

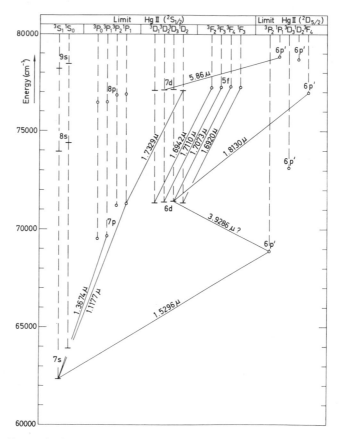

FIG. 9.26 Part of the term scheme of Hg I. Circles represent odd levels and horizontal lines, even levels. Transitions marked by lines correspond to observed laser lines. (After Bockasten et al. [71].)

Fig. 9.26. Such ions, however, play no part in observed laser transitions, therefore the corresponding levels are not shown in Fig. 9.27. That figure contains the levels of Hg II, with a closed $5d$ shell and one outside electron. The structure of this diagram is very similar to that encountered in the spectroscopy of alkalis.

The visible 5677-Å and 6150-Å lines of Hg II are very strong and probably quite suitable for many applications.* The lines are sharp because the Doppler effect causes only a small broadening, the velocities of the heavy ions being so small. A single laser line is known in Hg III.

* A peak output power of 40 W was measured at 6150 Å in the first experiments [47].

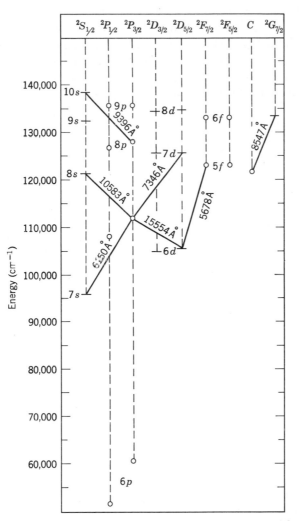

FIG. 9.27 Partial-energy-level diagram of Hg II showing observed laser transitions. Levels shown have the $5s^2 5p^6 5d^{10}$ core configuration with the exception of the level C, which is a $^2F_{5/2}$ level of unknown core configuration. Circles represent odd levels and horizontal lines even levels.

9.13 MOLECULAR LASERS

Spectroscopic Foundations. A systematic exposition of the elements of molecular spectroscopy is a more demanding task than the author is willing to undertake or the reader is likely to endure. Nevertheless, certain basic facts of this discipline must be brought into sharp focus in order to carry out the exposition of the working principles of molecular lasers. It is also desirable to introduce in an elementary manner the nomenclature used in the literature of molecular phenomena. The discussion here is a continuation of the exposition begun in Section 9.1. To avoid excessive complications, the analysis will be limited to the examination of *linear molecules*. These include, of course, all diatomic molecules and also triatomic molecules of the type of CO_2, but unfortunately not the most common of them all: H_2O.

The energy of a molecule (or of any mechanical system) may be thought of as consisting of two major parts. The first derives from the motion of the center of mass and from its location in an external force field, the second is the energy calculated in a system located at the center of mass of the molecule. The first part of the energy is simply related to the translational motion of the molecule. It will not be of concern here. The second part, the internal energy, must be examined further. As has already been stated in Section 9.1, this internal energy may be attributed to three sources: the configuration of nuclei and electrons, the vibrations of the nuclei, and the rotational motions of the entire molecule. Absence of interactions between these phenomena is not implied. What is meant is the following: The internal energy E_i of the molecule is thought of as being the sum of three terms:

$$E_i = E_e + E_v + E_r. \tag{13.1}$$

Here E_e is the energy of the nuclear and electronic configuration, that is, the energy the molecule would possess in the absence of all vibration and rotation, E_v is the additional energy that results from vibrations without rotation, and E_r is the balance that must be added when rotations are also present.

It is customary in molecular spectroscopy to replace energy values by *term* values that result by dividing the energy values by hc. This practice is equivalent to measuring energies in wavenumbers. As a consequence; the term differences give the reciprocal wavelengths of the transitions directly. These reciprocal wavelengths are often referred to as frequencies and are denoted by the letter ν. The terms of molecular spectroscopy are in their customary forms

$$T_e = \frac{E_e}{hc}, \quad G(v) = \frac{E_v}{hc}, \quad F(J) = \frac{E_r}{hc}.$$

These three types of terms are of different orders of magnitude. Consequently, when we examine the energy level structure of the N_2 molecule,

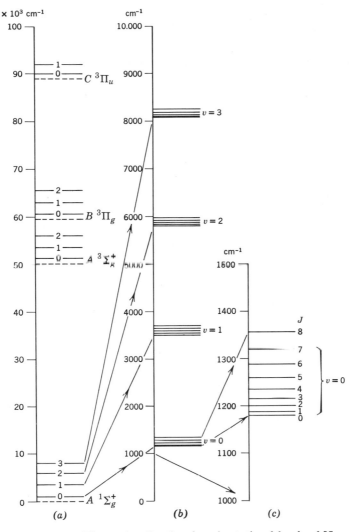

FIG. 9.28 Electronic, vibrational, and rotational levels of N_2.

for example, we must do so on three different scales. The first of these is shown at the left of Fig. 9.28. There the first four electronic levels are displayed with a few of the vibrational levels added. The rotational levels can not be shown on this scale. The center of the figure shows the ground electronic state with its first four vibrational states on a 10-fold enlarged scale. A few of the rotational levels are indicated. Another ten-fold enlargement on the right shows rotational levels of a single vibrational state.

In analogy to atomic spectroscopy, in which the letters S, P, and D are used to designate different electron configuration types, the corresponding Greek capital letters Σ, Π, and Δ are used in molecular spectroscopy. They indicate the value of a quantum number that specifies the component of orbital angular momentum along the molecular axis. The letter Σ indicates that this component is 0, Π that it is $\hbar$, and Δ that it is $2\hbar$. The multiplicity of the level is indicated by a superscript to the left, as in atomic spectroscopy. When there are two electrons participating in the formation of the molecule, the symbol $^1\Sigma$—pronounced singlet sigma—indicates that the spins of these electrons are opposite, while the symbol $^3\Sigma$—triplet sigma—indicates a sigma configuration with the two spins aligned parallel. The symbol $+$ is added as a superscript if the molecular wave function remains unchanged on reflection in a plane through the molecular axis. The symbol $-$ is used when such reflection changes the sign of the wave function. There is an additional pair of symbols, g and u, used to indicate the parity of the electronic state. It refers to the symmetry of the electronic state for reversal of all coordinates, that is, inversion about the center of the molecule. There are several $^3\Sigma_u^+$ states in N_2 for example, they are distinguished by an additional Latin character placed in front; for example, $A\,^3\Sigma_u^+$. The reason for the use of symmetry symbols in molecular spectroscopy is the same that motivates the use of the symbols S, P, D, etc. in atomic spectroscopy. If one knows the selection rules, it is relatively easy to make predictions for the intensities of transitions from the symmetry types of the states involved. Unfortunately, the selection rules for electronic transitions in molecules are rather complicated. They are different for molecules with similar nuclei, for example N_2, and for heteronuclear molecules such as CO. The exposition of these rules requires a rather lengthy analysis that we are not prepared to undertake. The interested reader will find this material in Herzberg's treatise [1].

The vibrational terms of a molecule are characterized by the vibrational quantum numbers. In the case of N_2 there is only one vibrational degree of freedom, therefore one quantum number v is sufficient. For any fixed electronic level, the vibrational levels associated with $v = 0$, 1, 2, etc., are nearly, but not accurately, equidistant. The vibrational term formula, which is the generalization of the harmonic oscillator formula (1.1), is usually written in the form

$$G(v) = \omega_e[(v + \tfrac{1}{2}) - \chi_e(v + \tfrac{1}{2})^2 + \cdots], \qquad (13.2)$$

where the higher order terms indicated by the dots will be omitted in our discussion. When χ_e is neglected, we regain the classical oscillator formula with $\omega_e = \nu_0/c$. For the ground electronic state of N_2, the values of the constants in (13.2) are as follows [1]:

$$\omega_e = 2359.6 \text{ cm}^{-1}, \quad \omega_e \chi_e = 14.46 \text{ cm}^{-1}.$$

These values indicate the preponderance of the first term in (13.2). The subscripts e remind us that the vibrational constants in another electronic state may be vastly different and result in different vibrational band spacings. It is interesting to note that the vibrational term is not zero for $v = 0$, but equals approximately to $\frac{1}{2}\omega_e$. For this reason the lowest energy state of the N_2 molecule is approximately 1180 cm^{-1} above the electronic energy of the molecule (see Fig. 9.28).

The rotational term in first approximation is given by

$$F(J) = B_v J(J + 1). \tag{13.3}$$

Here J is the rotational quantum number (not the total angular momentum) and

$$B_v = \frac{h}{8\pi^2 cI}, \tag{13.4}$$

I being the moment of inertia of the molecule about the axis of rotation. This quantity depends slightly on the vibrational motion of the molecule, and this dependence accounts for the appearance of v as a suffix. Again, using the ground electronic state of N_2 as an example, the value of B_v is 2.0 cm^{-1}.

In Section 9.1 we have already encountered the basic selection rule for the interaction of the radiation field with rotational transitions: If the rotational quantum number J changes at all it changes by $+1$ or -1.* The lines of the rotation—vibration spectrum are classified according to the change in J. When $\Delta J = +1$ the line is said to belong to the P-branch; when $\Delta J = 0$, to the Q-branch; when $\Delta J = -1$, to the R-branch. Here a transition is always counted in emission; that is, from the upper level to the lower. The lines are numbered by the rotational quantum number of the lower level. Thus the line P(1) designates one with a transition from a rotational level $J' = 0$ to a level with $J = 1$. (Naturally, the vibrational energy may also change.) The P-branch normally starts with 1, the Q-, and R-branches start with 0.† Figure 9.29 shows the structure of the branches of a single vibration—rotation band.

---

* This statement is not applicable to phenomena involving more than one photon, as is the case, for example, in the Raman effect.

† This situation applies when the electronic state is a Σ state, which is most commonly the case; for other cases, see Herzberg [1].

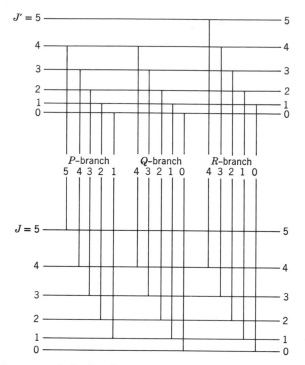

FIG. 9.29 Structure of the P-, Q-, and R-branches of a vibration—rotation band. (P-branch: $J - 1$ to J, Q-branch: J to J, R-branch: $J + 1$ to J.)

All these lines arise between two fixed vibrational levels of the same electronic state.

When the electronic state of the molecule does not change, the vibrational quantum number v may change by 0, $+1$, or -1. When it changes by $+1$, energy is absorbed. When v does not change, we have the *pure rotation spectrum*. When v decreases by one unit, emission of radiation takes place in one of the lines of the vibration-rotation spectrum.

When the electronic state changes, the change in v is not so restricted. Transitions are then possible into several different vibrational states and the relative probabilities of these transitions can be estimated from the Franck-Condon principle: The probability of a transition is large when it requires small readjustments of interatomic distances and momenta.

There are metastable levels in molecular spectroscopy, the $v = 1$ band of the ground electronic state of N_2 providing the best example. In the ground electronic state of N_2 the charge distribution is symmetrical about the main symmetry plane of the molecule. The molecule has no electric dipole moment. Oscillation of the N-atoms creates no electromagnetic field. Therefore a N_2 molecule, excited to one of the vibrational levels of the ground electronic state, cannot decay through electric dipole radiation. The effective lifetime of these states is determined by de-activation through collisions with other molecules and the walls. Although molecules in the $v = 2, 3$, etc., states may loose some of their energy by colliding with molecules in the $v = 0$ state, those in the $v = 1$ state are really stuck. Their population decays very slowly. Measurements indicate that the effective lifetime of this state is about 100 msec. In an electric discharge in N_2, a large number of molecules will accumulate in this $N_2(v = 1)$ level due to electron collisions which raise the molecules into various vibrational levels and also due to cascade processes from excited electronic levels. Accumulation of $N_2(v = 1)$ at a level 2331 cm^{-1} above ground level is of great importance in molecular laser technology. This level plays a role similar to that of the two metastable He levels in atomic gas laser technology.

So far the discussion was oriented toward molecules of the simplest type, exemplified by N_2. The rotation–vibration spectrum becomes a great deal more complicated in the presence of a third atom in the molecule. Because of the importance of CO_2 as a laser material, it is necessary to discuss the rotation—vibration spectrum of this molecule. Being a linear molecule, CO_2 has only one rotational degree of freedom, and its vibrational structure is relatively simple because of the symmetry plane σ perpendicular to the molecular axis. Carbon dioxide has four vibrational degrees of freedom. Its normal modes of vibration are illustrated in Fig. 9.30. In diagram (a) of this figure the atoms are shown frozen in their equilibrium positions. The most symmetrical vibration is illustrated in diagram (b). The carbon atom remains fixed at the center of mass, the oxygen atoms move toward and away from it, approaching the center at the same time, so that the plane σ remains a symmetry plane throughout the motion. Diagram (c) shows one of the bending modes of oscillation, with the oxygen atoms moving in the same direction and the carbon atom in the opposite direction, leaving the center of mass unchanged. The molecule loses its linear character during this vibration, but the plane σ remains a plane of symmetry as before. This bending mode is degenerate; another mode of equal frequency corresponds to motions in and out of the paper. The last, and least symmetric mode of vibration is shown in diagram (d). Here the carbon atom vibrates back and forth between the two oxygens, receding from one when approaching the other. More precisely, the center of mass remains fixed, and the two oxygens move in unison, with the carbon

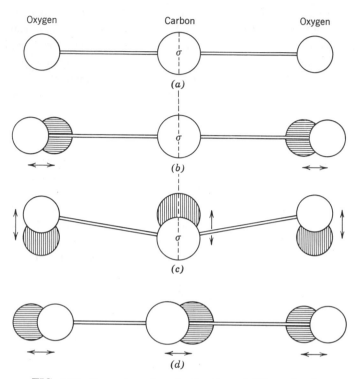

Oxygen   Carbon   Oxygen

(a)

(b)

(c)

(d)

FIG. 9.30 Normal modes of vibration of CO_2 molecule.

moving in phase opposition. In this oscillatory mode the plane σ is no longer a plane of symmetry. Four quantum numbers (integers) are associated with these four normal vibrational modes, and the vibrational state of the CO_2 molecule is described by a symbol of the type (v_1, v_2^l, v_3). The integers v_2 and l are associated with the degenerate bending mode, and v_1 and v_3 are associated with the normal modes illustrated in diagrams (b) and (d), respectively.

The above analysis applies to molecules similar to CO_2, such as CS_2, but not to linear molecules without central symmetry, such as N_2O, and certainly not to H_2O, which is not even linear.

Partial Inversion. When a gas is in thermal equilibrium at room temperature, only the lowest vibrational level ($v = 0$) of the ground electronic state is occupied to any significant extent. The concentration of molecules in the lowest vibrational level is the consequence of the fact that the thermal energy kT at $300°K$ is equivalent to only 208 cm^{-1}, and that it usually takes several times this energy to raise a molecule to the next vibrational level. The situation is different with respect to rotational levels that are more closely spaced. Quite a few of these levels (in the $v = 0$ vibrational level) will be occupied at room temperature, and it is of importance to determine the distribution of the gas molecules among them. This distribution is governed by Boltzmann's law: The number N_n at level n in thermal equilibrium at temperature T is proportional to $g_n \exp\ (-E_n/kT)$, where g_n is the multiplicity of the level (see Section 1.3.). This formulation of Boltzmann's law is valid not only for the entire gas in a vessel but also to any subsystem in thermal equilibrium. Such a subsystem may consist of all molecules on a particular vibrational level, provided that thermal equilibrium is established among all rotational sublevels of this level. When applying Boltzmann's law it is essential to distinguish between state and level. In other words, the multiplicities of the levels must be taken into account. Quantum theory teaches that vibrational levels are nondegenerate, rotational levels with quantum number J, however, have the multiplicity $2J + 1$. Thus, in thermal equilibrium, the distribution of N_v molecules among the rotational levels of a vibrational level (v) is given by

$$N_{v,J} - N_v P(T)(2J + 1) \exp\left[-\frac{F(J)hc}{kT}\right] \tag{13.5}$$

where P is the partition function defined by the requirement

$$\Sigma N_J = N_v. \tag{13.6}$$

A typical such distribution is illustrated in Fig. 9.31. There the energy is displayed on the vertical scale, and the lengths of the horizontal lines are drawn proportional to the population of the rotational levels in thermal equilibrium.

It is a very important fact that a redistribution of molecules by means of collisions can take place very rapidly if the redistribution involves only changes between rotational energy levels and if the vibrational and electronic quantum numbers of the molecules are not changed. The frequency with which molecules change their rotational quantum numbers, at gas pressures normally used in lasers, is between 10^7 and 10^8 per second. As a consequence of this fast exchange rate, a state of near thermal equilibrium is established within the rotational sublevels of each vibrational level in about 10^{-7} sec. The corresponding thermalization time between the vibrational levels of the same electronic state is orders of magnitude larger, about 10^{-3} sec. Because of this large difference between the rates at which the vibrational and rotational levels approach thermal

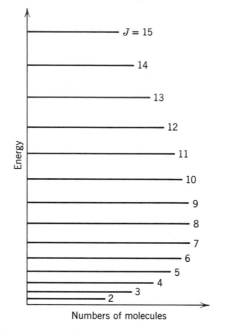

Numbers of molecules

FIG. 9.31 Distribution of molecules over rotational levels of a single vibrational level in thermal equilibrium.

equilibrium, a situation frequently arises in which there is near thermal equilibrium within each vibrational level, without such equilibrium prevailing between different vibrational levels. Laser technology creates and exploits such nonequilibrium situations. When an excitation process is sufficiently successful, so that it increases the total population of an entire vibrational level above that of the vibrational level immediately below, it is not difficult to see that optical gain will be available at least on some of the transitions between these vibrational levels. This is the case of *complete inversion*. It is not always easy to make the excitation quite so effective! Fortunately, it is possible to obtain optical gain on some transitions between vibrational levels even when the inversion is not complete. The general requirement for attaining optical gain on a transition from the state n to state m (see Section 1.3) is

$$\frac{N_n}{g_n} > \frac{N_m}{g_m}. \tag{13.7}$$

This condition must be satisfied for a particular pair of vibration—rotation transitions, not for the entire pair of vibrational levels. Figure 9.32

illustrates how such a situation, called *partial inversion,* may arise in the absence of complete inversion. In this figure the vibrational levels v and $v' = v + 1$ are drawn with nearly equal, total populations. It is assumed that within each of these vibrational levels thermal equilibrium is established, and that this equilibrium is appropriate to a common, relatively low, absolute temperature T_r. Inspection of the P-branch transitions above the most populated level shows that $N_{v',J-1}$ is greater than $N_{v,J}$ because of the exponential decay of populations with increasing J. It is then certainly true that

$$\frac{N_{v',J-1}}{2J-1} > \frac{N_{v,J}}{2J+1}. \tag{13.8}$$

This inequality is the condition for obtaining amplification on the $P(J)$

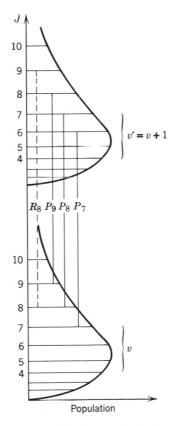

FIG. 9.32 Partial inversion with gain on the P_7, P_8, and P_9 transitions.

transition. It can be satisfied even when $N_{v'}$ is somewhat less than N_v. The exact condition for obtaining gain in partial inversion may be calculated in a simple form when it may be assumed that the rotational terms are adequately represented by the approximation given in (13.3), and when the same constant B is applicable to the upper and the lower vibrational level. Using the rotational level population as given by (13.5), the P-branch amplification condition takes the form

$$N_{v'} \exp\left[-\frac{F(J-1)hc}{kT_r}\right] > N_v \exp\left[-\frac{F(J)hc}{kT_r}\right]. \qquad (13.9)$$

Upon introducing the constant $\gamma = Bhc/kT_r$, this inequality becomes

$$N_{v'} \exp\left[-\gamma J(J-1)\right] > N_v \exp\left[-\gamma J(J+1)\right]. \qquad (13.10)$$

Hence

$$\exp 2\gamma J > \frac{N_v}{N_{v'}} \qquad (13.11)$$

and

$$J > \frac{1}{2\gamma} \log \frac{N_v}{N_{v'}}. \qquad (13.12)$$

This condition is satisfied for any J when there is complete inversion, but it may be satisfied for some J's without complete inversion. It is instructive to consider the consequences of (13.12) in the case of the lowest vibrational levels of nitrogen, when the rotational temperature is equivalent to 400°K. Then kT_r is equivalent to 280 cm^{-1} and $B = 2.0$ cm^{-1}, therefore $(2\gamma)^{-1} = 70$. Hence amplification will take place for $J > 70 \log N_v/N_{v'}$. A ten per cent excess population at the *lower* vibrational level would still permit amplification in the P-branch for $J > 7$.

The actual calculation of the gain as a function of J is more complicated. Curves have been computed giving this gain for various values of the ratio $N_v/N_{v'}$ [3].

The above analysis was carried out for the P-branch transitions. With respect to the other branches, the following conclusions may be reached: Q and R branches require complete inversion for gain, that is, $N_{v'} > N_v$. Even when such complete inversion exists, the gain on the P-branch transitions is greater than on the other branches, therefore P-branch transitions are expected to oscillate preferentially. This is actually observed.

Lasers on Electronic Transitions. Several common gases may be excited to laser oscillations involving transitions among different electronic levels of the molecule. These lasers produce the highest frequencies among the molecular lasers. Their spectrum consists of many closely spaced lines extending over bands, as one expects of a molecular spectrum. Lasers of this type are usually excited with high voltage pulses (10 to 30 kV). The pulses are of short duration (1 μsec or less) and carry sizeable current (50 to 100 A). The laser output is frequently self-terminating because of the filling up of the lower laser level.

NITROGEN. The nitrogen molecule is the basic material for a variety of lasers that operate on different transitions. The molecular electronic levels among which laser transitions have been observed are shown in Fig. 9.33. The most important nitrogen lasers are obtained in the first and second positive systems of the molecular spectrums. The lines arise on transitions among triplet levels of the molecule.

The first system corresponds to transitions from the $B^3\Pi_g$ to the $A^3\Sigma_u^+$

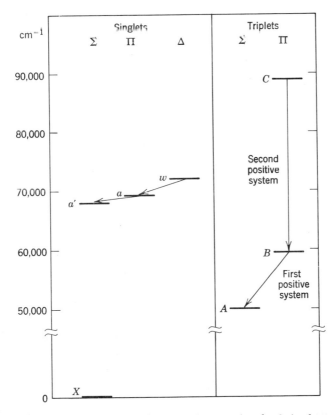

FIG. 9.33 Electronic levels of molecular nitrogen involved in lasers. (Arrows indicate laser transitions. Letters are the customary symbols used for the energy levels involved.)

electronic state. Mathias and Parker [74] discovered over 28 oscillating lines of this type in 1963, thereby commencing the development of molecular lasers. The early experiments soon revealed the power capabilities of this system, as more than 100-W peak output power was measured from oscillations in the wavelength ranges of 7700 to 7750 Å and 7580 to 7620 Å. Such power output from a tube 186 cm long and 1 cm in diameter was far in excess of the peak power levels attainable from gas lasers at that time. The detailed spectroscopy of the laser lines of the first positive system of N_2 was accomplished later by Kasuya and Lide [75], whose article contains extensive tables and diagrams. A condensed summary of these lines in our Table 9.7 indicates the complexity of this spectrum and the general spectral distribution of these laser lines.

TABLE 9.7

A Summary of Laser Lines Observed in
the First Positive System of N_2 [75]

Δv	Band $(v'-v)$	Wavelength Range in Air (Å)	No. of Lines Observed
-2	4–2	7482.7–7501.5	6
	3–1	7572.2–7625.1	15
	2–0	7712.1–7752.7	3
-1	2–1	8653.3–8722.2	34
	1–0	8833.7–8910.6	40
0	0–0	10435.8–10534.7	22
$+1$	0–1	12302.6–12346.3	7

The lines of the second positive system of N_2 lie in the near-ultraviolet. They result from the $C\Pi_u^+ - B^3\Pi_g$ transitions. At least 40 laser lines of this system were observed [76]. Most of these are 0–0 band transitions with 24 lines lying between 3370.44 and 3371.44 Å. All these are capable of very high gain and were observed in superradiant emission. Lines of the 0–1 band are found in the vicinity of 3576 Å. An extremely powerful pulsed nitrogen laser has been constructed operating at 3371 Å. In this laser the discharge occurs *across* the laser tube between long and sturdy electrodes extending about 180 cm along the optical path. Although the laser action is self-terminating in about 20 nsec, optical output power levels of 200 to 300 kw were attained [77].

Another group of laser lines of N_2 was observed by McFarlane [78]. The lines are in the infrared. The bands around 3.32, 3.47, and 8.25 μm were identified as belonging to the $a^1\Pi_g-a'^1\Sigma_u^-$ electronic transition and a band at 3.64 μm as belonging to the $w^1\Delta_u-a^1\Pi_g$ transition. The interesting fact about these lasers is not their practical potential, which is rather small, but the circumstance that these transitions were not known from ordinary spectroscopy, and that laser measurements on these lines made it possible to determine precisely the relevant characteristic of some uncommon electronic levels. With very few exceptions, the observed laser lines of N_2 belong to the P-branch.

CARBON MONOXIDE. Lasers based on an electronic transition in CO were also discovered by Mathias and Parker [79]. Oscillations occur on transition from the $B^1\Sigma-A^1\Pi$ electronic level. This system is called the *Angström system*. Laser oscillations were obtained on a number of transitions in the Q-branches of the following vibrational bands:

$v'-v$	Wavelength Range (Å)
0–3	5590.6–5603.8
0–4	6062.9–6074.2
0–5	6595.5–6613.5.

The oscillations are obtained under conditions similar to those favorable to nitrogen lasers with discharges excited by 30-kV, 80-A pulses in a gas of about 2-torr pressure. The oscillations are self-terminating in about 180 nsec. The optical output reported from these lasers does not compare with the high-intensity N_2 lasers.

HYDROGEN AND DEUTERIUM. Seven laser lines were observed on the $E^1\Sigma_g^+-B^1\Sigma_u^+$ electronic transition of H_2 [63, 80]. These are listed in Table 9.8. Some lines corresponding to these were also obtained in HD and in D_2. The highest gain is available on line 3. Apparently these lines of H_2 are comparatively easy to excite in discharges pulsed with 10 to 20 kV in a H_2 gas of 1 to 3-torr pressure.

Lasers on Vibration—Rotation Transitions. A number of lasers are known in which coherent radiation is obtained from molecules without a change in their electronic configurations. The technically most important laser of this type is the high-power CO_2 laser, but there are many other molecules including H_2O, CO, N_2O, and several cyanides which provide laser emission on transitions between vibration—rotation levels of their ground electronic state.

Carbon monoxide is the simplest of these molecules. Its structure is very similar to the N_2 molecule, with one important difference: the electric dipole moment of N_2 is zero in the ground electronic state because the two atoms that form the molecule are entirely alike. The situation is quite different for CO because the nuclear charges are not the same. While the vibration—rotation transitions of N_2 are not coupled to the electromagnetic field by electric dipole radiation, those of CO are.

TABLE 9.8
LASER LINES OF MOLECULAR HYDROGEN [63]

Line Number	Wavelength in Air (Å)	Transition $v'-v$	$P(J)$
1	8349.6	2–1	$P(2)$
2	8876.3	1–0	$P(4)$
3	8898.8	1–0	$P(2)$
4	11162.1	0–0	$P(4)$
5	11222.0	0–0	$P(2)$
6	13057.8	0–1	$P(4)$
7	13162.3	0–1	$P(2)$

Carbon monoxide has only one vibrational degree of freedom. Its vibrational structure is therefore considerably simpler than that of CO_2 and the other molecules mentioned above. Because of its structural simplicity, CO lends itself as a prototype for the description of laser phenomena on vibration—rotation transition. We will largely confine our discussion in this subsection to the results obtained on this gas. Carbon dioxide is the subject of the next subsection. The reader interested in vibration—rotation lasers in other gases is referred to specialized review articles [3, 81].

Several different excitation methods produce inversion in CO. Lasers have been made with direct pulse excitation in CO gas, with the excitation of N_2 in a discharge and subsequent transfer of excitation and, finally, by chemical reaction. These excitation processes do not in general produce the same laser lines, although some overlap does occur. The variation is not surprising in light of the fact that the different excitation processes need not favor the population of the same levels.

We defer the discussion of the chemically excited CO laser to the next section, which is dedicated to chemical lasers in general, and we shall concern ourselves now with the CO lasers whose energy of excitation is furnished by an electric discharge.

Direct excitation may be accomplished by 15-kV, 15-A direct current pulses of about 1-μsec duration passing through pure CO at about 0.8-torr pressure. These parameters were used by the discoverers of this laser, Patel and Kerl [82], who had a rather large tube (5 m long and 4.7 cm in diameter) to experiment with. It is likely that none of the parameters are too critical and that many of the laser lines are obtainable in smaller tubes. Laser oscillations occur in many vibration—rotation transitions, but only in the P-branches, which is not surprising considering the intrinsic advantage of these branches discussed earlier in this section. A summary of the observed spectral characteristics is given in Table 9.9. It is important to note that laser oscillations do not occur during the discharge, but that their onset is considerably delayed, the delay varying from one vibrational level to the other.

TABLE 9.9

SUMMARY OF THE SPECTRAL CHARACTERISTICS OF
CO VIBRATION-ROTATION LASERS OBTAINED IN
PULSED DISCHARGES [82]

Band $v'-v$	Wavelength Range [μm (vac)]	Range of J [in $P(J)$]	Time delay (μsec)
6–5	5.0375–5.1098	7–14	60
7–6	5.1041–5.1886	7–15	70
8–7	5.1722–5.2471	7–14	100
9–8	5.2419–5.3182	7–14	150
10–9	5.3241–5.3795	8–13	220

The evolution of the transitions in time is a very interesting phenomenon. The CO vibrational—rotational lasers are cascades, with oscillations starting at the lower end. Two cascade chains are given below, with their approximate starting times indicated. These times are reckoned from the beginning of the current pulse which lasts only about 1 μsec. The times are characteristics of the vibrational bands.

Two cascades in CO:

$$
\begin{array}{c}
v \\
J
\end{array}
\quad
\left.\begin{array}{c}10\\8\end{array}\right\}
\rightarrow
\left.\begin{array}{c}9\\9\end{array}\right\}
\rightarrow
\left.\begin{array}{c}8\\10\end{array}\right\}
\rightarrow
\left.\begin{array}{c}7\\11\end{array}\right\}
\rightarrow
\left.\begin{array}{c}6\\12\end{array}\right\}
\rightarrow
\left.\begin{array}{c}5\\13\end{array}\right\}
$$

t 220   150   100   70   60 μsec

$$
\begin{array}{c}
v \\
J
\end{array}
\quad
\left.\begin{array}{c}9\\8\end{array}\right\}
\rightarrow
\left.\begin{array}{c}8\\9\end{array}\right\}
\rightarrow
\left.\begin{array}{c}7\\10\end{array}\right\}
\rightarrow
\left.\begin{array}{c}6\\11\end{array}\right\}
\rightarrow
\left.\begin{array}{c}5\\12\end{array}\right\}
$$

The evolution of the cascade chain begins at the lower end of the energy scale. Oscillations spread upward as the laser terminal levels are depopulated. This type of laser action is naturally of rather limited duration. The oscillations are extinguished in about 30 to 50 μsec.

Continuous laser oscillation may be obtained in the vibration—rotation bands of CO by using nitrogen as an intermediary. This method of excitation was also developed by Patel [83]. It was originally intended for the study of the mechanism of the high-intensity CO_2 laser, but the experimental system has been used successfully for the production of laser oscillations in a number of other gases, including CO. The apparatus is illustrated in an abbreviated form in Fig. 9.34, the drawing being the representation of a tube approximately 2 m long. The discharge is excited continuously by means of a radio-frequency current—although the nature of the exciting current is of no great consequence. It is essential that in the experiment designed to explore the excitation process the discharge takes place in pure nitrogen, which is then pumped into the laser tube shown as the interaction region. In this region nitrogen is mixed with the other gas—in this case CO—whereupon transfer of

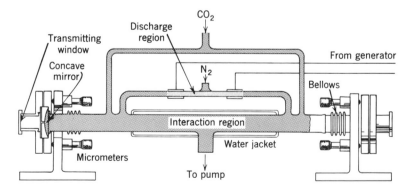

FIG. 9.34 Patel's apparatus for obtaining laser oscillations in CO_2 by vibrational energy transfer from $N_2{}^*(v = 1)$ to CO_2. (Drawing not to scale.) (Reprinted from *Physical Review Letters.*)

excitation takes place and the other gas becomes the active laser material. Naturally, a laser may be produced with discharge taking place in the already mixed gases. In the experiment described here, excitation of the active laser material is derived only from the products of discharge in N_2. What these are, and precisely how they excite the other molecules, is not uniformly clarified. In the case of the CO_2 laser the situation is relatively simple: An excitation to a level of CO_2 lying 2349 cm^{-1} above ground level is required, and this can easily be provided by N_2^* molecules in the metastable $v = 1$ band of the ground electronic state. The bottom of this band is at 2330 cm^{-1} above ground level. The situation is not so simple in the case of CO, where the results show that many vibrational levels are populated that have energies much higher than this metastable N_2 level. The $v = 18$ band of CO, for example, requires an energy of approximately 34500 cm^{-1}. This must come from more energetic by-products of the nitrogen discharge. What they are is uncertain. Experimental circumstances, described in the original publications [3, 84] pertaining to this type of CO laser, indicate that it is unlikely that vibrational levels of the N_2 ground state with large vibrational quantum numbers $(v \gg 1)$ are the carriers of the excitation energy.

About 160 laser lines were obtained in continuous oscillation, excited by means of an N_2 discharge. A summary catalog of these lines is Table 9.10. It is interesting to compare the material in this table with that in Table 9.9 because continuous wave laser oscillations do not occur on the same transitions that are listed in the earlier table. Even though there is overlap among the vibrational bands, the rotational laser levels are consistently higher in the continuous wave case. The lines represented in Table 9.9 do not all appear under identical conditions. They are the results of measurements made over the pressure range from 0.035 to 0.7 torr of CO. Within each band the low J-transitions were obtained at low pressures; with increasing J the optimum pressure for laser oscillation progressively increases. The low J-transitions are best obtained when the interaction region is cooled to −78°C, while the high J-transitions are favored at room temperature.

Patel's apparatus provides a large variety of laser lines in CO and permits the study of the underlying physical processes. It is not particularly suitable (or intended) for the production of high laser output. A high-power CO laser may be constructed by producing the discharge in a mixture of CO, N_2, and He. Helium is required in comparatively large quantities and serves to remove molecules from lower levels by cooling the discharge plasma. With a mixture of 0.3 torr CO, 1.4 torr air, and 8.8 torr He a laser output of 20 W was attained in a moderate-size tube cooled by liquid nitrogen [85].

TABLE 9.10

Summary of the Spectral Characteristics of
cw CO Vibration-Rotation Lasers Obtained
with Nitrogen-Assisted Excitation [3, 84]

Band $(v'-v)$	Wavelength Range [μm (vac)]	Range of J [in $P(J)$][a]
5–4	5.0869–5.1924	18–27
6–5	5.1316–5.2740	16–28
7–6	5.1886–5.3449	15–28
8–7	5.2471–5.4175	14–28
9–8	5.3296–5.4919	15–28
10–9	5.3911–5.5544	14–27
11–10	5.4540–5.6049	13–25
12–11	5.5790–5.6841	15–25
13–12	5.6330–5.7514	15–24
14–13	5.7136–5.8487	15–25
15–14	5.7835–5.9195	14–24
16–15	5.8945–5.9777	16–22
17–16	5.9818–6.0377	16–20
18–17	6.0575–6.1284	15–20
19–18	6.1490–6.2068	15–19
20–19	6.2432–6.2870	15–18
21–20	6.3260–6.3848	14–17
22–21	6.4252–6.4704	14–17
23–22	6.5120–6.5584	13–16
24–23	6.6476–6.6632	15–16

[a] A few isolated rotational transitions included in
the range were not observed.

Nitrogen-assisted excitation is effective for producing laser oscillations in a number of molecular gases. Many laser lines were observed by Patel using his original continuous gas-flow apparatus already described. Oscillations in N_2O are produced in the 10.77- to 11.04-μm region and are identified as transitions between 001 and 100 vibrational levels [86]. Laser oscillations in a CS_2—N_2 system occur between 11.48 and 11.54 μm. Their spectroscopic identification is not complete [87].

Carbon Dioxide Lasers. Laser oscillations may be obtained in pure CO_2, but such oscillations provide an output that is insignificant in comparison to the outputs of CO_2 lasers containing additional gases. One of these additives is N_2, which serves as a carrier of excitation, the other is a quenching gas, which serves to remove excitation from the terminal laser levels. This latter gas is usually He or H_2O. The role of N_2 was already briefly explored in connection with CO vibration—rotation lasers. Now we examine the entire cycle of the CO_2 laser in greater detail.

The vibrational levels of CO_2 involved in laser action are shown in Fig. 9.35, with the first vibrational level of N_2 shown on the right side. The various vibrational levels of these molecules form almost equally spaced ladders. The figure shows only the lowest rungs. The rotational levels are very closely spaced and are not shown in the figure.

The starting level of the CO_2 lasers is the $00°1$ level. It is the level of the lowest asymmetric-stretch vibrational mode. The terminal level is either the $10°0$ level (lowest symmetric vibration), or the $02°0$ level (second bending mode). Transitions take place between the rotational sublevels of these vibrational levels. The $00°1 \rightarrow 10°0$ transitions produce radiation around 10.6 μm, the $00°1 \rightarrow 02°0$ transitions around 9.6 μm. These two transition types are in competition with each other for molecules on their common starting level and the 10.6-μm oscillations win out, unless steps are taken to favor the shorter wavelength type.

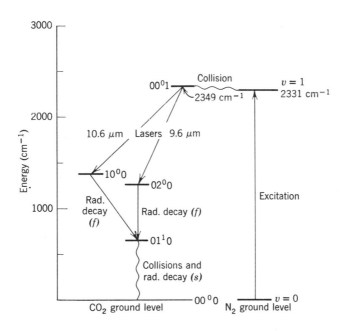

FIG. 9.35 Energy-level diagram showing pertinent vibrational levels of CO_2 and N_2. Rotational levels are omitted for simplicity; f = fast, s = slow radiative decay.

The $00^\circ 1$ level is a good starting level for several reasons, even in the absence of nitrogen. First, it is the lowest one of a series of equally spaced levels. Electron collision in the discharge causes the excitation of the CO_2 molecule to a number of levels of the type $00^\circ v_3$. When such a molecule excited to a $v_3 > 1$ state collides with a CO_2 molecule in its ground $(00^\circ 0)$ state, an energy exchange is likely to take place with the more energetic molecule losing one quantum of vibrational energy to the other molecule. In this manner a $(00^\circ v_3 - 1)$ and a $(00^\circ 1)$ molecule are produced, and ultimately most of the $00^\circ v_3$ type excitation is converted into $00^\circ 1$ type. A similar degradation process takes place on the other ladders, causing accumulation of molecules on the $10^\circ 0$ and $01^\circ 0$ levels. What is then the advantage of the $00^\circ 1$ level? First, it is the fact that direct electron collisions are more likely to excite the asymmetric than the symmetric stretch vibrations, as one can see intuitively by imagining the bombardment of a linear structure by fast, small particles from all sides. Second, the $10^\circ 0$ and the $02^\circ 0$ levels drain relatively fast to the $01^1 0$ level; so they are suitable as terminal laser levels. The bottleneck in the process is the draining of the $01^1 0$ levels which represent vibrational energy in the lowest bending mode. The energy of this vibrational mode must be removed by collisions, but this time the energy exchange is no longer a resonant one; the vibrational energy must be converted into translational kinetic energy. Such a conversion proceeds at a slower rate than resonant exchanges. The de-excitation may be greatly enhanced by the presence of He or water vapor and by keeping the walls of the laser tube cool.

Most important is the role of nitrogen, whose molecules accumulate in the $v = 1$ vibrational level, as a result of excitation by electron collision to levels of different v's and subsequent collisions with ground-state N_2 molecules. On collision with ground-state CO_2 molecules a resonant transfer of excitation takes place at a relatively high rate because the energy difference is only 18 cm^{-1}, less than the mean thermal energy of the molecules. Transfer may also take place at the $v = 2$ and 3 levels, but the contribution of these transfers is not significant.

Both the 10.6-μm and the 9.6-μm laser transitions may occur over a number of rotational levels. In the first group—the $00^\circ 1$ to $10^\circ 0$ band—the P-branch transitions extend from $P(12)$ to $P(38)$, covering the wavelength range 10.5135 μm to 10.7880 μm, the lines with the highest gain being around $P(22)$ with a wavelength of 10.6118 μm (in vacuo) [3]. For the second group the rotational levels extend from $P(22)$, 9.5691 μm, to $P(34)$, 9.6762 μm with strongest transitions in the middle of this range [3]. The oscillations usually occur in only one of the rotational transitions. This is the case at any rate in low-power CO_2 lasers operated in a steady state. Because of the rapid thermalization of the rotational sublevels, they replenish each other when one is depleted. Therefore, when radiation density builds up faster in one of the several competing rotational transitions than in the others, oscillation begins at a wavelength corresponding to that transition, and the other levels tend to lose their population by transferring to the favored level which started its oscillations. It is possible to influence the spectral output by adjusting the cavity length and by incorporating frequency-selective elements into the laser as a means of favoring one of the rotational levels. The situation is quite different in Q-switched operation, when the pulse develops in a time shorter than the thermal relaxational time of the rotational levels. In this case several rotational transitions will break into oscillation at the same time.

As was demonstrated by Patel [83], laser oscillations may be obtained in CO_2 even when there is no discharge in that gas and when all excitation is furnished by means of N_2^* molecules, excited in a side arm of an instrument shown in Fig. 9.34. This technique is not used in ordinary CO_2 lasers whose purpose is the production of large optical output. The gases are mixed in power lasers and the discharge is established through the mixture so that there is only one tube. For a CO_2 laser of relatively small power output, say 5 to 50 W, the structure may look like the one shown in Fig. 9.11. The typical gas mixture in such a laser is 3 torr CO_2, 5 torr N_2 and 14 torr He. The discharge is operated with a dc supply that provides a voltage drop of 5 to 8 kV over the tube with an approximately equal drop over a ballast resistor capable of relatively high dissipation to carry a current of 8 to 25 mA, which is a normal load for a small CO_2 laser. The conversion efficiency of a well-designed and well-operating laser of this type is 15 to 20% when related to the electric power dissipated in the discharge tube.

Carbon dioxide lasers may be sealed off. Those sold commercially are usually sealed for convenience. Sealing the laser creates new technical problems which manifest themselves in the deterioration and limited lifetime of these lasers. Several changes in the gas composition have been proposed to prolong the satisfactory operation of sealed-off lasers.

The reflectors of small CO_2 lasers may be glass or quartz concave lenses with their front surface gold-coated for reflection. The centers are drilled through to provide an exit port for the output. High-power CO_2 lasers must have mirrors that may be cooled, such as steel reflectors with gold coating. In fact, lasers with an output of 50 W and over present a variety of technological and safety problems that are beyond the scope of this book. They are potentially dangerous weapons that may cause fire and destruction in the laboratory. There are lasers with continuous wave outputs of several kilowatts. They are rather large structures requiring special installations.

Carbon dioxide lasers have been the subject of many investigations. The interested reader will find considerable additional information and references in review articles [3, 88].

Thermal Excitation of Molecular Lasers. We already noted (see p. 345) that thermal equilibrium is approached rapidly among the rotational sublevels of each vibrational level, but that the thermalization process between vibrational levels is approximately one thousand times slower. This great disparity between thermalization rates makes it possible to produce partial inversion by purely thermal means. A gas is first raised to a high temperature, thereby nearly equalizing the occupancy of two of its consecutive vibrational levels. It is then cooled so suddenly that very little rearrangement takes place between different vibrational levels, but the molecules thermalize among the rotational sublevels of each vibrational level. Population inversion may then occur for several of the higher P-branch transitions in the manner described.

The possibility of utilizing such a mechanism for the excitation of laser oscillations was pointed out in 1963 by Basov and Oraevskii, who later proposed specific laser systems utilizing N_2 as a carrier gas, and other gases—such as CO and CO_2—as working gases [89]. The scheme starts out with N_2 at high temperature rapidly cooled by adiabatic or nozzle-flow expansion, which removes a large fraction of its translational kinetic energy and which results in the production of N_2 gas at a moderate temperature enriched in molecules in the N_2^* ($v = 1$) vibrational level. These molecules then excite the CO or CO_2 molecules in the same manner as similar electrically excited N_2 molecules do in Patel's experiments.

An entirely thermally excited CO_2 laser built on this principle was constructed and successfully operated by Fein, Verdeyen, and Cherrington [90]. The nitrogen gas is first heated to 1200°C at 400-torr pressure, then allowed to expand through a nozzle into a region of much lower pressure where it is mixed with CO_2. The usual laser lines of CO_2 are obtained. So far only small amounts of laser power were obtained in this manner, in comparison to the power obtainable by electric excitation.

9.14 CHEMICAL LASERS

The term "chemical laser" describes a device in which population inversion is established as a direct result of a chemical reaction. Simple as this definition is, it leaves a gray area because it is not entirely clear how direct and how exclusive the chemical action must be before a laser is called a *chemical* laser. The oxygen lasers produced by electric discharges in Ne-O_2 and Ar-O_2 mixtures involve a chemical reaction in the excitation process, namely the process of disociation. Nevertheless, the energy for the excitation of these lasers derives from the electric discharge, therefore it would not be proper to include them among chemical lasers.

Closer to the boundary line is the photodissociation laser in iodine. This may be regarded as a precursor of truly chemical lasers, partly because it was discovered accidentally in the course of a search for truly chemical lasers. Irradiation of CF_3I by a powerful flash from a Xe flashtube causes photodissociation of the molecule producing I in the excited $^2P_{1/2}$ state. The transition to the $^2P_{3/2}$ state then results in laser oscillation at 1.30 μm [91].*

The next step in the evolution of chemical lasers is also based on photodissociation. A one-to-one mixture of CS_2 and O_2 is irradiated. As a result, CS_2 dissociates and CO forms. The CO molecules are formed in vibrationally excited states as demonstrated by the fact that under proper circumstances a large number of laser oscillations can be obtained on vibrational levels ranging from $v = 5$ to $v = 13$. Most of these lines are contained among the vibration—rotation lines of CO listed in Table 9.9. Successful operation of this type of a laser requires the presence of a large amount of He. Pollack [92], who discovered this laser, recommends 1 torr CS_2, 1 torr O_2, and 150 torr He. In a sense this CO laser is already chemical, the formation of the CO molecule provides the excitation energy, but naturally the system does not work without the light flash providing the energy necessary for dissociation.

The first truly chemical laser was created in 1965 by Kasper and Pimentel [93]. The population inversion in HCl is obtained as a result of the chemical reaction

$$H + Cl_2 \rightarrow HCl^* + Cl, \tag{14.1}$$

where * indicates the formation of hydrogen chloride in vibrationally excited states. Laser oscillations are obtained on several P-branch lines of the 2–1 and 1–0 vibrational transitions. The wavelength region is 3.70 to 3.84 μm. The chemical process takes place in a mixture of H_2 and Cl_2 and the reaction is initiated by light causing the photolysis of Cl_2:

$$Cl_2 + h\nu \rightarrow Cl + Cl. \tag{14.2}$$

While molecular chlorine does not interact with molecular hydrogen,

---

* The $^2P_{1/2}$-$^2P_{3/2}$ transition in I is a magnetic dipole transition.

atomic chlorine does, and the following reaction takes place

$$Cl + H_2 \rightarrow HCl + H. \tag{14.3}$$

No excited molecule has been produced so far, but atomic hydrogen has been made. The next reaction is (14.1), which produces the excited molecules. Since an atom of free chlorine is also produced, this atom can continue the (14.3) reaction and light is no longer needed. Thus light only serves as a trigger.

Another type of chemical laser involves the use of an electric discharge. The simplest materials involved in such lasers are the halides of hydrogen and deuterium formed in an electric discharge. With H_2 and CF_4 as a starting material, HF is formed in an excited vibrational state when CF_4 is dissociated in a discharge. The excitation is the result of the chemical reaction, not electron collisions. Excited molecules of HCl and HBr are formed in mixtures of H_2 and Cl_2, and H_2 and Br_2, respectively. Similar experiments can be carried out with deuterium instead of H_2, and a large variety of laser lines is obtained [3, 94]. The most important of these are the vibration—rotation lines of HCl which occur in the P-branches of the 3–2, 2–1 and 1–0 vibrational transitions covering the wavelength range 2.70 to 3.04 μm. Deutsch, who discovered this type of vibration—rotation laser, also discovered a HF laser based on purely rotational transitions [95]. The purely rotational transitions occur within the first few vibrational levels. They are summarized as follows:

Vibrational level	Rotational level range	Wavelength range
0	15 to 27	10.20 to 16.02 μm
1	11 to 22	12.26 to 21.70 μm
2	12 to 29	10.58 to 20.94 μm
3	12 to 27	11.54 to 21.79 μm

All these oscillations were found in HF formed from H_2 and CF_4, $CClF_3$, and CCl_3F. The operation of the laser requires a rather powerful discharge with 50– to 200-A pulses of about 1-μsec duration. Thus it is clear that this chemical laser still requires the use of some electrical power.

9.15 STABILITY IN WAVELENGTH AND FREQUENCY

Gas lasers are the finest sources of light as far as intensity, coherence, and spectral purity is concerned. They are, therefore, excellent tools in metrological investigations which involve interferometry and fringe counting. The stability and reproducibility of radiation from gas lasers is therefore of great concern. The question arises: Could lasers supplant ordinary, incoherent spectral standards of length? This question is still being actively studied. We shall now examine the main factors that determine the spectral content of laser radiation and describe some of the methods proposed for making the wavelength of a laser well defined, stable, and reproducible.

Research and development in this field centers around He-Ne lasers with concentration on the visible 6328-Å laser. We will accordingly confine our attention to the properties of this laser.

In the region of the electromagnetic spectrum accessible with ordinary electronic and radio techniques, one is accustomed to speak of precision and stability of *frequency* because all measurements are based on absolute frequency measurements or frequency comparisons. In the optical region, wavelength is directly measured, frequency is only calculated from the measured values of the wavelength and the velocity of light. Because of the uncertainty in the latter, frequencies are not as precisely known as wavelengths.

Although absolute measurements of laser frequencies in the visible range have not yet been made, frequency comparisons between similar stabilized lasers are made often. These measurements of frequency differences are accomplished by bringing radiation from both lasers together on a photodetector and detecting the frequency of the beat signal. Measurements of this type are used to explore the mode structure of lasers because it is relatively easy to measure the differences of frequencies emitted by the same laser in different modes. These differences, although in the MHz or GHz range, are still extremely small in comparison to the laser frequencies.

Frequency stability and wavelength stability are identical. They are numerically characterized by the ratio

$$S = \frac{\lambda}{\Delta\lambda} = \frac{\nu}{\Delta\nu},$$

where $\Delta\lambda$ and $\Delta\nu$ are wavelength and frequency fluctuations, or variations, under given circumstances. Following the established practice, we speak of frequency stability even when only wavelength comparisons are made.

Three characteristic stability figures are of interest: The short time stability S_s indicates the constancy of the frequency during a single measurement, say a fringe count of the interferometer. The long term stability S_l determines the constancy of the laser frequency over a whole series of measurements, over a period of hours perhaps. Finally, the absolute reproducibility, or resettability S_0 refers to the precision with which the frequency can be reproduced in a similarly constructed laser. The three stability figures differ, of course, in the conditions specifying the frequency spread, $\Delta\nu$.

As was explained in Chapter 3, the factor that broadly defines the spectral region in which a laser may emit is the amplification curve, which is the same thing as the absorption line reversed. Laser oscillations may take place at cavity-mode frequencies located in that region near the center of the line where amplification exceeds the minimum required by the threshold condition. This situation was graphically represented in Fig. 3.6, and it was noted that a laser will in general emit in more than one mode. When the laser is to be used in metrology, matters must be so arranged that only one mode of oscillation is present. The off-axis modes are eliminated by insuring a symmetrical structure and by a diaphragm, if necessary. Excitation of a single axial mode is insured by reducing excitation to a minimum and by spacing these modes out so that few of them will lie near the peak of the atomic line. Under these circumstances the precise frequency of the laser output is determined by the dimensions of the Fabry-Perot cavity and by the index of refraction of the materials within it. Even the smallest variations in the distance between the mirrors and in the refractive index of air, windows, and gas are reflected in frequency changes.

To reduce the fluctuations in frequency to a minimum, the laser structure must be stable—isolated from mechanical vibrations and thermal fluctuations. A simple and quite economical method for the construction of a stable He—Ne laser was developed at the Philips Research Laboratories in Eindhoven (The Netherlands) [96]. The cross section of such a laser is shown in Fig. 9.36. Its main element is a fused quartz rod of about 35-mm diameter with a thin coaxial bore. Side channels are bored into the center hole to connect side arms containing the electrodes and possibly a gas reservoir. The ends of the quartz rod are ground and polished strictly parallel to each other and perpendicular to the axis of the bore. Optical flats are provided with reflective, dielectric coating in their central areas and are wrung together with the quartz rod. The adhesion forces between these optically flat surfaces are sufficient to hold the structure rigidly together in proper alignment and to keep the system vacuum-tight. With the optical flats only about 12 cm from each other, the axial modes are 1250 MHz apart. Since this is about the same order of magnitude as the width of the 6328-A line, it is easy to excite only one mode of oscillation. Because of the rigidity of the structure and its relative insensitivity to thermal fluctuations, the influence of the environment on this laser is small. The stability of such lasers was measured by beating the radiation derived from two

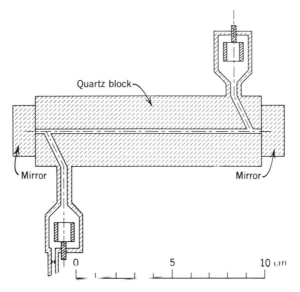

FIG. 9.36 Construction of short plane-mirror gas laser.

lasers. A short-time stability in excess of 10^9 has been measured for a period of several minutes. The reproducibility of such a laser is, of course, poor, a value of S_0 between 10^6 and 10^7 is indicated so that for precision measurements each of these lasers must be calibrated against a standard.

More elaborate and ingenious techniques are necessary when an absolute stabilization is desired, that is, when the laser frequency is made to coincide with the center of an atomic or molecular line, and adjustments are made in the cavity structure to insure that the cavity frequency is kept in coincidence with the center of the atomic line in spite of environmental fluctuations. The center of the atomic line may be located by the use of the Lamb dip. As the length of the cavity is slowly changed, the active mode frequency sweeps through the center of the Doppler-broadened line with a resultant dip in the output at the center. The process of searching for the center of the line can be made automatic as was done by Shimoda and Javan [97], who provided frequency modulation by driving one of the mirrors back and forth sinusoidally with a magnetostrictive drive. The resulting amplitude-modulated laser output was detected, and the fundamental, the second, and third harmonics of the modulation signal were used to control the mean position and the orientation of the mirrors in such a manner as to keep the oscillating mode at the center of the line. This method was applied to a 1.15-μm He—Ne laser produced in a neon gas containing only one isotope. The resettability of this laser was around 10^8 and the short-time stability over 10^9. A commercial laser (Spectra Physics Model 119) operating at 6328 Å is locked to the atomic line by means of the Lamb dip. It has a long-term stability of 5×10^8 per day and a resettability of 1×10^8. In the long run, pressure shifts degrade the precision of all these lasers.

Several ingenious tricks have been proposed to overcome the difficulty of precisely locating the top of the atomic line—a difficulty which arises from its flatness. One of these schemes involves the splitting of the atomic line by means of the Zeeman effect and the location of the intersection of two steep curves. This intersection occurs at the center frequency of the original line [98]. Another scheme stabilizes the 6328-Å line on a hyperfine transition of I_2 with the use of an iodine absorption cell [99].

The most precise measurements on the wavelengths of Lamb-dip-stablized He—Ne lasers were collated as a cooperative effort of American, British, and German national standards laboratories. Lasers of different construction and origin were consistent with themselves to better than one part in 10^8, but the overall wavelength measurements varied from 6329.91373 Å to 6329.91448 Å, indicating that an uncertainty of about one part in 10^7 must be associated with the wavelength of stabilized He—Ne lasers. The measurements represent wavelengths in vacuum. They were obtained by comparison with the Kr^{86} standard [100].

Additional information on stabilization and metrology may be found in the review articles of Birnbaum [101] and Polanyi and Tobias [102].

REFERENCES

1. G. Herzberg, *Spectra of Diatomic Molecules*, Van Nostrand, New York, 1950.
2. H. S. W. Massey and E. H. S. Burhop, *Electronic and Ionic Impact Phenomena*, Clarendon, Oxford, 1952.
3. C. K. N. Patel, *Gas Lasers, Lasers, Vol. 2*, A. K. Levine, Ed., Dekker, New York, 1968.
4. A. C. G. Mitchell and M. W. Zemansky, *Resonance Radiation and Excited Atoms*, Cambridge University Press, Cambridge, 1934, 1961.
5. D. C. Sinclair and W. E. Bell, *Gas Laser Technology*, Holt, Rinehart and Winston, New York, 1969.
6. W. R. Bennett Jr., Hole-burning effects in a He—Ne optical maser, *Phys. Rev.*, **126**, 580–593 (1962).
7. W. E. Lamb, Theory of an optical maser, *Phys. Rev.*, **134**, A 1429–1450 (1964).
8. R. A. McFarlane, W. R. Bennett Jr., and W. E. Lamb, Single-mode tuning-dip in the power output of an He—Ne optical maser, *Appl. Phys. Letters*, **2**, 189–190 (1963).
9. A. Szöke and A. Javan, Isotope shift and saturation behavior of the 1.15-μ transition of neon, *Phys. Rev. Letters*, **10**, 521–524 (1963).

10. A. L. Schawlow and C. H. Townes, Infrared and optical masers, *Phys. Rev.*, **112**, 1940–1949 (1958).
11. H. Z. Cummins, I. Abella, O. S. Heavens, N. Knable, and C. H. Townes, Alkali vapor infrared masers, *Advances in Quantum Electronics*, J. R. Singer, Ed., Columbia University Press, New York, 1961, pp. 12–17.
12. P. Rabinowitz, S. Jacobs, and G. Gould, Continuous optically pumped Cs laser, *Appl. Opt.*, **1**, 513–516 (1962).
13. C. H. Moore, *Atomic Energy Levels*, Natl. Bur. Std. (U.S.), Circ. 467, Vols. 1–3, U.S. Govt. Printing Office, Washington, D.C., 1949–1958.
14. G. F. Koster and H. Statz, Probabilities for the neon laser transitions, *J. Appl. Phys.*, **32**, 2054–2055 (1961).
15. W. L. Faust and R. A. McFarlane, Line strengths for noble-gas maser transitions, *J. Appl. Phys.*, **35**, 2010–2015 (1964).
16. W. R. Bennett, Radiative lifetimes and collision transfer cross sections of excited atomic states, *Advances in Quantum Electronics*, J. R. Singer, Ed., Columbia University Press, New York, 1961, pp. 28–43.
17. A. Javan, W. R. Bennet, Jr., and D. R. Herriott, Population inversion and continuous optical maser oscillation in a gas discharge containing a He-Ne mixture, *Phys. Rev. Letters*, **6**, 106–110 (1961).
18. A. D. White and J. D. Rigden, Continuous gas maser operation in the visible, *Proc. IRE*, **50**, 1697 (1962).
19. A. D. White and J. D. Rigden, The effect of super-radiance at 3.39μ on the visible transitions in the He-Ne maser, *Appl. Phys. Letters*, **2**, 211–212 (1963).
20. A. L. Bloom, W. E. Bell, and R. C. Rempel, Laser operation at 3.39μ in a helium-neon mixture, *Appl. Opt.*, **2**, 317–318 (1963).
21. W. R. Bennett, Jr., Gaseous optical masers, *Appl. Opt. Suppl.*, **1**, 24–61 (1962).
22. A. L. Bloom, *Gas Lasers*, Wiley, New York, 1968.
23. E. I. Gordon and A. D. White, Similarity laws for the effects of pressure and discharge diameter on gain of He-Ne lasers, *Appl. Phys. Letters*, **3**, 199–201 (1963).
24. A. D. White, E. I. Gordon, and J. D. Rigden, Output power of the 6328-Å gas maser, *Appl. Phys. Letters*, **2**, 91–93 (1963).
25. I. M. Belousova, O. B. Danilov, and I. A. Elkina, Optimum operating mode of a He-Ne laser, *Soviet Phys. JETP*, **17**, 748–749 (1963); **44**, 1111–1113 (1963).
26. P. W. Smith, On the optimum geometry of a 6328-Å laser oscillator, *IEEE J. Quant. Electr.*, **QE-2**, 77–79 (1966).
27. B. A. Lengyel, *Introduction to Laser Physics*, Wiley, New York, 1966.
28. C. K. N. Patel, W. R. Bennett, W. L. Faust, and R. A. McFarlane, Infrared spectroscopy using stimulated emission techniques, *Phys. Rev. Letters*, **9**, 102–104 (1962).
29. R. Cagnard, R. der Agobian, R. Echard, and J. L. Otto, L'émission stimulée de quelques transitions infrarouges de l'helium et du néon, *Compt. Rend.*, **257**, 1044–1047 (1963).
30. J. S. Levine and A. Javan, Far infrared continuous-wave laser oscillation in pure helium, *Appl. Phys. Letters*, **14**, 348–350 (1969).
31. C. K. N. Patel, W. L. Faust, R. A. McFarlane, and C. G. B. Garrett, Laser action up to 57.355 microns in gaseous discharges, *Appl. Phys. Letters*, **4**, 18–19 (1964).
32. W. L. Faust, R. A. McFarlane, C. K. N. Patel, and C. G. B. Garrett, Noble gas optical maser lines at wavelengths between 2 and 35 microns, *Phys. Rev.*, **133**, A 1466–1486 (1964).

33. C. K. N. Patel, W. L. Faust, R. A. McFarlane, and C. G. B. Garrett, Cw optical maser action up to 133μ in Ne discharges, *Proc. IEEE*, **52**, 713 (1964).
34. H. J. Gerritsen and P. V. Goedertier, A gaseous (He-Ne) cascade laser, *Appl. Phys. Letters*, **4**, 20–21 (1964).
35. R. Grudzinski, M. Paillette, and J. Becrelle, Étude de transitions laser couplées dans un mélange helium-néon, *Compt. Rend.*, **258**, 1452–1454 (1964).
36. D. Rosenberger, Oscillation of three 3p-2s transitions in a He-Ne laser, *Phys. Letters*, **9**, 29–31 (1964).
37. R. der Agobian, R. Cagnard, E. Echard, and J. L. Otto, Nouvelle cascade de transitions stimulées du néon, *Compt. Rend.*, **258**, 3661–3663 (1964).
38. D. Rosenberger, Laserübergänge and Superstrahlung bei 6143 Å und 5944 Å in einer gepulsten Neon-Entladung, *Phys. Letters*, **13**, 228–229 (1964).
39. D. M. Clunie, R. S. A. Thorn, and K. E. Trezise, Asymetric visible superradiant emission from a pulsed neon discharge, *Phys. Letters*, **14**, 28–29 (1965).
40. K. G. Ericsson and L. R. Lidholt, Superradiant transitions in argon, krypton and xenon, *IEEE J. Quant. Electr.*, **QE-3**, 94 (1967).
41. O. Andrade, M. Gallardo, and K. Bockasten, High-gain laser lines in noble gases, *Appl. Phys. Letters*, **11**, 99–100 (1967).
42. K. Bockasten, T. Lundholm, and O. Andrade, New near-infrared laser lines in Ar I, *Phys. Letters*, **22**, 145–146 (1966).
43. R. A. Paananen and D. L. Bobroff, Very high gain gaseous (Xe-He) optical maser at 3.5 μ, *Appl. Phys. Letters*, **2**, 99–100 (1963).
44. H. Brunet, Laser gain measurement in a xenon-krypton discharge, *Appl. Opt.*, **4**, 1354 (1965).
45. D. C. Sinclair, Near-infrared oscillation in pulsed noble-gas-ion laser, *J. Opt. Soc. Am.*, **55**, 571–572 (1965).
46. D. Rosenberger, Superstrahlung in gepulsten Argon-, Krypton-, und Xenon-Entladungen, *Phys. Letters*, **14**, 32 (1965).
47. W. E. Bell, Visible laser transitions in Hg⁺, *Appl. Phys. Letters*, **4**, 34–35 (1964).
48. W. B. Bridges, Laser oscillation in singly ionized argon in the visible spectrum, *Appl. Phys. Letters*, **4**, 128–130 (1964); err. **5**, 39 (1964).
49. W. B. Bridges and A. N. Chester, Visible and uv laser oscillation at 118 wavelengths in ionized Ne, Ar, Kr, Xe, O, and other gases, *Appl. Opt.*, **4**, 573–580 (1965).
50. J. P. Goldsborough, E. B. Hodges, and W. E. Bell, Rf-induction excitation of cw visible laser transitions in ionized gases, *Appl. Phys. Letters*, **8**, 137–139 (1966).
51. E. I. Gordon, E. F. Labuda, R. C. Miller, and C. E. Webb, Excitation mechanisms of the argon-ion laser, *Physics of Quantum Electronics*, P. L. Kelley, B. Lax and P. E. Tannenwald, Eds., McGraw-Hill, New York, 1966, pp. 664–687.
52. E. I. Gordon, E. F. Labuda, and W. B. Bridges, Continuous visible laser action in singly ionized argon, krypton, and xenon, *Appl. Phys. Letters*, **4**, 178–180 (1964).
53. W. B. Bridges and A. N. Chester, Spectroscopy of ion lasers, *IEEE J. Quant. Electr.*, **QE-1**, 66–84 (1965).
54. H. A. Statz et al., Transition probabilities, lifetimes and related considerations in ionized argon lasers, *Physics of Quantum Electronics*, P. L. Kelley, B. Lax and P. E. Tannenwald, Eds., McGraw-Hill, New York, 1966, pp. 674–687.
55. T. J. Bridges and W. W. Rigrod, Output spectra of the argon ion laser, *IEEE J. Quant. Electr.*, **QE-1**, 303–308, (1965).

56. W. E. Bell, A. L. Bloom, and J. P. Goldsborough, Visible laser transitions in ionized Se, As, and Br, *IEEE J. Quant. Electr.*, **QE-1**, 400 (1965).
57. W. B. Bridges and A. S. Halstead, New cw laser transitions in Ar, Kr, and Xe, *IEEE J. Quant. Electr.*, **QE-2**, 84 (1966).
58. R. H. Neusel, New laser oscillations in Kr and Xe, *IEEE J. Quant. Electr.*, **QE-2**, 334 (1966); **2**, 758 (1966).
59. H. G. Cooper and P. K. Cheo, Ion laser oscillations in sulfur, *Physics of Quantum Electronics*, P. L. Kelley, B. Lax, and P. E. Tannenwald, Eds., McGraw-Hill, New York, 1966, pp. 690–697.
60. C. B. Zarowin, New visible cw laser lines in singly-ionized chlorine, *Appl. Phys. Letters*, **9**, 241–242 (1966).
61. W. R. Bennett, Jr., W. L. Faust, R. A. McFarlane, and C. K. N. Patel, Dissociative excitation transfer and optical maser oscillation in Ne-O_2 and Ar-O_2 rf discharges, *Phys. Rev. Letters*, **8**, 470–473 (1962).
62. C. K. N. Patel, R. A. McFarlane, and W. L. Faust, Optical maser action in C, N, O, S, and Br on dissociation of diatomic and polyatomic molecules, *Phys. Rev.*, **133**, A1244–1248 (1964).
63. K. Bockasten, T. Lundholm, and O. Andrade, Laser lines in atomic and molecular hydrogen, *J. Opt. Soc. Am.*, **56**, 1260–1261 (1966).
64. S. M. Jarrett, J. Nunez, and G. Gould, Laser oscillation in atomic Cl in HCl gas discharges, *Appl. Phys. Letters*, **8**, 150–151 (1966).
65. R. A. Paananen, C. L. Tang, and F. A. Horrigan, Laser action in Cl_2 and He-Cl_2, *Appl. Phys. Letters*, **3**, 154–155 (1963).
66. K. Bockasten, On the classification of laser line in Cl and I, *Appl. Phys. Letters*, **4**, 118–119 (1964).
67. S. M. Jarrett, J. Nunez, and G. Gould, Infrared oscillation in HBr and HI gas discharge, *Appl. Phys. Letters*, **7**, 294–296 (1965).
68. P. K. Cheo and H. G. Cooper, Uv and visible laser oscillations in fluorine, phosphorus and chlorine, *Appl. Phys. Letters*, **7**, 202–204 (1965).
69. W. M. Keeffe and W. J. Graham, Observation of new Br II laser transitions, *Phys. Letters*, **20**, 643 (1966).
70. G. R. Fowles and R. C. Jensen, Visible laser transitions in ionized iodine, *Appl. Opt.* **3**, 1191–1192 (1964).
71. K. Bockasten, M. Garavaglia, B. A. Lengyel, and T. Lundholm, Laser lines in Hg I, *J. Opt. Soc. Am.*, **55**, 1051–1053 (1965).
72. A. L. Bloom, W. E. Bell, and F. O. Lopez, Laser spectroscopy of a pulsed mercury-helium discharge, *Phys. Rev.*, **135**, A578–579 (1964).
73. H. J. Gerritsen and P. V. Goedertier, Blue gas laser using Hg^{2+}, *J. Appl. Phys.*, **35**, 3060–3061, 1964.
74. L. E. S. Mathias and J. T. Parker, Stimulated emission in the band spectrum of nitrogen, *Appl. Phys. Letters*, **3**, 16–18 (1963).
75. T. Kasuya and D. R. Lide, Measurements on the molecular nitrogen pulsed laser. *Appl. Opt.*, **6**, 69–80 (1967).
76. V. M. Kaslin and G. G. Petrash, Rotational structure of ultraviolet generation in molecular nitrogen, *Soviet Phys.-JETP*, **3**, 55–58 (1966).
77. J. D. Shipman and A. C. Kolb, A high-power, pulsed nitrogen laser, *IEEE J. Quant. Electr.*, **QE-2**, 298 (1966).
78. R. A. McFarlane, Precision spectroscopy of new infrared emission systems of molecular nitrogen, *IEEE J. Quant. Electr.*, **QE-2**, 229–232 (1966).
79. L. E. S. Mathias and J. T. Parker, Visible laser oscillations from carbon monoxide, *Phys. Letters*, **7**, 194–196 (1963).

80. P. A. Bazhulin, I. N. Knyazev, and G. G. Petrash, Stimulated emission from molecular hydrogen and deuterium in the near infrared, *Soviet Phy.-JETP,* **22,** 11–16 (1966), [**49,** 16–23 (1965)].

81. W. S. Benedict, M. A. Pollack, and W. J. Tomlinson, The water vapor laser, *IEEE-J. Quant. Electr.,* **QE-5,** 108–124 (1969).

82. C. K. N. Patel and R. J. Kerl, Laser oscillations on $X^1\Sigma^+$ vibrational-rotational transitions of CO, *Appl. Phys. Letters,* **5,** 81–83 (1964).

83. C. K. N. Patel, Selective excitation through vibrational energy transfer and optical maser action in N_2-CO_2, *Phys. Rev. Letters,* **13,** 617–619 (1964).

84. C. K. N. Patel, Cw laser on vibrational-rotational transitions of CO, *Appl. Phys. Letters,* **7,** 246–247 (1965).

85. R. M. Osgood and W. C. Eppers, High-power CO-N_2-He laser, *Appl. Phys. Letters,* **13,** 409–411 (1968).

86. C. K. N. Patel, Cw laser action in N_2O, *Appl. Phys. Letters,* **6,** 12–13 (1965).

87. C. K. N. Patel, Cw laser oscillations in an N_2-CS_2 system, *Appl. Phys. Letters,* **7,** 273–274 (1965).

88. K. Gürs, Der CO_2 Laser, *Z. angew. Physik,* **25,** 379–386 (1968).

89. N. G. Basov, A. N. Oraevskii, and V. A. Shcheglov, Thermal methods for laser excitation, *Soviet Phys.-Techn. Phys.,* **12,** 243–249 (1967), [**37,** 339–348 (1967)].

90. M. E. Fein, J. T. Verdeyen, and B. E. Cherrington, A thermally pumped CO_2 laser, *Appl. Phys. Letters,* **14,** 337–340 (1969).

91. J. V. V. Kasper and G. C. Pimentel, Atomic iodine photodissociation laser, *Appl. Phys. Letters,* **5,** 231–233 (1964).

92. M. A. Pollack, Laser oscillation in chemically formed CO, *Appl. Phys. Letters,* **8,** 237–238 (1966).

93. J. V. V. Kasper and G. C. Pimentel, HCl chemical laser, *Phys. Rev. Letters,* **14,** 352–354 (1965).

94. T. F. Deutsch, Molecular laser action in hydrogen and deuterium halides, *Appl. Phys. Letters,* **10,** 234–236 (1966).

95. T. F. Deutsch, Laser emission from HF rotational transitions, *Appl. Phys. Letters,* **11,** 18–20 (1967).

96. H. G. Van Buren, J. Haisma, and H. De Lang, A small and stable continuous gas laser, *Phys. Letters,* **2,** 340–341 (1962).

97. K. Shimoda and A. Javan, Stabilization of the He-Ne maser on the atomic line center, *J. Appl. Phys.,* **36,** 718–726 (1965).

98. A. D. White, E. I. Gordon, and E. F. Labuda, Frequency stabilization of single-mode gas lasers, *Appl. Phys. Letters,* **5,** 97–98 (1964).

99. G. R. Hanes and K. M. Baird, I_2-controlled He-Ne laser at 633 nm, *Metrologia,* **5,** 32–33 (1969).

100. K. D. Mielenz et al., Reproducibility of He-Ne wavelengths at 633 nm, *Appl. Opt.,* **7,** 289–292 (1968).

101. G. Birnbaum, Frequency Stabilization of gas lasers, *Proc. IEEE,* **55,** 1015–1026 (1967).

102. T. G. Polanyi and I. Tobias, The frequency stabilization of gas lasers, *Lasers Vol. 2,* A. K. Levine, Ed., Dekker, New York, 1968, pp. 373–423.

Appendix A Basic physical constants and conversion factors

The values of the velocity of light (c), the electronic charge (e), Planck's constant (h), and Boltzmann's constant (k) are listed below. These are the value recommended by the U.S. National Bureau of Standards Technical News Bulletin, October 1963.

$$c = 2.997925 \times 10^8 \text{ m/sec},$$
$$e = 1.60210 \times 10^{-19} \text{ coulomb},$$
$$h = 6.6256 \times 10^{-34} \text{ J-sec},$$
$$k = 1.3806 \times 10^{-23} \text{ J/}^\circ\text{K}.$$

The last digits given are uncertain. From these constants the following useful conversion factors are derived:

$$\hbar = \frac{h}{2\pi} = 1.0545 \times 10^{-34} \text{ J-sec},$$

$$\frac{e}{ch} = 8065.7 \text{ cm}^{-1} \text{ (eV)}^{-1},$$

$$\frac{ch}{e} = 1.2398 \times 10^4 \text{ cm eV}.$$

The last two factors serve to convert energy level differences, expressed in electron volts, to wavenumbers expressed in reciprocal centimeters, and vice versa. The accuracy of these conversion factors is limited mainly by the inaccuracy in our knowledge of h.

Appendix B Conversion of electron volts into wavenumbers and wavelengths

The tables that follow are constructed using the conversion factors of Appendix A. They are meant for orientation, not for precise scientific calculations. The energy data in many publications dealing with semiconductors are given in electron volts, and one is often forced to convert energies of the order of 1 eV to wavenumbers. The precision of the conversion is limited, the conversion factor is not known with the same precision as the spectroscopic measurements of wavelength and wavenumber. The conversion in these tables is to wavelengths in vacuo. For wavelengths measured in air correction must be made according to Appendix C.

TABLE B

ΔE (eV)	σ (cm^{-1})	λ (μm)	ΔE (eV)	σ (cm^{-1})	λ (μm)
0.10	806.57	12.398	0.35	2823.0	3.5423
0.11	887.23	11.271	0.36	2903.7	3.4439
0.12	967.89	10.332	0.37	2984.3	3.3508
0.13	1048.5	9.5370	0.38	3065.0	3.2627
0.14	1129.2	8.8558	0.39	3145.6	3.1790
0.15	1209.9	8.2654	0.40	3226.3	3.0995
0.16	1290.5	7.7488	0.41	3306.9	3.0239
0.17	1371.2	7.2930	0.42	3387.6	2.9519
0.18	1451.8	6.8879	0.43	3468.3	2.8833
0.19	1532.5	6.5253	0.44	3548.9	2.8178
0.20	1613.1	6.1991	0.45	3629.6	2.7551
0.21	1693.8	5.9039	0.46	3710.2	2.6952
0.22	1774.5	5.6355	0.47	3790.9	2.6379
0.23	1855.1	5.3905	0.48	3871.6	2.5829
0.24	1935.8	5.1659	0.49	3952.2	2.5302
0.25	2016.4	4.9593	0.50	4032.9	2.4796
0.26	2097.1	4.7685	0.51	4113.5	2.4310
0.27	2177.7	4.5919	0.52	4194.2	2.3843
0.28	2258.4	4.4279	0.53	4274.8	2.3393
0.29	2339.1	4.2752	0.54	4355.5	2.2960
0.30	2419.7	4.1327	0.55	4436.2	2.2542
0.31	2500.4	3.9994	0.56	4516.8	2.2140
0.32	2581.0	3.8744	0.57	4597.5	2.1751
0.33	2661.7	3.7570	0.58	4678.1	2.1376
0.34	2742.3	3.6465	0.59	4758.8	2.1014

TABLE B (*continued*)

ΔE (eV)	σ (cm^{-1})	λ (μm)	ΔE (eV)	σ (cm^{-1})	λ (μm)
0.60	4839.4	2.0664	0.90	7259.2	1.3776
0.61	4920.1	2.0325	0.91	7339.8	1.3624
0.62	5000.8	1.9997	0.92	7420.5	1.3476
0.63	5081.4	1.9680	0.93	7501.1	1.3331
0.64	5162.1	1.9372	0.94	7581.8	1.3190
0.65	5242.7	1.9074	0.95	7662.4	1.3051
0.66	5323.4	1.8785	0.96	7743.1	1.2915
0.67	5404.0	1.8505	0.97	7823.8	1.2782
0.68	5484.7	1.8233	0.98	7904.4	1.2651
0.69	5565.4	1.7968	0.99	7985.1	1.2523
0.70	5646.0	1.7712	1.00	8065.7	1.2398
0.71	5726.7	1.7462	1.01	8146.4	1.2275
0.72	5807.3	1.7220	1.02	8227.0	1.2155
0.73	5888.0	1.6984	1.03	8307.7	1.2037
0.74	5968.6	1.6754	1.04	8388.4	1.1921
0.75	6049.3	1.6531	1.05	8469.0	1.1808
0.76	6130.0	1.6313	1.06	8549.7	1.1696
0.77	6210.6	1.6101	1.07	8630.3	1.1587
0.78	6291.3	1.5895	1.08	8711.0	1.1480
0.79	6371.9	1.5694	1.09	8791.6	1.1374
0.80	6452.6	1.5498	1.10	8872.3	1.1271
0.81	6533.2	1.5306	1.11	8953.0	1.1169
0.82	6613.9	1.5120	1.12	9033.6	1.1070
0.83	6694.6	1.4938	1.13	9114.3	1.0972
0.84	6775.2	1.4760	1.14	9194.9	1.0876
0.85	6855.9	1.4586	1.15	9275.6	1.0781
0.86	6936.5	1.4416	1.16	9356.2	1.0688
0.87	7017.2	1.4251	1.17	9436.9	1.0597
0.88	7097.8	1.4089	1.18	9517.6	1.0507
0.89	7178.5	1.3930	1.19	9598.2	1.0419

TABLE B (*continued*)

ΔE (eV)	σ (cm^{-1})	λ (μm)	ΔE (eV)	σ (cm^{-1})	λ (μm)
1.20	9678.9	1.0332	1.45	11695.3	0.8550
1.21	9759.5	1.0246	1.46	11776.0	0.8492
1.22	9840.2	1.0162	1.47	11856.6	0.8434
1.23	9920.8	1.0080	1.48	11937.3	0.8377
1.24	10001.5	0.9998	1.49	12017.9	0.8321
1.25	10082.2	0.9919	1.50	12098.6	0.8265
1.26	10162.8	0.9840	1.51	12179.3	0.8211
1.27	10243.5	0.9762	1.52	12259.9	0.8157
1.28	10324.1	0.9686	1.53	12340.6	0.8103
1.29	10404.8	0.9611	1.54	12421.2	0.8051
1.30	10485.4	0.9537	1.55	12501.9	0.7999
1.31	10566.1	0.9464	1.56	12582.5	0.7948
1.32	10646.8	0.9393	1.57	12663.2	0.7897
1.33	10727.4	0.9322	1.58	12743.9	0.7847
1.34	10808.1	0.9252	1.59	12824.5	0.7798
1.35	10888.7	0.9184	1.60	12905.2	0.7740
1.36	10969.4	0.9116	1.61	12985.8	0.7701
1.37	11050.1	0.9050	1.62	13066.5	0.7653
1.38	11130.7	0.8984	1.63	13147.1	0.7606
1.39	11211.4	0.8920	1.64	13227.8	0.7560
1.40	11292.0	0.8856	1.65	13308.5	0.7514
1.41	11372.7	0.8793	1.66	13389.1	0.7469
1.42	11453.3	0.8731	1.67	13469.8	0.7424
1.43	11534.0	0.8670	1.68	13550.4	0.7380
1.44	11614.7	0.8610	1.69	13631.1	0.7336

Appendix C

TABLE C
CORRECTION OF WAVELENGTHS FOR MEASUREMENTS IN AIR,[a] $\Delta\lambda = \lambda_0 - \lambda$
The wavelength in vacuum, λ_0, is the reciprocal of the wave number σ
3000 Å to 14,900 Å
Corrections in Ångströms

λ_0 (Å)	000	100	200	300	400	500	600	700	800	900
3000	0.87	0.90	0.92	0.95	0.98	1.00	1.03	1.05	1.08	1.10
4000	1.13	1.16	1.18	1.21	1.24	1.26	1.29	1.32	1.34	1.37
5000	1.39	1.42	1.45⁻	1.47	1.50	1.53	1.55	1.58	1.61	1.64
6000	1.66	1.69	1.72	1.74	1.77	1.80	1.83	1.85⁻	1.88	1.90
7000	1.93	1.96	1.98	2.01	2.04	2.07	2.09	2.12	2.15⁻	2.17
8000	2.20	2.23	2.25	2.28	2.31	2.34	2.36	2.39	2.42	2.44
9000	2.47	2.50⁻	2.52	2.55	2.58	2.61	2.63	2.66	2.69	2.71
10,000	2.74	2.77	2.80	2.82	2.85⁻	2.88	2.90	2.93	2.96	2.99
11,000	3.01	3.04	3.07	3.09	3.12	3.15⁻	3.18	3.20	3.23	3.26
12,000	3.28	3.31	3.34	3.37	3.39	3.42	3.45⁻	3.47	3.50	3.53
13,000	3.56	3.58	3.61	3.64	3.66	3.69	3.72	3.75⁻	3.77	3.80
14,000	3.83	3.85	3.88	3.91	3.94	3.96	3.99	4.02	4.04	4.07

1.0 μm to 9.9 μm
Corrections in Ångströms

λ_0 (μm)	0.0	0.1	0.2	0.3	0.4	0.5	0.6	0.7	0.8	0.9
1	2.7	3.1	3.3	3.6	3.8	4.1	4.4	4.6	4.9	5.2
2	5.5⁻	5.7	6.0	6.3	6.5	6.8	7.1	7.4	7.6	7.9
3	8.2	8.5	8.7	9.0	9.3	9.5	9.8	10.1	10.4	10.6
4	10.9	11.2	11.5⁻	11.7	12.0	12.3	12.5	12.8	13.1	13.4
5	13.6	13.9	14.2	14.4	14.7	15.0	15.3	15.5	15.8	16.1
6	16.3	16.6	16.9	17.2	17.4	17.7	18.0	18.3	18.5	18.8
7	19.1	19.4	19.6	19.9	20.2	20.4	20.7	21.0	21.3	21.5
8	21.8	22.1	22.4	22.6	22.9	23.2	23.4	23.7	24.0	24.3
9	24.5	24.8	25.1	25.3	25.6	25.9	26.2	26.4	26.7	27.0

Beyond the range of this table: $\Delta\lambda = 2.725 \times 10^{-4}\lambda_0$.

[a] The corrections are applicable for standard air at 15°C and 760-torr pressure.

Author Index

377

Subject Index